# Uncooled Infrared Imaging
# Arrays and Systems

SEMICONDUCTORS
AND SEMIMETALS
Volume 47

# Semiconductors and Semimetals

A Treatise

*Edited by R. K. Willardson*
CONSULTING PHYSICIST
SPOKANE, WASHINGTON

*Eicke R. Weber*
DEPARTMENT OF MATERIALS SCIENCE
AND MINERAL ENGINEERING
UNIVERSITY OF CALIFORNIA
AT BERKELEY

*In memory of Dr. Albert C. Beer, Founding Co-Editor in 1966
and Editor Emeritus of Semiconductors and Semimetals.
Died January 19, 1997, Columbus, OH.*

# Uncooled Infrared Imaging Arrays and Systems

SEMICONDUCTORS
AND SEMIMETALS

Volume 47

*Volume Editors*

PAUL W. KRUSE

INFRARED SOLUTIONS, INC.
MINNEAPOLIS, MINNESOTA

DAVID D. SKATRUD

DEPARTMENT OF THE ARMY
PHYSICS DIVISION
ARMY RESEARCH OFFICE
RESEARCH TRIANGLE PARK
NORTH CAROLINA

*ACADEMIC PRESS*

*San Diego   London   Boston*
*New York   Sydney   Tokyo   Toronto*

This book is printed on acid-free paper.

COPYRIGHT © 1997 BY ACADEMIC PRESS

All rights reserved.
NO PART OF THIS PUBLICATION MAY BE REPRODUCED OR TRANSMITTED IN ANY FORM OR BY ANY MEANS, ELECTRONIC OR MECHANICAL, INCLUDING PHOTOCOPY, RECORDING, OR ANY INFORMATION STORAGE AND RETRIEVAL SYSTEM, WITHOUT PERMISSION IN WRITING FROM THE PUBLISHER.

The appearance of the code at the bottom of the first page of a chapter in this book indicates the Publisher's consent that copies of the chapter may be made for personal or internal use of specific clients. This consent is given on the condition, however, that the copier pay the stated per copy fee through the Copyright Clearance Center, Inc. (222 Rosewood Drive, Danvers, Massachusetts 01923), for copying beyond that permitted by Sections 107 or 108 of the U.S. Copyright Law. This consent does not extend to other kinds of copying, such as copying for general distribution, for advertising or promotional purposes, for creating new collective works, or for resale. Copy fees for pre-1997 chapters are as shown on the title pages; if no fee code appears on the title page, the copy fee is the same as for current chapters. 0080-8784/97 $25.00

ACADEMIC PRESS
525 B Street, Suite 1900, San Diego, CA 92101-4495, USA
1300 Boylston Street, Chestnut Hill, Massachusetts 02167, USA
http://www.apnet.com

ACADEMIC PRESS LIMITED
24–28 Oval Road, London NW1 7DX, UK
http://www.hbuk.co.uk/ap/

International Standard Serial Number: 0080-8784
International Standard Book Number: 0-12-752155-0

PRINTED IN THE UNITED STATES OF AMERICA
97 98 99 00 01 BB 9 8 7 6 5 4 3 2 1

# Contents

LIST OF CONTRIBUTORS . . . . . . . . . . . . . . . . . . . . . . . . . . . . . . xi
PREFACE . . . . . . . . . . . . . . . . . . . . . . . . . . . . . . . . . . . . . . . xiii

## Chapter 1  Historical Overview
*Rudolph G. Buser and Michael F. Tompsett*

  I. Introduction . . . . . . . . . . . . . . . . . . . . . . . . . . . . . . . . . . 1
 II. History of Electronic Materials Research for Uncooled Imagers . . . . . . . 6
    1. Ferroelectric–Pyroelectric Materials . . . . . . . . . . . . . . . . . . . . 6
    2. Resistive Materials . . . . . . . . . . . . . . . . . . . . . . . . . . . . . 8
III. Uncooled Imaging Arrays Using Silicon Read-Out . . . . . . . . . . . . . . 9
    1. Ferroelectric–Pyroelectric Arrays . . . . . . . . . . . . . . . . . . . . . 9
    2. Resistive Bolometric Arrays . . . . . . . . . . . . . . . . . . . . . . . . 11
IV. Future . . . . . . . . . . . . . . . . . . . . . . . . . . . . . . . . . . . . . . 12
    References . . . . . . . . . . . . . . . . . . . . . . . . . . . . . . . . . . . 14

## Chapter 2  Principles of Uncooled Infrared Focal Plane Arrays
*Paul W. Kruse*

  I. Importance of the Thermal Isolation Structure . . . . . . . . . . . . . . . 17
 II. Principal Thermal Detection Mechanisms . . . . . . . . . . . . . . . . . . . 23
    1. Resistive Bolometers . . . . . . . . . . . . . . . . . . . . . . . . . . . . 23
    2. Pyroelectric Detectors and Ferroelectric Bolometers . . . . . . . . . . . 25
    3. Thermoelectric Detectors . . . . . . . . . . . . . . . . . . . . . . . . . . 29
III. Fundamental Limits . . . . . . . . . . . . . . . . . . . . . . . . . . . . . . . 31
    1. Temperature Fluctuation Noise Limit . . . . . . . . . . . . . . . . . . . 31
    2. Background Fluctuation Noise Limit . . . . . . . . . . . . . . . . . . . 33
IV. Discussion . . . . . . . . . . . . . . . . . . . . . . . . . . . . . . . . . . . . 37
    References and Bibliography . . . . . . . . . . . . . . . . . . . . . . . . . 40

## Chapter 3  Monolithic Silicon Microbolometer Arrays
### R. A. Wood

| | |
|---|---|
| I. Background | 45 |
| II. Responsivity of Microbolometers | 47 |
|    1. Microbolometer Model | 47 |
|    2. Resistance Changes in Microbolometer Materials | 51 |
|    3. Microbolometer Heat Balance Equation | 56 |
|    4. Solutions of the Heat Balance Equation | 57 |
|    5. Heat Balance with No Applied Bias | 57 |
|    6. Heat Balance with Applied Bias | 59 |
|    7. Calculations of V–I Curves | 61 |
|    8. Load Line | 64 |
|    9. Low-Frequency Noise in Microbolometer with Applied Bias | 68 |
|    10. Microbolometer Responsivity with Pulsed Bias or Large Radiation Signals | 70 |
|    11. Numerical Calculation of Microbolometer Performance | 71 |
| III. Noise in Bolometers | 75 |
|    1. Bolometer Resistance Noise | 75 |
|    2. Noise from Bias Resistors | 79 |
|    3. Thermal Conductance Noise | 80 |
|    4. Radiation Noise | 81 |
|    5. Total Electrical Noise | 83 |
|    6. Preamplifier Noise | 85 |
| IV. Microbolometer Signal-to-Noise | 86 |
|    1. Noise Equivalent Power (NEP) | 86 |
|    2. Noise Equivalent Temperature Difference (NETD) | 86 |
|    3. Detectivity | 87 |
|    4. Comparison with the Ideal Bolometer | 89 |
|    5. Johnson Noise Approximation | 91 |
| V. Electric Read-Out Circuits for Two-Dimensional Microbolometer Arrays | 91 |
| VI. Offset Compensation Schemes | 95 |
| VII. Gain Correction | 97 |
| VIII. Modulation Transfer Function (MTF) | 98 |
| IX. Microbolometer Physical Design, Fabrication, and Packaging | 98 |
|    1. One-Level Microbolometers | 100 |
|    2. Two-Level Microbolometers | 102 |
|    3. Packaging | 109 |
| X. Practical Camera Development | 116 |
| References | 119 |

## Chapter 4  Hybrid Pyroelectric–Ferroelectric Bolometer Arrays
### Charles M. Hanson

| | |
|---|---|
| I. Introduction | 123 |
| II. Principles of Pyroelectric Detectors | 124 |
|    1. Pyroelectricity and Ferroelectric Materials | 124 |
|    2. Modes of Operation | 139 |
|    3. Signal and Noise | 144 |
| III. Practical Considerations and Designs | 154 |

|   |   |   |
|---|---|---|
| | 1. Ferroelectric Material Selection | 154 |
| | 2. Thermal Isolation | 156 |
| | 3. Modulation Transfer Function (MTF) | 158 |
| | 4. Read-out Electronics | 159 |
| | 5. System Electronics | 161 |
| | 6. Choppers | 162 |
| IV. | Systems Implementations | 169 |
| | References | 173 |

## Chapter 5  Monolithic Pyroelectric Bolometer Arrays
*Dennis L. Polla and Jun R. Choi*

|   |   |   |
|---|---|---|
| I. | Introduction | 175 |
| II. | Detector Design Methodology | 176 |
| | 1. Materials Processing | 178 |
| | 2. Materials Characterization | 181 |
| | 3. Thermal Isolation Structures | 183 |
| | 4. Micromachined Sensor Process Design | 184 |
| | 5. Integrated Circuits | 186 |
| III. | Process Design | 187 |
| IV. | Silicon-Based Integrated Pyroelectric Detector Arrays | 189 |
| | 1. Cell Structure | 190 |
| | 2. Circuit Operation | 191 |
| | 3. Silicon-Based $PbTiO_3$ Array Performance | 195 |
| V. | Gallium Arsenide-Based Integrated Pyroelectric Detectors | 197 |
| VI. | Summary | 199 |
| | References | 200 |

## Chapter 6  Thermoelectric Uncooled Infrared Focal Plane Arrays
*Nobukazu Teranishi*

|   |   |   |
|---|---|---|
| I. | Introduction | 203 |
| II. | Thermopile Infrared Detector | 204 |
| | 1. Mechanism for Uncooled Infrared Detector | 204 |
| | 2. Comparison Among Uncooled Infrared Detector Schemes | 205 |
| | 3. The Seebeck Effect | 206 |
| | 4. Various Thermopile Infrared Detectors | 209 |
| III. | A $128 \times 128$ Pixel Thermopile Infrared Focal Plane Array | 210 |
| | 1. Polysilicon Thermopile Infrared Detector | 210 |
| | 2. Characteristics of a Thermopile Infrared Detector | 211 |
| | 3. Signal Read-Out Circuit | 211 |
| | 4. Charge-Coupled Device Scanner | 213 |
| | 5. Package | 214 |
| | 6. Performance | 215 |
| | 7. Future Improvements | 217 |
| IV. | Summary | 217 |
| | References | 218 |

## Chapter 7  Pyroelectric Vidicon
*Michael F. Tompsett*

|  |  |
|---|---|
| I. History | 219 |
| II. Performance Analysis | 223 |
| References | 225 |

## Chapter 8  Tunneling Infrared Sensors
*T. W. Kenny*

|  |  |
|---|---|
| I. Introduction | 227 |
| II. Sensor Modeling | 229 |
|    1. Sensor Thermal Model | 229 |
|    2. Sensor Mechanical and Electrical Model | 232 |
|    3. Noise Model and Considerations | 236 |
| III. Tunneling Transducer Background | 239 |
|    1. Comparison of Tunneling and Capacitive Transducers | 241 |
|    2. Tunneling Transducer Design Considerations | 243 |
| IV. Tunneling Infrared Sensor Design and Fabrication | 245 |
| V. Tunneling Transducer Operation | 253 |
| VI. Infrared Sensor Operation and Testing | 259 |
| VII. Future Prospects for the Tunneling Infrared Sensor | 264 |
| VIII. Conclusion | 266 |
| References | 266 |

## Chapter 9  Application of Quartz Microresonators to Uncooled Infrared Imaging Arrays
*John R. Vig, Raymond L. Filler, and Yoonkee Kim*

|  |  |
|---|---|
| I. Introduction | 269 |
| II. Quartz Microresonators as Infrared Sensors | 271 |
| III. Quartz Thermometers and Their Temperature Coefficients | 272 |
| IV. Oscillator Noise | 273 |
| V. Frequency Measurement | 274 |
| VI. Thermal Isolation | 275 |
| VII. Infrared Absorption of Microresonators | 277 |
| VIII. Predicted Performance of Microresonator Arrays | 279 |
| IX. Producibility and Other Challenges | 281 |
| X. Summary and Conclusions | 283 |
| Appendix. Performance Calculations | 284 |
| References | 294 |

## Chapter 10  Application of Uncooled Monolithic Thermoelectric Linear Arrays to Imaging Radiometers
*Paul W. Kruse*

|  |  |
|---|---|
| I. Introduction | 297 |
| II. Identification of Incipient Failure of Railcar Wheels | 298 |

|  |  |
|---|---|
| 1. Technical Description of the Model IR 1000 Imaging Radiometer | 298 |
| 2. Performance of the Model IR 1000 Imaging Radiometer | 300 |
| 3. Initial Application | 304 |
| 4. Summary | 309 |
| III. Imaging Radiometer for Predictive and Preventive Maintenance | 309 |
| 1. Description | 310 |
| 2. Operation | 312 |
| 3. Specifications | 317 |
| 4. Summary | 317 |
| References | 318 |
| Index | 319 |
| Contents of Volumes in This Series | 327 |

# List of Contributors

Numbers in parenthesis indicate the pages on which the authors' contribution begins.

RUDOLPH G. BUSER (1), *US Army Communications and Electronics Command, Fort Monmouth, New Jersey 07703-5601*

JUN R. CHOI (175). *Microtechnology Laboratory, University of Minnesota, Minneapolis, Minnesota 55455*

RAYMOND L. FILLER (269). *US Army Communications and Electronics Command, Fort Monmouth, New Jersey 07703-5601*

CHARLES M. HANSON (123). *Uncooled Infrared Systems Department, Texas Instruments, Dallas, Texas 75243*

T. W. KENNY (227). *Department of Mechanical Engineering, Stanford University, Stanford, California 94305-4021*

YOONKEE KIM (269). *US Army Research Laboratory, Fort Monmouth, New Jersey 07703-5601*

PAUL W. KRUSE (17, 297). *Infrared Solutions, Inc., Minneapolis, Minnesota 55442*

DENNIS L. POLLA (175). *Microtechnology Laboratory, University of Minnesota, Minneapolis, Minnesota 55455*

NOBUKAZU TERANISHI (203). *Microelectronics Research Laboratories, NEC, Corporation, Sagamihara, Kanagawa 229-11, Japan*

MICHAEL F. TOMPSETT (1, 219). *US Army Research Laboratory, Fort Monmouth, New Jersey 07703-5601*

JOHN R. VIG (269). *US Army Communications and Electronics Command, Fort Monmouth, New Jersey 07703-5601*

R. A. WOOD (43). *Honeywell Inc., Plymouth, Minnesota 55441-4799*

# Preface

A tremendous effort has been spent on developing the capability to see during conditions of darkness and obscured visibility. The motivations are obvious and manifold; there has been a natural desire to extend all daylight activities to periods of darkness, including work, play, commerce, and especially activities involving safety and defense. There are three possible approaches to "seeing" in the dark: to provide artificial illumination; to greatly amplify the small amounts of natural illumination that may be present; or to detect and image the electromagnetic radiation emitted by objects—infrared thermal imaging. For various technological and financial reasons, infrared thermal imaging has been the least used of these approaches. However, advances in uncooled detectors for infrared thermal imaging are revolutionizing its practical utility, providing a level of performance and cost that will enable a wide range of new applications and capabilities. The information available in the open literature on uncooled infrared imaging arrays has been quite limited because of military classification of much of the original work. This volume describes in detail the scientific and technological breakthroughs that enabled this revolution; outlines expected advances that will further enhance uncooled infrared imaging; and describes existing and anticipated applications.

To appreciate the significance of the developments in uncooled thermal imaging, it is helpful to understand the limitations of the other approaches to nighttime viewing. Of the three approaches, artificial illumination has been the most successful and ubiquitous, ranging from the fires of cavemen to highly efficient street lamps, automobile headlights, and radar systems. Nevertheless, artificial illumination suffers from several shortcomings. Artificial illumination with visible light is not effective during conditions of obscured visibility such as is present with fog, smoke, and dust. Also, the fraction of the scene that can be illuminated is typically much less than can be seen during daylight. For example, automobile headlights cover a width and range that is much less than the eye can accommodate. In addition, artificial illumination is an active system and hence has several deleterious attributes. For example, the glare from a vehicle's headlights can blind an oncoming driver. There are also problems with the use of active systems for security and military applications. The police car's spotlight alerts burglars to the presence of the police, and the burglars can attempt to avoid illumination by the spotlight. Similarly for the military, spotlights, flares, and radar act as beacons, alerting the enemy to the presence and location of the observer. Although artificial illumination provided by radar systems is much less limited by obscurants than visible light, radar systems typically have poor resolution because of their long wavelength. Consequently, radar provides limited information regarding the shape or image of a detected object, and radar also suffers from the beacon effect of all active systems.  Amplification of the existing low level of natural illumination has been very successfully provided by image intensifiers. They found early and widespread use in the military. Attractive for passive viewing, image intensifiers evolved through significant research and development investments in photoelectronic detection and vacuum electronic amplification. With reductions in price arising from advances in solid-state electronics, image intensifiers are becoming commonly used by law enforcement and security agencies, and they are even affordable to sportsmen such as hunters and boaters. A significant limitation of image intensifiers is the need for existing, albeit low-level, light. For satisfactory performance, even the best image intensifiers require lighting equivalent to a bright starlit night. Also, since image intensifiers amplify visible and near-visible light, they cannot see through clouds, fog, smoke, or thick dust, even if the ambient light level is high.

Infrared thermal imaging does not suffer from the limitations of techniques that use artificial illumination or amplification of existing low levels of visible light. Thermal imaging systems detect the electromagnetic radiation emitted by all objects. The so-called black body, or thermal, emission described by the Planck relationship is a function of the body's temperature and emissivity. The intensity of the radiation and its peak frequency increase with temperature. The sun, with a temperature of about 6000 K, has a peak emission wavelength near 500 nm, which is close to the middle of the visible region of the electromagnetic spectrum, whereas terrestrial objects, with temperatures that are typically about 300 K, have a peak in the infrared with wavelengths near 10 $\mu$m. This wavelength is fortuitously in the 8–14 $\mu$m atmospheric absorption window. Infrared imagers can also work in the 3–5 $\mu$m atmospheric window by detecting the radiation emitted in the tail of the Planck curve, or by detecting reflected infrared sunlight during daytime operation. Since the wavelength of detected radiation is typically more than 15 times that of visible light, the infrared radiation is scattered by water vapor and particulates much less than is visible light. Consequently, infrared thermal imaging can provide superior performance under conditions of fog, clouds, smoke, and dust.

There are several technological challenges to infrared imaging. Since many objects in a scene will tend to have nearly the same temperature, the difference in the radiation emitted by the objects is typically quite small, often less than 1%. This low contrast requires very sensitive detectors with fast optics. Infrared thermal imaging has the attractive feature that the images appear similar to those in the visible (unlike some alternative imaging technologies such as synthetic aperture radar); however, objects may switch from appearing hot to appearing cold with the diurnal cycle or other changing heat loads. This can cause problems with the automatic processing of infrared images, but it is usually not a significant problem for humans.

The two principal types of infrared detectors are photon detectors and thermal detectors. In photon detectors the absorbed photons directly produce free electrons or holes. In thermal detectors the absorbed photons produce a temperature change, which is then indirectly detected by measuring a temperature-dependent property of the detector material. Photon detection-based infrared imaging arrays provide outstanding performance for imaging scenes at 300 K. However, they are very expensive and require cooling to cryogenic temperatures. The development of uncooled thermal detector arrays potentially provides much lower-cost, more compact, and lower-power thermal infrared imagers. This enables a wide range of civilian applications, in addition to greatly increased military uses. The Army was a primary stimulus for the development of uncooled imaging as a cost-effective means of extending infrared imaging from expensive weapons platforms, where the high performance and high unit cost of cooled imaging was justified, to much broader applications such as use by individual soldiers. It should be noted that much of the impetus to develop uncooled imagers was a result of the historic utility of its sophisticated and expensive predecessor, cooled HgCdTe imagers. The experience of using HgCdTe imagers, albeit at high unit costs, did much to demonstrate the usefulness of thermal infrared imaging.

The development of uncooled infrared arrays capable of imaging scenes at 300 K has been an outstanding technical achievement. Much of the technology was developed under classified military contracts in the United States, so the public release of this information in 1992 surprised many in the worldwide infrared community. There had been an implicit assumption that only cryogenic photon detectors operating in the $3-5$ or $8-12\,\mu$m atmospheric window had the necessary sensitivity to image room-temperature objects. Uncooled intrinsic photon detectors, including PbS and PbSe photoconductors, lacked the required spectral response. Uncooled InSb photoconductor and photoelectromagnetic detectors operating out to 7 $\mu$m lacked adequate sensitivity. Extrinsic photon detectors require even more cooling than intrinsic detectors. Thermal detectors such as bolometers, pyroelectric and thermoelectric, have the required spectral response but were believed to lack sensitivity and speed of response. The pyroelectric vidicon had been found to have adequate performance for some less demanding, nonmilitary applications, but it did not find widespread use.

Although the desirability of uncooled thermal imaging systems was obvious, there existed the following thought process, which was not enunciated explicitly:

- A photon detector having either a small forbidden energy gap or a small donor or acceptor activation energy is required to image infrared scenes at 300 K.
- High-sensitivity, background-limited operation requires that most of the free carriers in the photon detector be excited by the radiating background, not by thermal excitation associated with the operating temperature of the detector.
- The number of photoexcited free carriers is proportional to the free carrier lifetime. To be 300 K background-limited requires a free carrier lifetime in the millisecond range. However, the lifetime of small-gap semiconductors at 300 K is in the microsecond or less range due to the nature of the recombination processes (Auger or radiative).
- Therefore, it is not possible to prepare a background-limited photon detector operating at 300 K having a spectral response in the 3–5 or 8–12 $\mu$m region.

The fallacy in the above arguments is the initial assumption that photon detectors are required to image thermal scenes with useful frame rates. In the United States it was the vision of the Defense Advanced Research Projects Agency (DARPA), the Army Night Vision and Electronic Sensors Directorate (NVESD), and later the Naval Air Warfare Center (NAWC) that an approach could be found to providing an uncooled thermal imaging system. Further, in the United States it was the technical excellence of the research staffs at Honeywell and Texas Instruments, who, under contract with the above agencies, developed the following approach:

- The focus will be on staring focal plane arrays based on thermal detection mechanisms.
- The detector response time requirement for staring focal plane arrays is the video frame time, which in the United States is 33 msec.

- Excellent imagery can be obtained from staring focal plane arrays even if the pixel $D^*$ values are low. An NETD (noise equivalent temperature difference) of 0.1°C, which is better than that of the first-generation (Ha, Cd)Te common modular FLIR, can be obtained with a $D^*$ less than $1 \times 10^9$ cm Hz$^{1/2}$/Watt and a fast ($f/1$) lens.
- Staring arrays of bolometers (the Honeywell approach) and unbiased and electrically biased pyroelectric detectors (the Texas Instruments approach) can readily meet the NETD and frame rate requirements for uncooled thermal imagers.
- Modern integrated circuit processing technology can enable the development of very large, low-cost, two-dimensional arrays.

Thus, a paradigm shift in focal plane array development arose from the vision of the above organizations. Similar approaches were underway elsewhere, especially in the United Kingdom, where the Plessey Company, later part of GEC Marconi, successfully developed pyroelectric staring focal plane arrays with funding from the Defence Research Agency.

This volume describes a variety of approaches to uncooled thermal imaging. The bolometric and pyroelectric approaches have received the most emphasis and success, but there are other novel approaches that may be successful, at least in niche applications. The book starts with two overview chapters. In Chapter 1, Rudolph Buser and Michael Tompsett give a historical overview of the development of uncooled thermal imaging, with an emphasis on pyroelectric and bolometric detectors. They also describe possible future developments in the field. In Chapter 2, Paul Kruse describes the most widely used thermal detection mechanisms and provides an analysis of the fundamental performance limits common to all of them.

The remaining chapters describe specific approaches to uncooled thermal imaging. In Chapter 3, Andrew Wood describes the development by Honeywell and others of micromachined silicon microbolometer arrays. Due in large part to the inherent excellent thermal isolation of the micromachining approach, microbolometer-based systems have been demonstrated with NETDs better than 40 mK for $240 \times 336$ arrays with $f/1$ optics. In Chapter 4, Charles Hanson discusses the hybrid pyroelectric approach pioneered by Texas Instruments. The relatively poor thermal isolation of the bump-bonded, hybrid pyroelectric structure is offset by the high thermal response of the ferroelectric material. This approach has also achieved system NETDs of less than 40 mK for $328 \times 245$ arrays with $f/1$ optics. In Chapter 5, Dennis Polla and Jun Choi present a monolithic approach to uncooled pyroelectric arrays that combines micromachining with thin-film deposition of a pyroelectric material. Although still in the development stage, the combination of the excellent thermal isolation and high material responsivity should provide significantly improved NETDs. Thermoelectric infrared detector arrays are covered by Nobukazu Teranishi in Chapter 6. Thermoelectric detectors do not require temperature stabilization or chopping, and they are relatively easy to fabricate. Consequently, they provide a relatively inexpensive, but lower-performance, alternative to other uncooled thermal imagers. In Chapter 7, Michael Tompsett describes the development of pyroelectric vidicons, which provide good spatial resolution and modest NETDs. The following two chapters present novel approaches to uncooled thermal imaging. In Chapter 8, Thomas Kenny describes the extension to imaging of Golay detectors, which measure the thermal expansion with temperature of a volume of trapped gas. Tunneling displacement transducers are used to measure the membrane deflection of miniature Golay cells fabricated by silicon micromachining. Although this approach will be difficult to extend to large arrays, it may be useful for applications such as space-based, very-long-wavelength infrared detection. In Chapter 9, John Vig, Raymond Filler, and Yoonkee Kim describe the application of quartz microresonators to uncooled thermal imaging. There are several challenging technical issues to this approach, including thermal isolation and the construction of large arrays. Nevertheless, it is intriguing because of the extreme sensitivity of these oscillators to temperature and the extraordinary precision to which the frequency deviations can be measured. The final chapter, by Paul Kruse, describes two applications of uncooled silicon microstructure thermoelectric linear arrays to imaging radiometers. In contrast to thermal imagers, imaging radiometers provide not only a thermal image of a scene, but also the means to determine the temperature of every point in the scene.

With funding from DARPA, the Army NVESD, the Army Research Office, the National Science Foundation, industry, and other organizations, extensive research and development continues on uncooled thermal imagers. Most of the programs are following the monolithic bolometric approach, the monolithic pyroelectric electric approach, or the monolithic thermoelectric approach. Since most of the fabrication process is similar to that for silicon integrated circuits, very-low-cost thermal imagers can be manufactured with NETD values satisfying almost all commercial requirements and many military requirements. It will be exciting to watch the rapid growth in the market and applications of uncooled thermal infrared imaging systems as they mature.

PAUL W. KRUSE<br>DAVID D. SKATRUD

CHAPTER 1

# Historical Overview

*Rudolph G. Buser*

US ARMY COMMUNICATIONS AND ELECTRONICS COMMAND
FORT MONMOUTH, NEW JERSEY

*Michael F. Tompsett*

US ARMY RESEARCH LABORATORY
FORT MONMOUTH, NEW JERSEY

I. INTRODUCTION . . . . . . . . . . . . . . . . . . . . . . . . 1
II. HISTORY OF ELECTRONIC MATERIALS RESEARCH FOR UNCOOLED IMAGERS . . . . 6
   1. *Ferroelectric–Pyroelectric Materials* . . . . . . . . . . . . . 6
   2. *Resistive Materials* . . . . . . . . . . . . . . . . . . . . 8
III. UNCOOLED IMAGING ARRAYS USING SILICON READ-OUT . . . . . . . . . . 9
   1. *Ferroelectric–Pyroelectric Arrays* . . . . . . . . . . . . . . 9
   2. *Resistive Bolometric Arrays* . . . . . . . . . . . . . . . . 11
IV. FUTURE . . . . . . . . . . . . . . . . . . . . . . . . . . 12
   *References* . . . . . . . . . . . . . . . . . . . . . . . . 14

## I. Introduction

The historical basis for thermal imaging was laid when man first identified the position of a recently extinguished fire by placing his hand over the black but radiating embers and felt the warmth. It was many millennia before this one-pixel bolometric sensor was improved. This required not only the theoretical underpinnings that involved understanding the physics of blackbody radiation but a vast basis of material science related to the chemistry and crystallography of thermally sensitive materials and the technology of microelectronic image scanning and displays. From the point of view of physics there are only a limited number of effects that provide a quantitative pathway to a sensitive infrared (IR) imaging device, although the properties of most materials show temperature sensitivity. Indeed a number of these materials have been used in thermometers over the centuries. Conceptually, thermal imaging is a form of remote thermometry, but with simultaneous measurements over many neighboring regions. The materials issues are primarily of sensitivity, uniformity, and adaptability to available read-out mechanisms. These aspects and their history will be explored in this chapter.

There are two steps to infrared (IR) imaging: First, the thermal radiation of the scene must be focused onto a sensor to cause a physical effect, such as photoconductivity, which generally requires cooled devices, or a change in a physical property caused by an increase in temperature, as in bolometric devices. Second, the resulting physical effect must be read and displayed. This latter has been realized dramatically for visible imaging with the television technology that was developed from the late 1920s and provided powerful leverage for the development of IR imaging. Many efforts were devoted to making mechanically scanned imagers and electron-beam scanned IR camera tubes. However, it was the development of metal-oxide semiconductor (MOS) silicon (Si) technology beginning in the early 1960s and particularly the concept of X-Y addressed imaging arrays first described in 1968 by Noble that has been adapted to create the read-out integrated circuits (ROICs) necessary for making large solid-state imaging arrays, which has enabled the development of uncooled solid-state imaging arrays that are the primary topic of this book.

The modern development of IR imaging really started with the early work in the late 1920s on the understanding of photoemission from which it was known that the silver–oxygen–caesium (Ag–O–Cs) photocathode system (S-1) had a response in the near-IR range and led to the first night-vision image intensifier devices used by the military. The early work was empirical but led to progressive improvements that continue today, and the development of multi-alkali photocathodes such as the latest extended-red S-25 photocathode. Another approach was triggered by the discovery in 1965 of the negative electron affinity of caesiated gallium arsenide (GaAs) photocathodes. This led eventually to the very high sensitivity third generation of image intensifiers (Roaux, Richard, and Piaget, 1985) that are in use by military forces today, providing near-IR night operation capability using starlight illumination out to wavelengths of $2\,\mu\mathrm{m}$.

A natural extension of the work in low light level image-intensification for the military, which started in the 1940s, led to research on how to use the thermal differences between an object and its background to image in the IR spectrum. Thus the mid-IR (3 to $5\,\mu\mathrm{m}$) and far-IR (8 to $14\,\mu\mathrm{m}$) spectral regions defined by the atmospheric windows having been ignored by science as a means of night vision began to receive attention in the 1950s (Kruse *et al.*, 1962). There were several generic approaches that were pursued. One approach that was crucial to the establishment of the viability of true thermal IR imaging was to optomechanically scan the image across a single IR-sensitive detector. This approach was developed in the 1960s and commercialized. The AGA Company in Sweden (Borg, 1968) used a cooled single-element indium antimonide intrinsic photoconductor. Barnes Engineering in the United States (Wormser, 1968) initially used a thermistor bolometer (DeWaard and Wormser, 1959) but then adopted a pyroelectric detector (Astheimer and Schwartz, 1968).

Cooled IR detectors really began with the work by Lawson *et al.*, (1959) at the then Royal Signals and Radar Establishment in the United Kingdom, who identified mercury cadmium telluride (MCT) as an intrinsic semiconducting photoconductor that only required cooling to 77 K. The French pursued a photovoltaic approach; however, the British and Honeywell in the United States initially developed mechanically scanned, hybridized linear arrays of MCT detectors cooled to 77 K that operated in the photoconductive mode (Kruse, 1981). This first generation of imaging systems operating in the 8 to 12 $\mu$m range and, in the United States built from common module elements defined by the US Army Night Vision Laboratory, are today the mainstay of battlefield night-vision imaging systems. They have also spawned a range of military, civilian, and research applications, as shown in Table I. The new second generation of devices operating in the photovoltaic mode can have superb performance with very low noise equivalent temperature differences (NETD) of 0.02°C. All-electronically scanned large MCT staring arrays are just becoming viable, but are expensive and require cryogenic systems.

Because detection systems make extensive use of the figure-of-merit term specific detectivity D*, before going further we will define this most widely used method of comparing the performances of individual detectors (Jones, 1959). Specific detectivity is derived from the concept of noise equivalent power (NEP), which is the power of the input signal in watts required to give a response, in a 1-Hz bandwidth, that is equal to the total root mean square noise voltage of the detector and amplifier. Since some noise sources are proportional to $A^{1/2}$, where A is the area of the detector, detectors of different sorts and sizes can be compared using the specific detectivity D* defined as $D^* = A^{1/2}/NEP$ in units of $cmHz^{1/2}W^{-1}$. The D* of MCT is particularly high, which gives the cooled MCT detectors an edge in critical applications.

Mechanically scanned systems have several obvious disadvantages that are overcome in electronically scanned television systems, such as those using electron-beam scanned camera tubes and solid-state imagers for optical to electronic conversion. The initial step therefore was to consider

TABLE I
APPLICATIONS OF INFRARED IMAGING SYSTEMS

| Military, Law Enforcement, and Rescue | Industrial | Medical | Scientific |
|---|---|---|---|
| Night operations | Industrial surveillance and crime prevention | Early detection and identification of cancer | Satellite earth resource Surveying |
| Reconnaissance and surveillance | Process control | Optimum site for amputation determination | Gulf Stream location and mapping |
| Fire fighting and rescue in smoke | Nondestructive inspection of manufactured items such as thermal insulators, photographic film and infrared materials | Placental site location | Volcano studies |
| Submarine detection | | Efficiency of arctic clothing studies | Water pollution detection and studies |
| Detect underground missile sites, personnel, vehicles, weapons, mines and encampments | Hidden piping location | Early diagnosis of incipient stroke and vein blockage | Crevass and sea-ice Reconnaissance studies |
| | Microwave fields display | Wound healing monitoring | Crop detection |
| Damage assessment | Diseased Tree and crop detection | Onset of infection detection without removing bandages | Forgery detection |
| Night-time landing aids | Hot spot detection, e.g., hot boxes on railroad cars and power lines | | Nocturnal animal studies |
| Missile guidance and proximity fusing | Brake linings, cutting tool, weld, and ingot temperature measurement | | Remote sensing of weather conditions |
| Earthquake victim locations | Clear-air turbulence detection | | Heat transfer in plant studies |
| Hidden law violator location | Organic chemicals and gas analysis | | Earth's heat balance measurements |
| Forest fire detection | Pipeline leak detection | | Farming techniques improvements |
| | Oil spill detection | | |

using electron-beam scanning for thermal imaging. The first suggestion came in 1963 from Hadni at the University of Nancy in France, who proposed a ferroelectric bolometric tube operating in dielectric mode (Hadni et al., 1965). In 1969, Tompsett (1969a, b) at EEV Company in the United Kingdom patented a thermal-imaging camera tube operating in the pyroelectric mode and predicted (Tompsett, 1971) a performance that was successfully realized following developments through the 1970s, reaching 400 lines of resolution and an NETD of 0.2°C (Goss, 1987). Pyroelectric vidicon camera tubes are still manufactured today and are used primarily for fire-fighting applications. A more detailed description of the tube and its history will be given in chapter 7.

At the same time as he proposed the pyroelectric vidicon in 1969, Tompsett (1969c) also made the first proposal for solid-state thermal-imaging arrays. These devices consisted of an array of field-effect transistors (FETs) with a thin layer of a pyroelectric–ferroelectric material attached or deposited onto the gates of the FETs but thermally insulated from them. This concept has evolved into the hydridized single-crystal-slice imaging devices available today (described in chapter 4) and the devices using deposited ferroelectric layers that are currently under development for the next generation of monolithic uncooled ferroelectric imagers. Another approach based on monolithic devices using a thermistor material deposited onto air-bridges was proposed by Johnson (1978) and are described in chapter 3.

The development of the technology of uncooled arrays really took off in the 1980s, particularly in the United States where the US Army Night Vision Laboratories recognized the significance of low-cost, lightweight applications and jointly funded with the Defence Advanced Projects Agency (DARPA) the development of both ferroelectric and resistive arrays. These programs have culminated with arrays having resolutions of over 300 lines and NETDs as low as 0.04°C, which are now manufactured for military and civilian applications. Today, both types of arrays have comparable performances and are the main emphasis of this book. The history of the materials' research that enabled these devices will be reviewed in Part II of this chapter, and that of the imaging arrays in Part III. Chapters 3, 4 and 5 will be devoted to the detailed science and technology of these imagers.

Thermoelectric junctions, or thermopiles, were first used to detect IR radiation a few years after the discovery of the Seebeck effect in 1824 (Putley, 1966). The use of thermopiles for an imaging array has been under development by NEC (Kanno et al., 1994). Thermal voltages produced across doped polysilicon junctions are used to control the flow of charge into a frame-transfer charge-coupled device (Tompsett, 1971). The arrays have some potential advantages but the performance so far is a modest 0.5°C. They will be described in Chapter 6.

Another early detector technique using thermal expansion of a gas in a Golay cell, named after its inventor (Golay, 1947), has recently been made much more sensitive by using microstructure techniques to fabricate small cells whose expansion is sensed with a tunneling displacement transducer (Kenny et al., 1991). The technique will be described by Chapter 8. This technique is very sensitive; however, because electrical feedback is required to control the tunneling distance to fractions of a nanometer at each cell, the problems of integrating it into an imaging array are significant. A similar structure was proposed as an accelerometer so that micophonics can be expected to be a problem in such arrays (Waltman and Kaiser, 1989).

Another sensitive bolometric device first proposed in 1962 (Wade and Slutsky, 1962) has been revised to make use of quartz microresonators and will be described in Chapter 10. Since the resonant frequency of a quartz crystal changes with temperature, and frequency can be measured with extreme accuracy, individual detectors promise very high sensitivity. However, the task of accessing an array of resonators and making $10^7$ precision multi-MHz frequency measurements per second is daunting. There are also other approaches discussed in section IV using micro-electro-mechanical structures that lead to high resolution ultra sensitive uncooled imaging arrays.

## II. History of Electronic Materials Research for Uncooled Imagers

Since 1940 many materials and devices have been proposed seriously for IR detection and are excellently reviewed by Putley (1966). Two classes have stood the test of time for uncooled imaging arrays: one uses ferroelectric–pyroelectric bolometry; the other uses resistive bolometry with materials having very large temperature coefficients of resistivity. Neither requires cooling, although both require temperature stabilization, and both are compatible with the high electrical impedance of today's integrated read-out circuits. The histories of these two types are outlined in the next two sections.

### 1. FERROELECTRIC–PYROELECTRIC MATERIALS

Lang (1974) has reviewed the 2300-year history of pyroelectricity, the phenomenon where complementary electrical charges appear on opposite sides of certain classes of polar crystals when subjected to a change of temperature. The first author to use the term *pyroelectricity* was D. Brewster in 1824. In the 19th century, many experiments were conducted that established a basic understanding of the pyroelectric effect in crystals such as quartz and tourmaline. A rudimentary way of determining the charge distribution, which could have been used as a direct-view display concept, was experimentally demonstrated in the 19th century by Kundt (1883). He used a powdered mixture of sulfur and red lead oxide blown through a cloth screen. The sulfur particles were charged negatively by friction and attracted to the positive parts of the crystal, while the red lead particles became positively charged and migrated to the negative areas. An important step was the discovery in 1920 by Valasek, who showed that the dielectric properties of ferroelectric crystals were in many respects similar in nature to the ferromagnetic properties of iron, namely, that there is a hysteresis effect in the field polarization curve, that a Curie temperature exists, and that large dielectric changes can be observed around this temperature where the transition to a high-temperature nonferroelectric form occurs.

In the 1930s a series of ferroelectric crystals were discovered and grown (Busch and Scherrer, 1935; Busch, 1938). Barium titanate was found to have a very high dielectric constant of approximately 3000, and displayed ferroelectricity. Further success in material research came quickly (Matthias, 1949; Matthias and Remeika, 1949; Shirane et al., 1950). In 1941, Slater developed the microscopic model of ferroelectricity, which he extended in 1960 to include quantum modeling (Slater, 1960). A good review of the field of ferroelectric crystals up to 1962 is available (Jona and Shirane, 1962). As recently as 1994 Resta developed the latest and perhaps definitive explanation of macroscopic electric polarization.

The first proposal that a pyroelectric crystal, tourmaline in this case, could be used as an IR detector was made in 1938 by Ta. A patent for an "Energy Translation Device" using a pyroelectric to detect a modulated infrared beam was issued in 1942 (Sivian, 1942). In 1956, Chynoweth used the pyroelectric effect to study ferroelectricity. He showed that an unbiased crystal subjected to a change in temperature develops a change in charge along the polar axis of the crystal. If this axis is made normal to the crystal surface, a voltage and charge displacement, which can be measured, will then appear across the crystal. Hanel (1961) also suggested using the temperature-sensitive dielectric constant of ferroelectrics in a discrete bolometer. In this case, a dc bias voltage is applied across the crystal. A change in temperature causes a change in dielectric constant and hence a charge displacement or current, which can be measured. Alternatively, if the target is charged to a voltage and then electrically isolated, this change in dielectric constant is converted to a change in voltage across the crystal, which can be sensed. Cooper (1962) showed that the pyroelectric approach is preferable to the dielectric approach and has significantly more sensitivity. The concept was further developed (Beerman, 1967) into a commercial product (Astheimer and Schwartz, 1968). A simple figure of merit for pyroelectric materials in a detecting configuration is the pyroelectric coefficient (Coulombs $°C^{-1}cm^2$) divided by the dielectric constant. On this basis,

Beerman showed that an organic crystal triglycine sulfate (TGS) (Mathias et al., 1956; Putley, 1966) has a figure of merit significantly higher than some of the better-known inorganic ferroelectric materials. Putley (1968) discussed the principal noise sources and performance limitations of pyroelectric detectors within the thermal and electrical environments in which they would be applicable. In practice, both modes contribute to the charge–voltage signal output (Watton, 1989). An excellent review of pyroelectric materials and their measured properties and appropriate figures of merit relative to their use in pyroelectric vidicons and imagers has been published by Whatmore (1986).

2. RESISTIVE MATERIALS

The second practical path to uncooled imaging arrays and systems has been to use the long-known dependence of material resistance on temperature. In 1880, Langley made the first bolometer of platinum strips in a

FIG. 1. Contemporary photograph of cows grazing taken at a distance of 500 m with an uncooled image sensor made by Texas Instruments, Inc. (Dallas, Texas) 90 years after Langley's 1901 experiment with a platinum strip and a meter.

Wheatstone bridge. He worked on this for 20 years and in 1901 showed (Langley, 1901) that he could detect a cow at a quarter mile as a deflection on a meter. This should be compared with the contemporary image shown in Fig. 1 of a field of cows taken with a modern resistive bolometric imager. Since then, the concept has become well established and expressions for responsivity and noise limitations of thin-films resistance bolometers have been published (Putley, 1966). Thermsistor bolometers were also made (Putley, 1966) from sintered mixtures of various oxides proposed in 1946 (Brattain and Becker, 1946). Interest in these increased as the microfabrication technology of the 1980s allowed thermally insulated microstructures to be fabricated, as is discussed in Part III, § 2 and chapter 3. Several materials have been explored for modern devices. Vanadium oxide (VO) has a particularly high temperature coefficient of resistance (Moffatt et al., 1986; Umadevi et al., 1991; Jerominek et al., 1993). Both platinum (Pt) and vanadium oxide have been incorporated into today's manufactured devices that have specific detectivities, D*'s of $1 \times 10^9 \, \text{cmHz}^{1/2}\text{W}^{-1}$.

## III. Uncooled Imaging Arrays using Silicon Read-out

1. FERROELECTRIC–PYROELECTRIC ARRAYS

A variety of all–solid-state, that is, non–electron-beam, approaches to uncooled detector technology have been proposed, researched, and developed since Tompsett (1969b, c) first suggested using an array of FETs as the read-out circuit on which was placed a layer of a pyroelectric material with contacts to the gates of the FETs but thermally insulated from them. The first work in this area proceeded under a US government program at RCA Laboratories between 1972 and 1973 (Boornard et al., 1973). This program investigated several approaches to the basic concept of using TGS attached to a silicon ROIC. Both an X-Y addressed array and a bucket brigade array were considered. Reticulation of the TGS was developed as a way to improve the resolution. Since thermal isolation from the substrate is important for good temperature sensitivity, a novel technique using stress in deposited thin films to make thin-film spring microfingers was developed to provide low thermal impedance electrical contacts from the ROIC to the back of the TGS crystal. The results of calculations for an X-Y addressed array showed that a very thin, 10-$\mu$m-thick, TGS slice with 100-$\mu$m square elements on 200-$\mu$m centers could be expected to resolve 0.4°C. However, there is a trade-off because halving the spatial resolution would more than quadruple the temperature resolution.

The next step in the historical sequence was a mock-up of a solid-state array made by Plessey Ltd. Thirty-two discrete TGS detectors on a 250-$\mu$m pitch were mounted on a rotating assembly, with the associated amplifiers driving a similar array of light-emitting diodes (LEDs) (Watton, 1976). This was packaged as a handheld, direct-view instrument.

Texas Instruments (TI) began work on uncooled imaging in the mid-1970s that led to hybridized devices patented by Hopper in 1979. In 1987, TI demonstrated the first 100 × 100 element uncooled imaging array using a barium strontium titanate (BST) crystal-slice bump-bond hydridized to a ROIC with an NETD of 0.5°C. TI reported a 245 × 328 pixel array in 1992 (Hanson et al., 1992) with much higher performance. At the same time, the British were publishing extensively in this area (Watton et al., 1983, 1984; Watton, 1986; Manning et al., 1985; Whatmore and Ainger, 1985; Mansi and Liddicoat, 1986; Watton and Manning, 1987; Mansi et al., 1987; Watton and Mansi, 1987) and showed imagery from a 100 × 100 array in 1994 (Watton, 1994).

In 1987, the US Army Night Vision Laboratories with DARPA sponsored the High-Density Array Development (HIDAD) program which led to the delivery in 1990 of the first HIDAD brassboard imager for evaluation. The focal plane array more than met the promised 0.30°C specification, with a measured NETD of 0.08°C. The 245 × 328 pixel array (Hanson et al., 1992) used an X-Y addressed array. The devices operated near the phase transition of the BST detector material and used a mechanical chopper for field difference processing. Additional refinements included target reticulation to reduce thermal cross talk between pixels and increase the modulation transfer function (MTF), and an IR absorbing coating (Advena et al., 1993). The TI array used dual-mode detection (Watton, 1989), that is, a combined dielectric–pyroelectric mode of operation in which both changes in polarization and dielectric constant with temperature are used to generate the output signal. The dielectric mode requires having a dc-bias on the slice but this has other advantages (Hanson et al., 1992). In order to obtain the high dielectric sensitivity a low-power, single-stage thermoelectric cooler was used to stabilize the focal plane array close to the Curie temperature. The ROIC had an amplifier at each pixel to optimize the signal-to-noise-ratio.

Based on the success of the HIDAD program, a Low Cost Uncooled Sensor Prototype Program (LOCUSP) was implemented to develop low-cost lightweight fieldable IR sensor hardware. Performance of the arrays was improved to 0.047°C with f/1 optics (Hanson et al., 1993). TI is now providing night-vision sensors for law enforcement purposes (Hanson and Beratan, 1994).

Loral Inc. also described a 192 × 128 pixel array in 1992 (Butler and Iwasa, 1992). This array used lithium tantalate in a true pyroelectric mode without thermal stabilization of the substrate. The hydrid array was reticulated with an absorbing layer that had an NETD of 0.07°C.

Similar work was proceeding in Britain over this same time period that migrated to using lead scandium tantalate (Watton and Mansi, 1987; Watton, 1989; Shorrocks and Edwards, 1990; Shorrocks et al., 1990; Watton et al., 1990). Imagery was shown from a 100 × 100 array, with a 100-µm pitch that included reticulation for improved resolution (Todd and Watton, 1990; Watton, 1994). Work also has been done (Watton, 1992, 1994; Patel et al., 1994) to deposit ferroelectrics onto a microbridge structure in order to realize monolithic arrays.

## 2. Resistive Bolometric Arrays

In 1983, in the United States, Honeywell Inc. (Plymouth, Minnesota) began to develop room tempeature thermal detectors using silicon micromachining technology, because this offered the highest possible thermal isolation and the promise of low-cost monolithic arrays (Wood et al., 1988, 1989, 1990, 1991, 1992; Cole et al., 1994). This prompted the US Army Night Vision Laboratories and the Defense Advanced Research Projects Agency (DARPA) to fund Honeywell in 1985 to make a device with 80,000 pixels and an NETD of 0.3°C using resistive bolometric detection. A 64 × 128 device was demonstrated in 1989 (Wood et al., 1989). The approach adopted by Honeywell used a two-dimensional array of VO resistors whose resistance changes approximately 2% per °C change in temperature. The VO is deposited on silicon nitride ($Si_3N_4$) bridges that are supported on thin arms fabricated on the ROIC. This structure ensures thermal isolation from the ROIC and is designed to be a very efficient absorber. The focal plane temperature is at room temperature via a low-power, single-stage thermoelectric stabilized cooler onto which the ROIC package is mounted. The ROIC is an array of gate-controlled switches that sequences a bias current through all of the detectors in a row, and the read-out is in parallel via the column leads to a high-speed multiplexer at one end of the columns. In 1991, Honeywell provided 336 × 240 arrays with 50-µm pixel spacing that were integrated into a HIDAD brassboard. Using an f/1.0 germanium lens at 30 frames/sec, this brassboard had an NETD of 0.1°C and impressive imagery (Wood et al., 1991).

Because the approach used by Honeywell does not require any AC coupling of the incoming radiation, there is no shutter required for normal operation. This doubles the integration time for the signal and therefore the responsivity of the system. The narrower bandwidth and lack of spatial and time-dependent noise from the shutter are also advantages for improved performance. However, a shutter must still be provided because it is needed for intermittent system calibration. The use of freestanding bridges also eliminates thermal cross talk between pixels and allows the modulation transfer function for the system to be optics-limited, which is seen in practice. Another important consideration for this technology is that the monolithic Si structure used for these arrays does not require hybrid bump-bonding of the detector array to a ROIC, which eliminates a major yield and cost factor. Arrays based on this approach are now available from several manufacturers.

## IV. Future

The ability of the Si industry to deliver large high-resolution low-cost ROICs, continuing progress in microstructure fabrication, and advances in material science are providing the technical underpinnings for creating extremely high-performance low-cost arrays that can be mass-produced. These devices can be expected routinely to have a NETD of approximately 50 mK, with performance comparable with first generation cooled MCT systems. These devices are already militarily and commercially viable and are now going into mass production. Two of the major advantages of the uncooled arrays over the MCT photonic detectors are the linearity and uniformity of their response and the small number of defective pixels, both of which are a major advantage at the system level because of the need for less correction circuitry. At a more theoretical level Kruse (1995) has recently published a study of the performance limits of thermal and photon detector imaging arrays and noise in pyroelectric detectors is thoroughly reviewed by Whatmore (1986).

The relative merits of monolithic ferroelectric and resistance arrays can be compared. Resistance devices require bias current for maximum response, which leads to undesirable power dissipation on the array. An advantage of ferroelectric arrays is that there is no dc current or power from the dc-bias voltage. A detectivity $D^*$ of $5 \times 10^9 \, \text{cmHz}^{1/2}/\text{W}$ has been reported for resistive bolometric detectors (Butler and Iwasa, 1992), which is within a factor of four of the ultimate limit for any detector operating at room temperature. However, this is not the true figure of merit for performance, which is better defined by the ratio of the detectivity to the square root of the detector noise bandwidth (Watton and Mansi, 1987). Unlike MCT and resistive arrays, the smoothing and smaller bandwidth is automatically provided by the high capacitance of the ferroelectric capacitance. Watton (1992, 1994) claims that ferroelectric arrays have the potential for the highest performance based on signal-to-noise considerations, particularly for larger array sizes. The use of a chopper for pyroelectric arrays is inconvenient, but it does provide an ac-coupled system with spatial high-pass filtering that eliminates dc-offsets, nonuniformity, and low-frequency spatial noise, which thereby enhances fine detail.

A simple argument would say that since today's hybrid ferroelectric array has essentially the same performance as do resistive arrays, any solution to the materials problem of growing thin ferroelectric layers with close to bulk properties directly onto the ROIC would enable monolithic ferroelectric arrays to be made having significantly better performance than do today's resistive arrays. Developments that significantly increase the temperature coefficient of resistance of resistive bolometer materials would, of course, change this comparison.

The challenge for technologists currently working in this area is to realize the 0.005°C NETD performance of the best low-temperature cooled photonic arrays in a monolithic chip with high-definition television (HDTV) resolution of 1900 × 1100 and pixels fabricated with high yield. Thin films of ferroelectric material grown onto thermally isolating structures will not only increase the performance, as long as bulk properties can be preserved, but will make the arrays monolithic, enabling higher resolution and lower cost. This should also enable future uncooled imaging arrays to perform close to their theoretical NETD performance at 0.36 K.

A recent concept that could allow closure to this theoretical limit to be realized is currently in development (Amantea *et al.*, 1997) at the Sarnoff Laboratories. This device uses an array of micro-bimaterial arms one per pixel that flex when heated. The flexure is detected electrostatically between the arm and the substrate by low noise feedback amplifiers on the ROIC on which these arms are fabricated. Micro-electromechanical technology is used to fabricate the structure but with one important new aspect. Silicon carbide (SiC) is used not only as one of the bimaterials but as supports, which are very thin and more thermally insulating by an order of magnitude than the silicon nitride (SiN) that would otherwise be used. The expected performance of this approach is for a NETD under 5 mK for 50 $\mu$m square pixels. It is monolithic and 100% compatible with silicon IC technology. It has the potential to yield an uncooled imager with the performance comparable to today's best cooled imagers and with similar resolution and cost to today's active pixel visible sensors. All the elements of the technology have been demonstrated but working arrays have not yet been fabricated.

Another intriguing concept is to develop solid-state pixels having the sensitivity in the visible of "GEN3" image intensifiers and combine them with solid-state IR-sensitive pixels of the types we have described. With appropriate fusion of the images in a high-resolution lightweight helmet-mounted display, the user would truly be able to see in the dark under most operational conditions.

## References

Advena, D. J., Bly, V. T., and Cox, T. J. (1993). *Appl. Opt.* **32,** 1136–1144.
Amantea, R., Knoedler, C. M., Pantuso, F. P., Patel, V. K., Sauer, D. J. and Tower, J. R. (1997). *Proc. IRIS Symp. on Passive Sensors*, to be published.
Astheimer, R. W., and Schwartz, F. (1968). *Appl. Opt.* **7,** 1687–1695.
Beerman, H. P. (1967). *Am. Ceram. Bull.* **46,** 737–749.
Boornard, A., Hall, D., Herrmann, E., Larrabee, R. D., Morren, W., Southgate, P. D., and Stephens, W. L. (1973). "Pyroelectric/Integrated Infrared Imaging Array Development," Tech. Rep. AFAL-TR-73-258. U.S. Department of Defense, Washington DC.
Borg, S.-B. (1968). *Appl. Opt.* **7,** 1697.
Brattain, W. H., and Becker, J. A. (1946). *J. Opt. Soc. Am.* **37,** 354.
Busch, (1938). *Helv. Phys. Acta* **11,** 269.
Busch, and Scherrer, (1935). *Naturwissenschaften* **25,** 737.
Butler, N., and Iwasa, S. (1992). *SPIE* **1685,** 146–154.
Chynoweth, A. (1956). *J. Appl. Phys.* **27,** 78.
Cole, B., Horning, R., Johnson, B., Nguyen, K., Kruse, P. W., and Foote, M. C. (1994). *IEEE Int. Symp. Appl. Ferroelectr.* pp. 653–656.
Cooper, J. (1962). *Rev. Sci. Instrum.* **33,** 92–95.
DeWaard, R., and Wormser, E. M. (1959). *Proc. IEEE* **47,** 1508.
Golay, M. J. E. (1947). *Rev. Sci. Instrum.* **18,** 357.
Goss, A. J. (1987). *Proc. SPIE* **807,** (Passive Infrared Syst. Technol.), 25–32.
Hadni, A. (1963). *J. Phys. (Paris)* **24,** 694.
Hadni, A., Henniger, Y., Thomas, R., Vergnat, P., and Wynche, B. (1965). *J. Phys. (Paris)* **26,** 345–360.
Hanel, R. A. (1961). *J. Opt. Soc. Am.* **51,** 220–224.
Hanson, C., and Beratan, H. (1994). *IEEE Int. Symp. Appl. Ferroelectr.* pp. 657–661.
Hanson, C., Beratan, H., Owen, R., Corbin, M., and McKenney, S. (1992). *SPIE* **1735,** 17–26.
Hanson, C., Beratan, H., Owen, R., and Sweetser, K. (1993). *Proc. Detect. IRIS Meet.* Bedford, MA.
Hopper, G. S. (1979). U.S. Pat. 4,162,402; reissued as 4,379,232 (1983).
Jerominek, H., Picard, F., and Vincent, D. (1993). *Opt. Eng.* **32,** 2092–2099.
Johnson, R. G. (1978). Honeywell Internal Report.
Jona, and Shirane, G. (1962). "Ferroelectric Crystals." Macmillan, New York.
Jones, R. C. (1959). *Proc. IRE* **47,** 1495.
Kanno, T., Saga, M., Matsumoto, S., Fujimoto, N., and Teranishi, N. (1994). *SPIE* **2269,** 450–459.
Kenny, T. W., Kaiser, W. J., Waltman, S. B., and Reynolds, J. K. (1991). *Appl. Phys. Lett.* **59.**
Kruse, P. W. (1981). In "Semiconductors and Semimetals" (R. K. Willardson and A. C. Beer, eds.), Vol. 18, Chapter 1. Academic Press, New York.
Kruse, P. W. (1995). *Infrared Phys. Technol.* **36,** 862–882.
Kruse, P. W., McGlauchlin, L. D., and McQuistin, R. B., eds. (1962). "Elements of Infrared Technology." Wiley, New York.
Kundt, A. (1883). "Vorlesungen uber Experimentalphysik." Friedrich Vieweg und Sohn, Berlin.
Lang, S. B. (1974). *Ferroelectrics* **7,** 231–234.
Langley, S. P. (1880).
Langley, S. P. (1901).
Lawson, W. D., Nielson, S., Putley, E. H., and Young, A. S. (1959). *J. Phys. Chem. Solids* **9,** 325.
Manning, P., Burgess, D., and Watton, R. (1985). *Proc. SPIE* **590** (Infrared Technol. Appl.), 2–10.

Mansi, M. V., and Liddicoat, T. J. (1986). *SPIE Int. Tech. Symp. Opt. Optoelectron. Appl. Sci. Eng. 30th,* San Diego.
Mansi, M. V., Liddicoat, T. J., and Richards, L. J., (1987). *Proc. SPIE* **807** (Passive Infrared Syst. Technol.).
Mathias, (1949). *Phys. Rev.* **75,** 171.
Mathias, and Remeika, J. P. (1949). *Phys. Rev.* **76,** 1886.
Mathias, Miller, C. E., and Remeika, J. P. (1956). *Phys. Rev.* **104,** 849–850.
Moffatt, D., Runt, J, Safari, A., and Newnham, R. E. (1986). *Proc. IEEE Int. Symp. Appl. Ferroelectr. 6th,* 673–676.
Noble, P. (1968). *IEEE Trans. Electron Devices* **ED-15,** 196–201.
Patel, A., Osbond, P. C., Shorrocks, N. M., Twiney, R. C., Whatmore, R., and Watton, R. (1994). *Proc. IEEE Int. Symp. Appl. Ferroelectr.* **9,** 647–652.
Putley, E. H. (1968). "The Hall Effect and Semiconductor Physics." Constable, London.
Putley, R. (1966). *J. Sci. Instrum.* **43,** 857–868.
Resta, R. (1994). *Ferroelectrics* **151,** 49–58.
Roaux, Richard, J. C., and Piaget, C. (1985). *Adv. Electron. Electron Phys.* **64A.**
Shirane, Hoshino, and Suzuki, (1950). *Phys. Rev.* **80,** 1105.
Shorrocks, N. M., and Edwards, I. M. (1990). *Proc. IEEE Int. Symp. Appl. Ferroelectr. 7th,* pp. 58–62.
Shorrocks, N. M., Porter, S. G., Whatmore, R. W., Parsons, A. D., Gooding, J. N., and Pedder, D. J. (1990). *SPIE* **1320,** 88.
Sivian, L. J. (1942). Patent for an "energy translating device."
Slater, J. C. (1941). *J. Chem. Phys.* **9,** 16.
Ta, Y. (1938). Action of radiation on pyroelectric crystals. See S. D. Lang, *Sourcebook of Pyroelectricity,* Gordon and Breach Science Publishers, New York, 1974.
Todd, M. A., and Watton, R. (1990). *SPIE* **1320,** 95–1920.
Tompsett, M. F. (1969a). U.S. Pat. 3,646,267.
Tompsett, M. F. (1969b). U.K. Pat. 1,239,243.
Tompsett, M. F. (1969c). U.K. Pat. 1,266,529.
Tompsett, M. F. (1971). *IEEE Trans. Electron Devices* **ED-18,** 1070–1074.
Tompsett, M. F., Amelio, G. F., Bertram, W. J., Buckley, R. R., McNamara, W. J., Mikkelsen, J. C., and Sealer, D. A. (1971). *IEEE Trans. Electron Devices* **ED-18,** 992–996.
Umadevi, P., Nagendra, C. L., Thutupalli, G. K. M., Mahadevan, K., and Yadgiri, G. (1991). *Proc. SPIE.* **1484,** 125–135.
Valasek, (1920). *Phys. Rev.* **15,** 537–538.
Wade, W. H., and Slutsky, L. J. (1962). *Rev. Sci. Instrum.* **33,** 212–213.
Waltman, S. B., and Kaiser, W. J. (1989). *Sens. Actuators* **19,** 201.
Watton, R. (1976). *Ferroelectrics* **10,** 91–98.
Watton, R. (1986). *Proc. Int. Symp. Appl. Ferroelectr. 6th,* pp. 172–181.
Watton, R. (1989). *Ferroelectrics* **91,** 87–108.
Watton, R. (1992). *Ferroelectrics* **133,** 5–10.
Watton, R. (1994). *Integr. Ferroelectr.* **4,** 175–186.
Watton, R., and Manning, P. A. (1987). "The Design of Low-noise Arrays of MOSFETs for *Proc. SPIE* **807** (Passive Infrared Syst. Technol.).
Watton, R., and Mansi, M. V. (1987) *SPIE* **865,** 78–84.
Watton, R., Smith, C., Gillham, J., Manning, P. A., and Burgess, D. (1983). *Int. Conf. Adv. Infrared Detect. Syst. 2nd, Publ.* **228,** pp. 49–53.
Watton, R., Ainger, F., Porter, S., Pedder, D., and Gooding, J. (1984). *Proc. SPIE* **510** (Infrared Technol. X), 139–148.
Watton, R., Todd, M. A., and Gillham, J. P. (1990). *Int. Conf. Adv. IR Detect. Syst. 4th,* pp. 69–77.

Whatmore, R. W. (1986). *Rep. Progr. Phys.* **49,** 1335–1386.
Whatmore, R. W., and Ainger, F. W. (1985). *SPIE* **395,** 261–266.
Wood, R. A., Carney, J., Higashi, R. E., Ohnstein, T., and Holmen, J. (1988).
Wood, R. A., Carlson, R. A., Higashi, R. E., and Foss, N. A. (1989). *Proc. IRIS DSG.*
Wood, R. A., Cole, B. E., Foss, N. A., Han, C. J., Higashi, R. E., and Lubke, E. (1990). *Proc. IRIS DSG.*
Wood, R. A., Cole, B. E., Han, C. J., and Higashi, R. E. (1991). *Proc. IRIS DSG.*
Wood, R. A., Han, C. J. and Kruse, P. W. (1992). *IEEE Solid State Actuator Workshop.*
Wormser, E. M. (1968). *Appl. Opt.* **7,** 1667–1670.

CHAPTER 2

# Principles of Uncooled Infrared Focal Plane Arrays

*Paul W. Kruse*

INFRARED SOLUTIONS, INC.
MINNEAPOLIS, MINNESOTA

I. IMPORTANCE OF THE THERMAL ISOLATION STRUCTURE . . . . . . . . . . . 17
II. PRINCIPAL THERMAL DETECTION MECHANISMS . . . . . . . . . . . . . . . . 23
   1. *Resistive Bolometers* . . . . . . . . . . . . . . . . . . . . . . . . 23
   2. *Pyroelectric Detectors and Ferroelectric Bolometers* . . . . . . . . . . . 25
   3. *Thermoelectric Detectors* . . . . . . . . . . . . . . . . . . . . . . . 29
III. FUNDAMENTAL LIMITS . . . . . . . . . . . . . . . . . . . . . . . . . . 31
   1. *Temperature Fluctuation Noise Limit* . . . . . . . . . . . . . . . . . . 31
   2. *Background Fluctuation Noise Limit* . . . . . . . . . . . . . . . . . . 33
IV. DISCUSSION . . . . . . . . . . . . . . . . . . . . . . . . . . . . . . 37
   *References and Bibliography* . . . . . . . . . . . . . . . . . . . . . . 40

This chapter establishes principles and limits that are fundamental to all types of infrared (IR) arrays that operate by means of thermal detection mechanisms.

## I. Importance of the Thermal Isolation Structure

All thermal IR detectors exhibit a change in some measurable property that accompanies a change in temperature of the sensitive element, that is, the picture element, or pixel, caused by the absorption of IR radiation by the pixel. The analysis of all types of thermal IR detectors therefore begins with a heat flow equation that describes the temperature increase in terms of the incident radiant power. The manifestation of the temperature increase depends on the detection mechanism. The most common detection mechanisms are the resistive bolometer; the pyroelectric detector that is operated

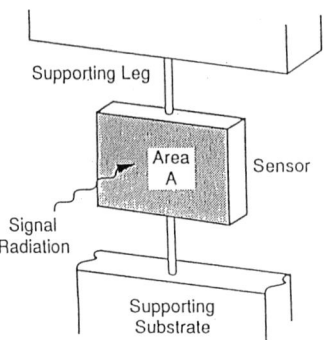

- Sensor is heated by incoming IR radiation
- Temperature change is measured by
  - Resistance change (bolometer)
  - Thermoelectric junction (TE sensor)
  - Pyroelectric effect
  - Oil-film evaporation (evaporagraph)
  - Semiconductor absorption-edge shift
  - Thermoelastic effect
  - Liquid crystal color change
  - Gas pressure change (Golay cell)

FIG. 1. Principle of thermal infrared (IR) sensors. (From R. A. Wood, Honeywell Technology Center, Plymouth Minnesota, personal communication. Reprinted with permission.)

unbiased; the pyroelectric detector under bias, known as the ferroelectric bolometer or the field-enhanced pyroelectric detector; and the thermoelectric detector, also referred to as a radiation thermocouple.

Consider a focal plane array consisting of a two-dimensional assembly of thermally sensitive pixels. Each pixel consists of a sensitive area connected to the substrate (Fig. 1). Infrared radiation falling on a pixel is absorbed by the sensitive area, causing its temperature to increase. Heat flows from the sensitive area to its surroundings. There are three potential mechanisms of heat transfer: conduction, convection, and radiation. Conduction can occur in three ways within the array. (1) Heat can flow from the sensitive area along its support to the substrate. (2) Heat may be able to flow directly from the sensitive area of one pixel to a neighboring pixel if the sensitive areas are contiguous; this is known as lateral heat flow. It is to be avoided because it reduces the resolution of the image. (3) Heat can flow through the surrounding atmosphere if the array is not mounted in an evacuated package.

Convection is the second method of heat transfer. It requires the presence of a surrounding atmosphere. In general, it is not an important heat transfer mechanism in thermal arrays. If the array package is not evacuated, heat loss from the sensitive element through the atmosphere is usually by conduction rather than by convection.

The third heat transfer mechanisms is radiation. Here the sensitive element radiates to its surroundings, and the surroundings radiate to it. This is the ideal case for a thermal IR array. If the principal heat loss mechanism is radiative, then the array is at the background limit, the limit fundamental to its performance.

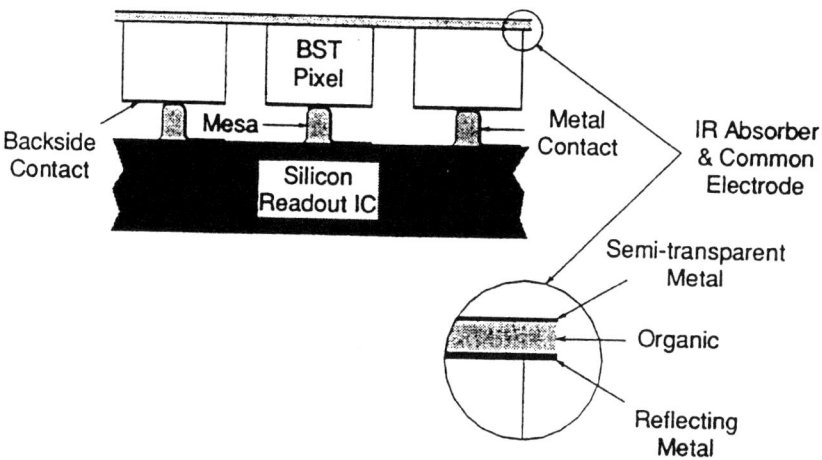

FIG. 2. Pyroelectric detector device structure. BST, barium strontium titanate; IR, infrared; IC, integrated circuit. (Reprinted from Hanson, *et al.* (1992). Uncooled thermal imaging at Texas Instruments. *Proc. SPIE 1735, Infrared Detectors: State-of-the-Art* **17**; with permission from the SPIE.)

To obtain high performance, thermal arrays should be in evacuated packages incorporating IR transmitting windows. The sensitive element should not be contiguous with the sensitive elements of adjacent pixels to avoid loss of image resolution due to thermal spreading. Assuming the array is not background limited, the principal heat loss mechanism should be by conduction down the support structure to the substrate.

The support structure is the key to high-performance thermal arrays. It must provide three functions: mechanical support, a thermally conducting path, and an electrically conducting path (assuming that the read-out is electrical rather than, say, optical). In practice, there are two types of support structure. The older type is a bump-bond such as has been employed in the Texas Instruments (Dallas, Texas) hybrid pyroelectric/ferroelectric bolometer array, (Fig. 2). Here the detecting layer is a relatively thick (approximately $25 \mu m$) layer of barium strontium titanate, which is prepared separately in the form of a wafer several inches in diameter. The silicon (Si) substrate incorporates multiplexer electronics at each pixel. The two are joined by a bump-bond at each pixel which provides mechanical support, high electrical conduction, and low thermal conduction.

The other principal type of support structure is the membrane type such as has been employed in the Honeywell monolithic bolometer array (Honeywell Inc. Plymouth, Minnesota) (Fig. 3). Here the detecting layer is

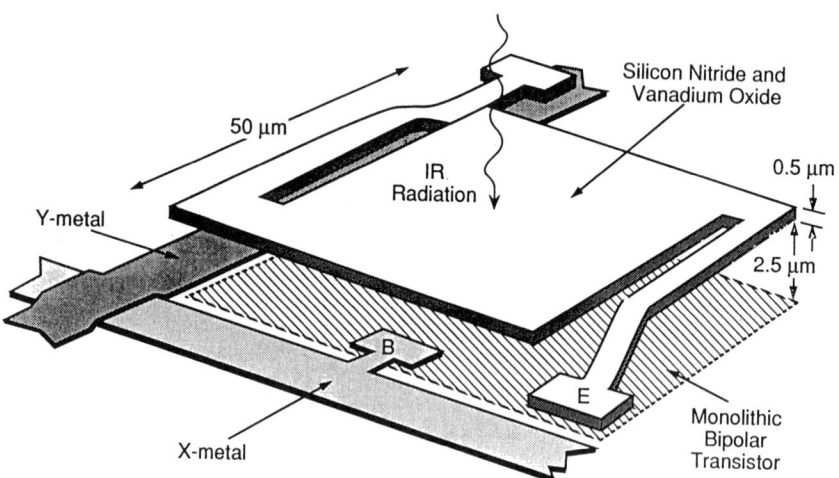

FIG. 3. Microbolometer pixel structure. IR, infrared. (From R. A. Wood, Honeywell Technology Center, Plymouth Minnesota, personal communication. Reprinted with permission.)

a thin film (less than 1 μm) deposited upon a silicon nitride ($Si_3N_4$) membrane. The membrane is raised above the Si substrate, supported by two legs at diagonally opposite corners. Electronics at each pixel are embedded in the substrate. This structure is built up by depositing thin films of selected materials including metals, silicon dioxide ($SiO_2$), $Si_3N_4$, and vanadium oxide ($VO_x$) (the thermally sensitive resistive material) on Si wafers that incorporate a transistor switch at each pixel. The Si wafers are then processed in a manner similar to that employed in preparing integrated circuits. Certain unique steps are employed, including one that removes the $SiO_2$ (known as the sacrificial layer) between the substrate and the $Si_3N_4$ membrane. In this structure, care must be taken to prevent the sensitive layer from tilting and touching the substrate; the process to prevent this is known as *stress balancing*. Furthermore, the structure must be designed to be rugged and free from mechanical vibrational modes that can be excited inadvertently.

A simplified analysis of the temperature increase at the sensitive area of a pixel when IR radiation falls on the pixel does not require knowledge of the detection mechanism (Kruse et al., 1962). Let the sensitive area of a pixel have a heat capacity of $C$. Let the thermal conductance of the principal heat loss mechanism, which usually is the thermal conductance of the support structure, be $G$. Let temporally modulated IR radiation of power amplitude $P_0$ fall on the pixel. Let the fraction of the incident radiation that is absorbed be $\eta$. Let the angular frequency of modulation of the radiation be $\omega$. Let the temperature increase of the sensitive area of the pixel be $\Delta T$. Then the heat flow equation describing the pixel is

$$C\frac{d(\Delta T)}{dt} + G(\Delta T) = \eta P = \eta P_0 \exp(j\omega t) \tag{1}$$

where $j = \sqrt{-1}$ and $t$ is time. This simplified equation assumes that the power dissipation in the sensitive area due to applied electrical bias (in either the resistive bolometer or ferroelectric bolometer mode of operation) can be neglected. The exact solution for a resistive bolometer is given in Kruse, et al., 1962. The solution of Eq. (1) is

$$\Delta T = \frac{\eta P_0 \exp(j\omega t)}{G + j\omega C} = \frac{\eta P_0}{G(1 + \omega^2 \tau^2)^{1/2}} \tag{2}$$

where $\tau$ is the thermal response time, defined as

$$\tau = \frac{C}{G} \tag{3}$$

Equation (2) is basic to thermal IR arrays. It describes the temperature increase of the sensitive area of the pixel when radiation of power amplitude $P_0$ sinusoidally modulated with angular frequency $\omega$ falls on the sensitive area. The pixel temperature increases and decreases as the input radiant power rises and falls in an oscillatory manner. At low frequencies for which $\omega t \ll 1$ the temperature increase is

$$\Delta T = \frac{\eta P_0}{G} \qquad (\omega \tau \ll 1) \tag{4}$$

At high frequencies for which $\omega \tau \gg 1$, the temperature increase falls off inversely with frequency and is described by

$$\Delta T = \frac{\eta P_0}{\omega C} \qquad (\omega \tau \gg 1) \tag{5}$$

The transition between the low- and high-frequency regions is characterized by the thermal time constant $\tau$.

This equation is generally applicable to all thermal detectors. To determine the responsivity of, say, a bolometer, we express the signal amplitude in terms of a temperature coefficient of the signal parameter; for a bolometer this is the temperature coefficient of resistance. We then determine the signal amplitude by multiplying the temperature coefficient by $\Delta T$ given in Eq. (2). The responsivity is determined by dividing both sides by the power $P_0$.

Of utmost importance in the design of thermal detectors is the need to minimize the thermal conductance $G$. Thermal isolation is the key to high-performance IR arrays. Of paramount importance is the design of the support structure, which determines $G$. Of secondary importance is the selection of the detection mechanism and material. It will be seen later that the fundamental limits to the array performance—temperature fluctuation noise and background fluctuation noise—also demand high thermal isolation.

How then is a thermal detector array properly designed? First, a support structure is designed having excellent thermal isolation, that is, a very small value of $G$ expressed in Watts/°C (W/°C). Second, a detection mechanism and material are selected that, in combination with the support structure, provide a high responsivity. Third, the heat capacity $C$ of the sensitive element, expressed in Joules/°C (J/°C), must be made low enough to meet the response time requirement $\tau$, expressed in seconds. For example, assume a two-dimensional (staring) array operating at 30 frames/sec (30 Hz). Then the pixel response time must be somewhat shorter than 1/30 sec, since in 1/30 sec the signal amplitude will be only 71% of the maximum. Common practice is to make the pixel response time one-third of the reciprocal of the frame time, say, 10 msec. Assume a structure with excellent thermal isolation, say, $G = 1 \times 10^{-7}$ W/°C, is to be designed. Then $C$ must be $1 \times 10^{-9}$ J/°C to meet the response time requirement. Because the pixel area is defined by the system requirement, the only adjustable parameter is the thickness. Assuming, say, 50 $\mu$m square pixels, to attain a value of $C = 1 \times 10^{-9}$ J/°C will require a thickness of the sensitive layer of no greater than 1 $\mu$m for most materials. Thus a monolithic thin-film approach is seen to be most desirable.

The monolithic approach such as that exemplified by the Honeywell bolometer array design has other advantages. The response time can be adjusted by selecting appropriate values of the dimensions of the "legs" of the support structure down which heat flows to the substrate. For example, increasing the leg length or reducing its thickness and width reduces the value of $G$, thereby increasing $\tau$ if $C$ remains the same.

In comparing the hybrid with the monolithic approach, there are many practical issues. The hybrid approach allows preparation of the detecting material before bump-bonding it to the substrate. This eases process compatibility issues that must be addressed in the monolithic approach. Thermal spreading of the image ("optical cross talk") is a problem to be solved in the hybrid approach, usually by laser scribing to isolate each pixel from its neighbors. Thermal spreading is not an issue in the monolithic approach; the substrate acts as an infinite heat sink that provides ideal thermal isolation between pixels. The monolithic approach employing Si process technology is in principle a lower cost approach than the hybrid, since it relies to a large part on technology developed for the integrated circuit industry.

## II. Principal Thermal Detection Mechanisms

### 1. Resistive Bolometers

The term *bolometer* is usually reserved for resistive devices for which the temperature increase accompanying the absorption of IR radiation causes a change in resistance. To distinguish these devices from field-enhanced pyroelectric devices, which are also called *ferroelectric bolometers*, the term *resistive bolometers* is used herein (Kruse, 1977).

The analysis of resistive bolometers assumes that the temperature increase $\Delta T$ of the bolometer due to the absorption of IR radiation is small enough so that the resistance change $\Delta R$ is linear with $\Delta T$, that is,

$$\Delta R = R\Delta T \tag{6}$$

so that it is possible to express the change in resistance in terms of $\alpha$, the temperature coefficient of resistance. Thus

$$\Delta R = \alpha R \Delta T \tag{7}$$

where

$$\alpha = \frac{1}{R}\frac{dR}{dT} \tag{8}$$

The temperature coefficient of resistance can be either positive or negative. For metals at room temperature it is positive, that is, the resistance increases with increasing temperature. For semiconductors at room temperature it is usually negative. For superconductors at their transition edge (*superconducting bolometers*)—which even for high-temperature superconductors is well below 300 K—it is positive. Typical values are

Metals  $\alpha = 0.002 \ (°C)^{-1}$
Semiconductors  $\alpha = -0.02 \ (°C)^{-1}$
Superconductors  $\alpha = 2.0 \ (°C)^{-1}$

The responsivity $\Re$ of an IR pixel is defined as the output signal (voltage or current) divided by the input radiant power falling on the pixel. Assume that the output signal is the voltage $V_S$. Then

$$V_S = i_b \Delta R = i_b \alpha R \Delta T \tag{9}$$

where $i_b$ is the bias current through the pixel. The temperature increase $\Delta T$ is given by Eq. (2). Thus the signal voltage is

$$V_S = \frac{i_b \alpha R \eta P_0}{G(1 + \omega^2 \tau^2)^{1/2}} \tag{10}$$

so that the responsivity is given by

$$\Re = \frac{i_b \alpha R \eta}{G(1 + \omega^2 \tau^2)^{1/2}} \tag{11}$$

Equation (11) shows that the responsivity is directly proportional to the temperature coefficient of resistance and inversely proportional to the thermal conductance associated with the principal heat loss mechanism. Both parameters are important; however, for uncooled IR resistive bolometers, values of $G$ can range over several orders of magnitude, whereas the range of possible values of $\alpha$ is far less. As stated previously, the primary focus should be on the thermal isolation structure. The choice of resistive material, which frequently has been considered the most important decision, is really of secondary importance. Given an excellent thermal isolation structure, a resistive material is chosen that has the highest possible value of $\alpha$ consistent with the requirements for depositing the material on the structure and processing the arrays.

The previous derivation assumes that the heating effect of the bias current can be ignored. In reality it can be of major importance in resistive bolometers incorporating semiconductor layers exhibiting a nonlinear and negative dependence of resistance on temperature at room temperature. See Kruse, et al. (1962) for an analysis of this issue.

Now consider the noise equivalent temperature difference (NETD). The NETD is defined as the change in temperature of a large blackbody within the field of view of an imaging system incorporating the focal plane array that will cause the signal-to-noise ratio at the output of the array and its read-out electronics to change by unity. It is given by

$$\text{NETD} = \frac{4F^2 V_N}{\tau_o A \Re (\Delta P/\Delta T)_{\lambda_1 - \lambda_2}} \qquad (12)$$

where $F$ is the focal ratio of the optics, $V_N$ is the total electrical noise within the system bandwidth, $\tau_o$ is the transmittance of the optics, $A$ is the area of a pixel, $\Re$ is the responsivity of a pixel and $(\Delta P/\Delta T)_{\lambda_1 - \lambda_2}$ is the change in power per unit area radiated by a blackbody at temperature $T$ with respect to $T$, measured within the spectral band from $\lambda_1$ to $\lambda_2$ (Lloyd, 1975). The values of $(\Delta P/\Delta T)_{\lambda_1 - \lambda_2}$ for a 295 K blackbody in the 3 to 5 $\mu$m and 8 to 14 $\mu$m spectral intervals (Lloyd, 1975) are as follows:

$$(\Delta P/\Delta T)_{3-5} = 2.10 \times 10^{-5} \text{ W/cm}^2 \text{ K}$$
$$(\Delta P/\Delta T)_{8-14} = 2.62 \times 10^{-4} \text{ W/cm}^2 \text{ K}$$

The literature includes expressions for the NETD that have $(4F^2 + 1)$ in the numerator and others that have $4F^2$. These expressions are equivalent if, in that containing $(4F^2 + 1)$, $F$ is defined as $f/D$, where $f$ is the focal length of the lens and $D$ is the diameter of the lens, and in that containing $4F^2$, $F$ is defined as the focal ratio. Here the focal ratio equals the reciprocal of $2 \sin U$, where $U$ is the slope angle of the marginal ray exiting the lens. The expression that includes $(4F^2 + 1)$, where $F$ is defined as $f/D$, is valid only when the scene being imaged is effectively at infinity, so that the image is at the focal length of the lens. The expression that includes $4F^2$ is valid for all object distances; however, the slope angle of the marginal ray is with respect to the focal point of the image, which is at the focal length of the lens only for scenes effectively at infinity.

The expression for the NETD given by Eq. (12) is applicable to any detection mechanism. Thus the NETD of a resistive bolometer can be obtained by substituting the expression for the responsivity, Eq. (11), into Eq. (12), then inserting appropriate values of the parameters.

## 2. Pyroelectric Detectors and Ferroelectric Bolometers

The pyroelectric effect (Putley, 1970) is exhibited by certain materials, among which are some ferroelectric crystals that exhibit spontaneous electric polarization. That is, opposite faces of certain crystallographic

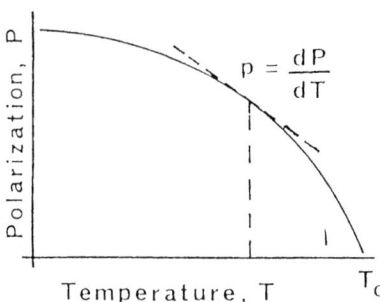

FIG. 4. Pyroelectric effect.

orientations exhibit opposite electrical charges that, at a constant temperature, are neutralized by free internal charges. The polarization, defined as the electric dipole moment per unit volume, exhibits a temperature dependence, as illustrated in Fig. 4, becoming zero above a characteristic temperature known as the Curie temperature, $T_c$. Below the Curie temperature, a change in temperature $\Delta T$ products a transient change in the surface charge, thereby causing a transient current $I_S$ to flow in an external circuit connected to the pyroelectric material. The magnitude of $I_S$ is given by

$$I_S = pA \frac{d(\Delta T)}{dt} \tag{13}$$

where $A$ is the sample area (the pixel area) and $p$, the pyroelectric coefficient, is the slope of the polarization versus temperature curve at the operating temperature (Fig. 4).

The responsivity of a pyroelectric detector is found by taking the time derivative of $\Delta T$ found in Eq. (2), then substituting it into Eq. (13) to find $I_S$:

$$I_S = \frac{\eta p \omega A P_0}{G(1 + \omega^2 \tau^2)^{1/2}} \tag{14}$$

where

$$\tau = C/G \tag{15}$$

Because pyroelectric detectors are capacitive, their properties include their capacitance $C_e$ and their loss resistance $R$. The loss resistance can be expressed in terms of a loss tangent, tan δ. The pyroelectric signal voltage $V_S$ is

$$V_S = \frac{I_S R}{(1 + \omega^2 R^2 C_e^2)^{1/2}} \tag{16}$$

The responsivity $\mathfrak{R}$ of a pyroelectric detector is

$$\mathfrak{R} = \frac{\eta \omega p A R}{G(1 + \omega^2 \tau_e^2)^{1/2}(1 + \omega^2 \tau^2)^{1/2}} \tag{17}$$

where

$$\tau_e = RC_e = (\omega \tan \delta)^{-1} \tag{18}$$

Thus pyroelectric detectors have both an electrical response time and a thermal response time. The electrical response time at low modulation frequencies for materials with a small, that is, desirable, loss tangent is usually longer than is the thermal response time. For example, at 30 Hz a material with a loss tangent of 0.01 has an electrical time constant of 0.53 sec, compared with a typical thermal response time of 10 msec.

Equation (17) shows that pyroelectric detectors have no dc response; they must employ radiation modulators (called *choppers*) or must be continuously panned across the scene. Figure 5 illustrates the dependence of responsivity on modulation frequency.

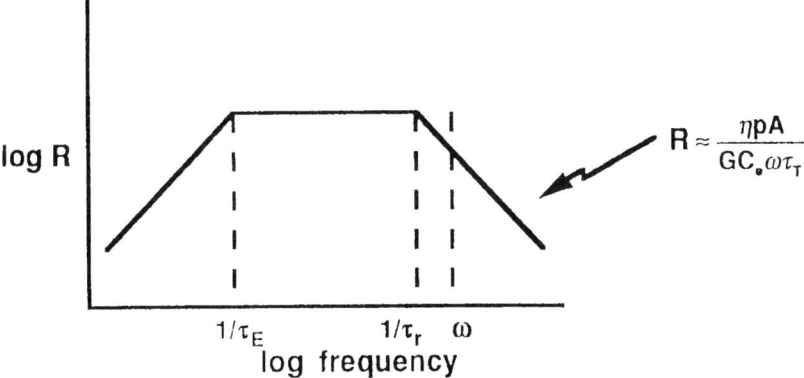

FIG. 5. Pyroelectric detector responsivity as a function of frequency.

An appropriate figure of merit (FOM) for comparing pyroelectric materials when the fundamental noise is Johnson noise in the loss resistance is

$$\text{FOM} = \frac{p}{c(\kappa \tan \delta)^{1/2}} \quad (19)$$

where $c$ is the volume specific heat of the pyroelectric material and $\kappa$ is the relative permittivity.

The NETD of a pyroelectric pixel or array is obtained by substituting Eq. (17) into Eq. (12) and inserting appropriate values of the parameters.

A variant of the pyroelectric effect is the field-enhanced pyroelectric effect, also known as the ferroelectric bolometer effect. The pyroelectric effect is exhibited in the absence of an applied voltage. In the presence of an electric field due to an applied voltage, the pyroelectric material exhibits a polarization that extends past the normal Curie temperature. By applying an electric field, it becomes possible to obtain an additional component of the

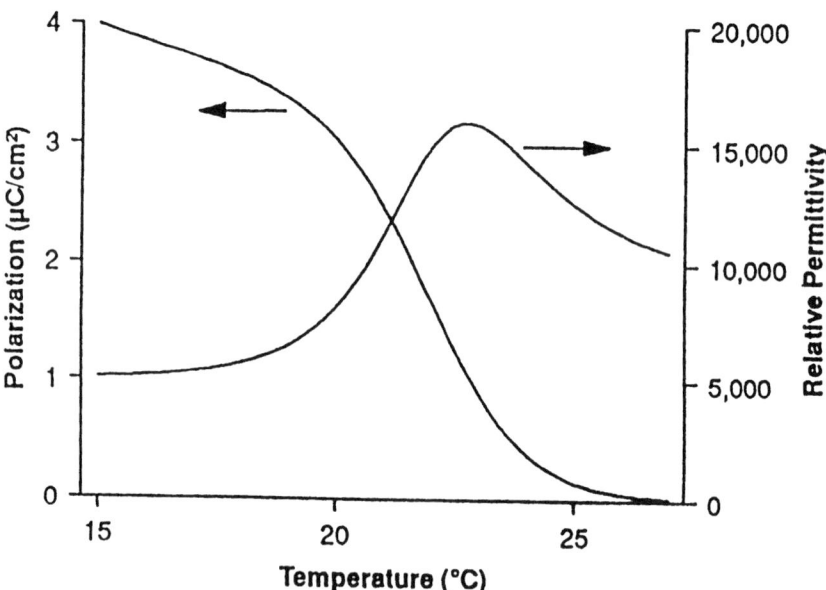

FIG. 6. Temperature dependence of spontaneous polarization and dielectric permittivity for ferroelectric ceramic barium strontium titanate (BST). (Reprinted from Hanson et al. (1992). Uncooled thermal imaging at Texas Instruments. *Proc. SPIE 1735, Infrared Detectors: State-of-the-Art* **17**; with permission from the SPIE.)

signal arising from the temperature dependence of the dielectric permittivity. Figure 6 illustrates the temperature dependence of the polarization and dielectric permittivity of barium strontium titanate, the pyroelectric material employed by Texas Instruments in their field-enhanced pyroelectric detectors. The pyroelectric coefficient under electrical bias is given by

$$p = p_0 + \int_0^E \frac{\partial \varepsilon}{\partial T} \, dE' \qquad (20)$$

where $p_0$ is the pyroelectric coefficient in the absence of electric bias, $\varepsilon$ is the dielectric permittivity, and $E$ is the applied electric field (Hanson, 1992). The second term on the right in Eq. (20), which has no simple analytic solution, represents the field-enhanced component.

Note that for all detection mechanisms, the responsivity of a pixel is defined as the signal voltage divided by the radiant power incident on the pixel. Because pyroelectric detectors and ferroelectric bolometers require modulated radiation, they suffer a $\sqrt{2}$ disadvantage with respect to detectors such as resistive bolometers and radiation thermocouples, which do not.

3. THERMOELECTRIC DETECTORS

The thermoelectric effect (Stevens, 1970) takes place in a circuit consisting of two different electrically conducting materials that are joined at two points (Fig. 7). If the junctions are at different temperatures, a voltage—termed the *thermoelectric voltage*—will be detected when the circuit is opened, as shown in the figure. The magnitude of the voltage depends on the type of materials and the temperature difference between the junctions. The voltage can be increased by connecting in electrical series pairs of hot and cold junctions, an arrangement known as a *thermopile*.

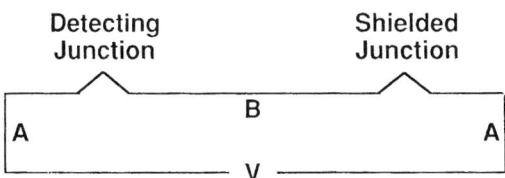

FIG. 7. Radiation thermocouple. A, material A; B, material B; V, thermoelectric voltage.

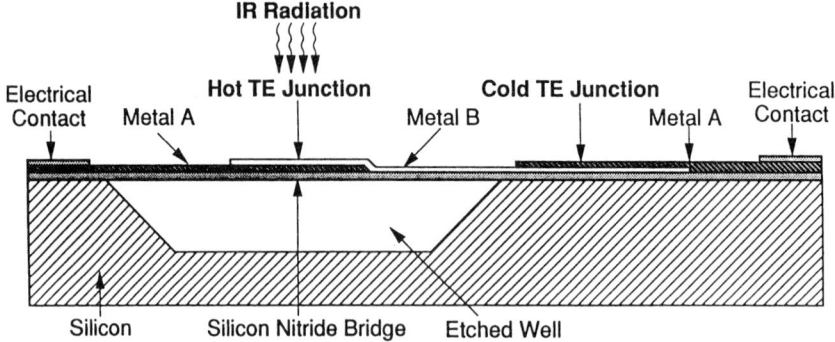

FIG. 8. Structure of Honeywell thermoelectric (TE) detector. (Reprinted from Wood et al. (1995). IR SnapShot™ camera. *Proc. SPIE 2552, Infrared Technology* **XXI**, 654; with permission from the SPIE.)

To employ the thermoelectric effect as an IR detector, thermocouples or thermopiles are prepared in the form of thin films deposited on thermally isolated substrates (Fig. 8). The thermoelectric signal voltage $V_S$ is given by

$$V_S = N(S_1 - S_2)\Delta T \tag{21}$$

where $S_1$ and $S_2$ are the thermoelectric coefficients (Seebeck coefficients) and their difference is known as the thermoelectric power (*thermopower*) of a junction. The hot junctions are deposited on the thermally isolated part of the pixels, whereas the cold junctions are deposited on the surrounding substrate; that is, the cold junctions are heat-sunk to the substrate. From Eqs. (2) and (21), the responsivity $\Re$ of a thermoelectric pixel is

$$\Re = \frac{\eta N(S_1 - S_2)}{G(1 + \omega^2 \tau^2)^{1/2}} \tag{22}$$

where $\eta$ is the optical absorptance and

$$\tau = C/G \tag{23}$$

The NETD is obtained by combining Eqs. (12) and (22), then inserting appropriate values of the parameters.

When compared with resistive bolometers and pyroelectric detectors or ferroelectric bolometers, thermoelectric detectors have low responsivity. On

TABLE I

COMPARISON OF THE PRINCIPAL TYPES OF UNCOOLED THERMAL ARRAYS

|  | Resistive Bolometer | Pyroelectric Detector | Ferroelectric Bolometer | Thermoelectric Detector |
|---|---|---|---|---|
| Responsivity | High | High | High | Low |
| Bias required | Yes | No | Yes | No |
| DC response | Yes | No | No | Yes |
| Chopper required | No | Yes | Yes | No |
| Reponse time | $C/G$ | $C/G, rC_e$ | $C/G, rC_e$ | $C/G$ |

$C/G$, pixel heat capacity/thermal conductance; $rC_e$, pixel electrical loss resistance multiplied by electrical capacitance.

the other hand, they operate unbiased, show no excess (current induced) noise and are relatively simple to prepare.

Table I compares the relative advantages and disadvantages of these principal detection mechanisms.

### III. Fundamental Limits

1. TEMPERATURE FLUCTUATION NOISE LIMIT

Any thermodynamic system, including thermal IR detectors in contact with its environment, exhibits random fluctuations in temperature, known as temperature fluctuation noise, arising from the statistical nature of the heat interchange with its environment (Kruse, 1995). The mean square temperature fluctuations $\overline{\Delta T^2}$ of the system (detector) over all frequencies are given by

$$\overline{\Delta T^2} = \frac{kT^2}{C} \tag{24}$$

where $k$ is Boltzmann's constant; $T$ is the system temperature, assumed to be the same as the environment temperature; and $C$ is the harmonic mean of the heat capacities of the system and its environment. When the system is a thermal IR detector, the harmonic mean of the heat capacities reduces to the heat capacity of the detector, or the pixel of an array.

To determine the frequency dependence of the temperature fluctuations, a heat flow equation identical to Eq. (2) is employed, except that $P$ now represents the heat power exchanged with the pixel environment and $\Delta T$ is the square root of $\overline{\Delta T^2}$ of Eq. (24). Assuming that fluctuations in the power interchange with the environment are independent of frequency, described by

$$\overline{P_f^2} = P_0 B \qquad (25)$$

where $\overline{P_f^2}$ is the mean square fluctuations in power, $P_0$ is a constant independent of frequency, and $B$ is the measurement bandwidth, then

$$\overline{\Delta T_f^2} = \frac{\overline{P_f^2}}{G^2(1 + \omega^2\tau^2)} = \frac{P_0 B}{G^2(1 + \omega^2\tau^2)} \qquad (26)$$

where $\overline{\Delta T_f^2}$ is the frequency dependence of the mean square temperature fluctuations and $\tau$ is the thermal time constant

$$\tau = \frac{C}{G} \qquad (27)$$

By integrating over all frequencies the value of $P_0$ is found to be

$$P_0 = 4GC\overline{\Delta T^2} \qquad (28)$$

By equating the expressions for $\overline{\Delta T^2}$ found in Eqs. (24) and (28), thereby solving for $P_0$ and substituting it into Eq. (26), the expression for the frequency dependence of the mean square temperature fluctuations is found to be

$$\overline{\Delta T_f^2} = \frac{4GkT^2B}{G^2(1 + \omega^2\tau^2)} \qquad (29)$$

The mean square fluctuation in heat power exchanged between the pixel and its surroundings, determined from Eqs. (26) and (29), is

$$\overline{P_f^2} = 4kT^2GB \qquad (30)$$

Now consider the temperature fluctuation noise-limited $D^*$. The defining equation for $D^*$ is

$$D^* = \frac{(AB)^{1/2}}{P_N} \qquad (31)$$

where $P_N$ is the noise equivalent power. In this example,

$$P_N = \left(\frac{\overline{P_f^2}}{\eta^2}\right)^{1/2} \tag{32}$$

so that the temperature fluctuation noise-limited $D^*$, that is, $D_{TF}^*$, is

$$D_{TF}^* = \left(\frac{\eta^2 A}{4kT^2G}\right)^{1/2} \tag{33}$$

It is now possible to calculate the temperature fluctuation noise-limited NETD. The relationship between $D^*$ and the responsivity $\Re$ is defined as

$$D^* = \frac{(AB)^{1/2}\Re}{V_N} \tag{34}$$

Combining Eqs. (12), (33), and (34), the temperature fluctuation noise-limited NETD, that is, NETD$_{TF}$, is found to be

$$\text{NETD}_{TF} = \frac{8F^2 T(kBG)^{1/2}}{\eta \tau_0 A(\Delta P/\Delta T)_{\lambda_1 - \lambda_2}} \tag{35}$$

Note that NETD$_{TF}$ is proportional to $G^{1/2}$. This is of paramount importance in the design of a thermal IR array. Depending on the design, the value of $G$ can vary by orders of magnitude. Arrays with poorly designed thermal isolation structures can be severely limited in their attainable NETD by temperature fluctuation noise.

2. BACKGROUND FLUCTUATION NOISE LIMIT

Background fluctuation noise is the manifestation of temperature fluctuation noise when radiative exchange is the dominant mode of heat exchange between the thermal detector (pixel) and its surroundings. If the detector and its surroundings are at the same temperature, then the background fluctuation noise-limited $D^*$ can be determined directly from Eq. (33). Remembering that the thermal conductance $G$ is the proportionality constant between the rate of heat flow and the temperature increment, the expression for $G$ when radiation exchange dominates, $G_{\text{rad}}$, is the derivative with respect to the temperature of the Stefan-Boltzmann expression

$$G_{\text{rad}} = 4\eta A\sigma T^3 \tag{36}$$

where $\sigma$ is the Stefan-Boltzmann constant. Substituting this value into Eq. (33) results in the expression for $D^*_{\text{BF}}$, the background fluctuation noise-limited $D^*$:

$$D^*_{\text{BF}} = \left(\frac{\eta}{16k\sigma T^5}\right)^{1/2} \tag{37}$$

Equation (37) is the background fluctuation noise-limited $D^*$ when the detector and the background are at the same temperature and make equal contributions. When they are at different temperatures, the background fluctuation noise-limited $D^*$ is given by

$$D^*_{\text{BF}} = \left[\frac{\eta}{8k\sigma(T_D^5 + T_B^5)}\right]^{1/2} \tag{38}$$

where $T_D$ is the detector temperature and $T_B$ is the background temperature. Figure 9 illustrates $D^*_{\text{BF}}$ from Eq. (38), assuming $\eta$ is unity.

The NETD for background-fluctuation noise-limited detectors, that is $\text{NETD}_{\text{BF}}$, is obtained by combining Eqs. (12), (34), and (38):

$$\text{NETD}_{\text{BF}} = \frac{8F^2[2k\sigma B(T_D^5 + T_B^5)]^{1/2}}{\tau_0(\eta A)^{1/2}(\Delta P/\Delta T)_{\lambda_1-\lambda_2}} \tag{39}$$

Figure 10 illustrates the temperature fluctuation noise and background fluctuation noise limits for a background temperature of 300 K and detector temperatures of 85 K and 300 K. Values of other parameters that are assumed are listed in Fig. 10. With increasing thermal isolation, $\text{NETD}_{\text{TF}}$ decreases as $G^{1/2}$, until $\text{NETD}_{\text{BF}}$ is reached.

Up to this point the discussion has assumed that the optical absorptance $\eta$ is wavelength-independent. If $\eta = 0$ for all wavelengths except between $\lambda_1 = c/v_1$ and $\lambda_2 = c/v_2$, where $c$ is the speed of light and $v_1$ and $v_2$ are optical frequencies, then Eq. (38) is replaced by (Low and Hoffman, 1963)

$$D^*_{\text{BF}} = \left[\frac{\eta}{8k\sigma T_D^5 + F(v_1, v_2)}\right]^{1/2} \tag{40}$$

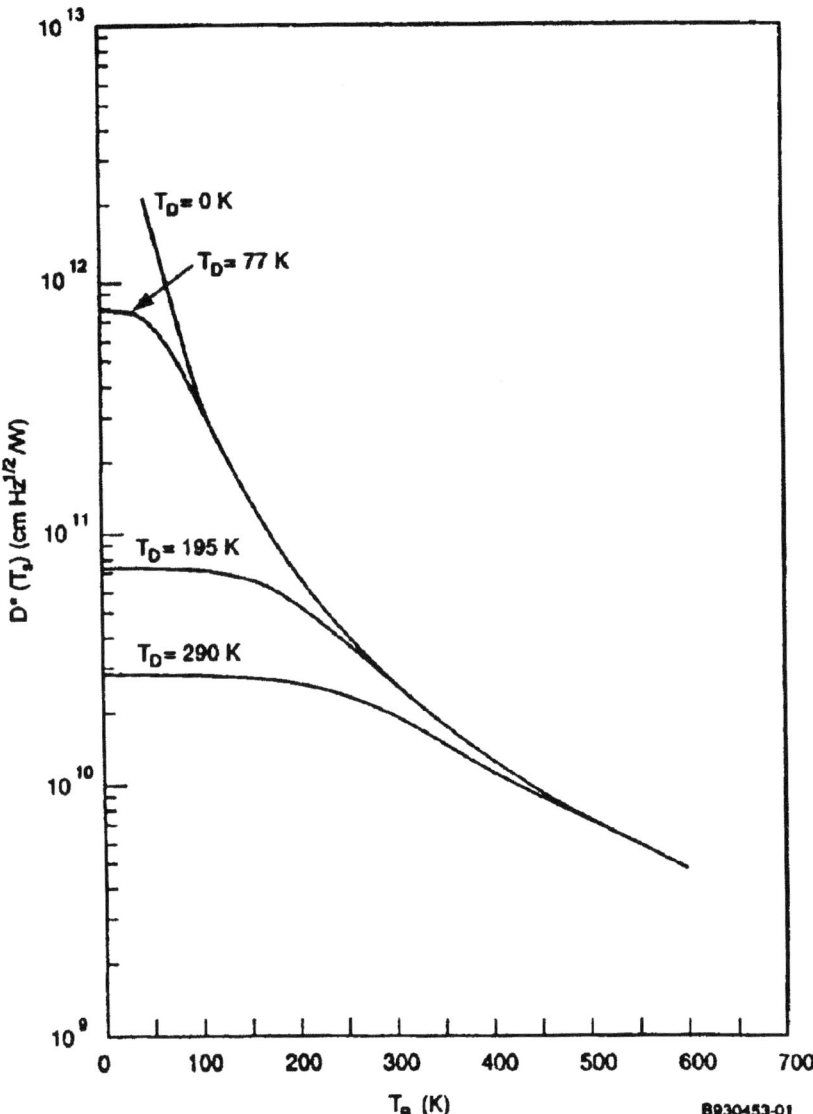

FIG. 9. Background fluctuation noise limit of thermal detectors as a function of detector temperature $T_D$ and background temperature $T_B$ for viewing angle of $2\pi$ steradians and unity absorptance. (Reprinted from Kruse et al. 1962). *Elements of Infrared Technology*. John Wiley & Sons, New York, Chapt. 9; with permission from Wiley.)

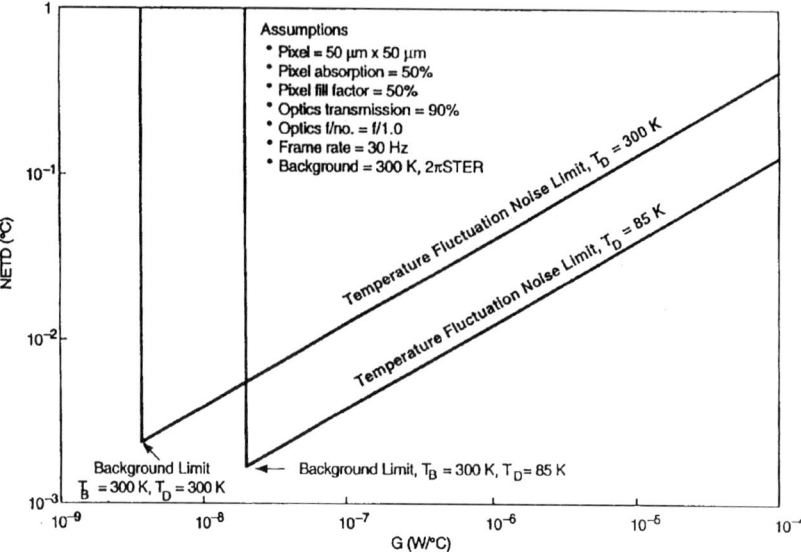

FIG. 10. Temperature fluctuation noise limit and background fluctuation noise limit for the assumed values of the parameters. $T_B$, background temperature; $T_D$, detector temperature; NETD, noise equivalent temperature dependence. (Reprinted from Kruse (1995). A comparison of the limits to the performance of thermal and photon detector imaging arrays. *Infra. Phys. Tech.* 36: 869; with the kind permission of Elsevier Science–NL.)

where

$$F(v_1, v_2) = 2 \int_{v_2}^{v_1} hv M(v, T_B) \frac{\exp(hv/kT_B)\,dv}{[\exp(hv/kT_B) - 1]} \qquad (41)$$

where $h$ is Planck's constant

$$M(v, T_B) = \frac{2\pi hv^3/c^2}{[\exp(hv/kT_B) - 1]} \qquad (42)$$

Figure 11 illustrates Eq. (40) for the case of a long wavelength cutoff $\lambda_0$ in which $\eta = 1$ for $\lambda < \lambda_0$ and $\eta = 0$ for $\lambda > \lambda_0$, and a short wavelength cutoff $\lambda_0$ in which $\eta = 0$ for $\lambda < \lambda_0$ and $\eta = 1$ for $\lambda > \lambda_0$, assuming $T_B = 300$ K.

Thermal detectors also benefit from cold shielding, just as photon detectors do, although it may not make sense to cold shield an uncooled

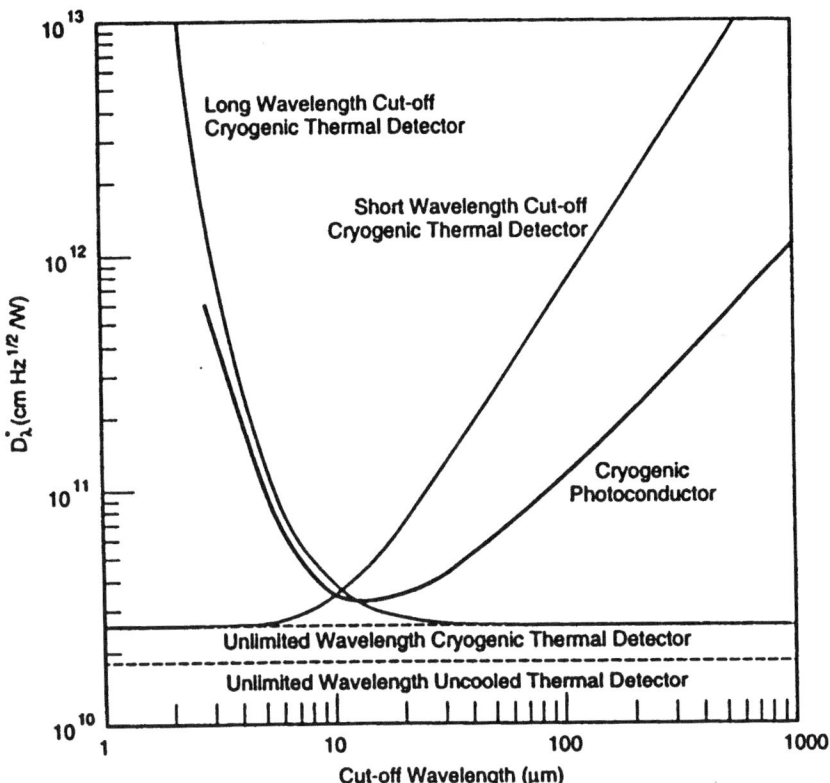

FIG. 11. Dependence $D_\lambda^*$ and $D^*$ ($T_S$) on long wavelength limit for thermal and photon detectors. (Reprinted from Low and Hoffman (1963). *Appl. Optics.* **2**: 649; with permission from the Optical Society of America.)

detector. Figure 12 illustrates the relative increase in the background fluctuation noise-limited performance accompanying cold shielding.

## IV. Discussion

The best way to design a high-performance uncooled thermal-imaging focal plane array based on a thermal detection mechanism is to begin with the design of the structure, for it is the structure that establishes the

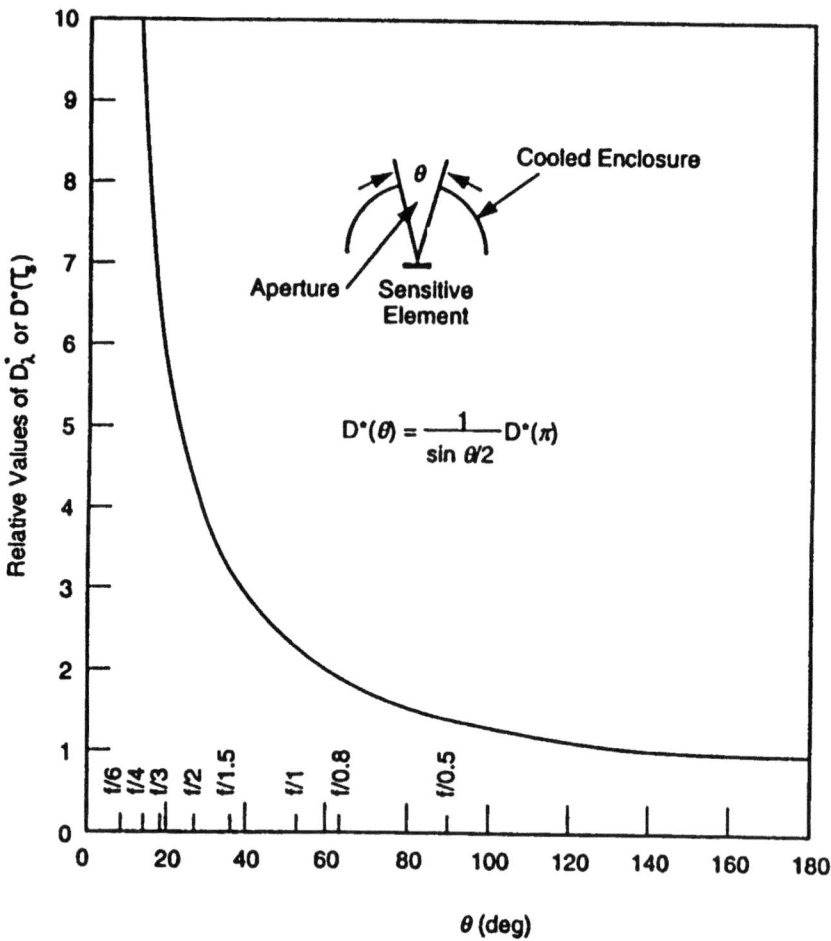

FIG. 12. Relative increase in background fluctuation noise-limited $D_\lambda^*$ and $D^*(T_S)$ for photon and thermal detectors achieved by using cold aperture. (Reprinted from Kruse et al. (1962). *Elements of Infrared Technology.* John Wiley & Sons, New York, Chapt. 9; with permission from Wiley.)

temperature fluctuation noise limit to the array performance. Furthermore, much of the array processing cost is structure-dependent. A monolithic thin-film approach based on Si technology appears to be the best.

Having selected the structure, it is then necessary to select a mode of operation and a material that will best allow operation at the temperature fluctuation noise limit imposed by the structure, and that also will allow meeting the response time $\tau$ requirement. Given the thermal conductance $G$ value imposed by the structure, the pixel heat capacity $C$ must be no greater than the $G\tau$ product. This is a function of the thickness and specific heat of the detecting layer, including the support membrane, electrical contacts, detecting material, absorbing coating (if necessary), and protective coating (if necessary).

To attain the temperature fluctuation noise limit imposed by the structure requires that temperature fluctuation noise be greater than Johnson noise and $1/f$ power law noise (if the detection mode requires electric bias). Minimizing $1/f$ power law noise is an art that is not amenable to exact analysis and will not be discussed here.

Thus, assuming no $1/f$ power law noise, to attain the temperature fluctuation noise limit imposed by the structure requires that temperature fluctuation noise exceed Johnson noise from the detecting layer. This, in turn, sets a lower limit to the pixel responsivity. Equation (24) shows that the mean square temperature fluctuation noise $\overline{\Delta T^2}$ is

$$\overline{\Delta T^2} = \frac{kT^2}{C} \tag{24}$$

The response mechanism acts on the root mean square temperature fluctuation noise, just as it does on the temperature increase $\Delta T$ associated with the absorption of radiant power. For example, if the response mechanism is a resistive bolometer, then the temperature fluctuation noise voltage $V_{TF}$ is given by

$$\begin{aligned} V_{TF} &= i_b \alpha R \left( \overline{\Delta T^2} \right)^{1/2} \\ &= i_b \alpha R \left( \frac{kT^2}{C} \right)^{1/2} \end{aligned} \tag{43}$$

This value needs to exceed the Johnson noise voltage $V_J$ associated with the system bandwidth:

$$V_J = (4kTRB)^{1/2} \tag{44}$$

Thus a large bias current (limited by joulean heating and the onset of $1/f$ power law noise), large temperature coefficient of resistance, large resistance, and small bandwidth are desirable in order to attain the temperature fluctuation noise limit of a resistive bolometer.

A similar line of reasoning holds for pyroelectric detectors. Analysis of pyroelectric detectors and ferroelectric bolometers is more difficult, in that Johnson noise follows a more complex behavior than that described by Eq. (44). Thermoelectric detectors, which have a low responsivity, are less likely than are resistive bolometers, pyroelectric detectors, and ferroelectric bolometers to be temperature fluctuation noise-limited.

The approach described in this chapter is a simplified synopsis of an approach to designing uncooled IR arrays based on thermal detection mechanisms. The linear process described is in reality an iterative one, with choices made, evaluated, modified, re-evaluated, and so on. However, the overall approach presented in this chapter is believed by the author to be best.

## REFERENCES

Hanson, C. M. (1992). *Proc. SPIE* **1735** Infrared Detectors: State-of-the-Art, 17.
Hanson, C. et al. (1992). *Proc. SPIE* **1735** (Infrared Detectors: State-of-the-Art), 17.
Kruse, P. W. (1977). *In* "Optical and Infrared Detectors", (R. D. Keyes, ed.). Springer-Verlag, New York.
Kruse, P. W. (1995). *Infrared Phys. Technol.* **36**, 869.
Kruse, P. W., McGlauchlin, L. D., and McQuistan, R. B. (1962). "Elements of Infrared Technology," Chapter 9. Wiley, New York.
Lloyd, J. M. (1975). "Thermal Imaging Systems." Plenum, New York.
Low, F. J. and Hoffman, A. R. (1963). *Appl. Opt.* **2**, 649.
Putley, E. H. (1970). *In* "Semiconductors and Semimetals" (R. K. Willardson and A. C. Beer, eds.), Vol. 5. Academic Press, New York.
Stevens, N. B. (1970). *In* "Semiconductors and Semimetals" (R. K. Willardson and A. C. Beer, eds.), Vol. 5. Academic Press, New York.
Wood, R. A. (1993). *Proc. SPIE* **2020** (Infrared Technol. XIX), 329.
Wood, R. A. et al. (1995). *Proc. SPIE* **2552** (Infrared Technol. XXI), 654.

## BIBLIOGRAPHY

### Overview

Kruse, P. W. (1994). Uncooled infrared focal plane arrays. *In* "Proceedings of the IEEE Ninth International Symposium on the Applications of Ferroelectrics". Pennsylvania State University, State Park.
Kruse, P. W. (1995). Uncooled IR focal plane arrays. *Proc. SPIE* **2552** (Infrared Technology XXI, 556.

### Resistive Bolometer Arrays

Flannery, R. E., and Miller, J. E. (1992). Status of uncooled IR imagers. *Proc. SPIE* **1689**, 379.
Gallo, M. A., Willits, D. S., Lubke, R. A., and Thiede, E. C. (1993). Low cost uncooled IR sensor for battlefield surveillance. *Proc. SPIE* **2020** (Infrared Technol. XIX), 351.
Liddiard, K. C. (1993). Thin film monolithic detector arrays for uncooled thermal imaging. *Proc. SPIE* **1696**, 206.

Liddiard, K. C. (1993). Application of interferometric enhancements to self-absorbing thin-film thermal IR detectors. *Infrared Phys.* **34**, 379.

Liddiard, K. C. (1995). Staring focal plane arrays for advanced ambient temperature infrared sensors. *Proc. SPIE* **2552** (Infrared Technol. XXI), 564.

Liddiard, K. C., Unewisse, M. H., and Reinhold, O. (1994). Design and fabrication of thin film monolithic uncooled infrared detector arrays. *Proc. SPIE* **2225** (Infrared Detect. Focal Plane Arrays III).

Ueno, M. *et al.* (1995). Monolithic uncooled infrared image sensor with 160 × 120 pixels. *Proc. SPIE* **2552** (Infrared Technol. XXI), 636.

Unewisse, M. H., Passmore, S. J., Liddiard, K. C., and Watson, R. J. (1994). Performance of uncooled semiconductor film bolometer infrared detectors. *Proc. SPIE* **2269** (Infrared Technol. Sens. XX).

Wood, R. A. (1993). Uncooled thermal imaging with monolithic silicon focal plane arrays. *Proc. SPIE* **2020** (Infrared Technol. XIX), 329.

Wood, R. A., and Foss, N. A. (1993). Micromachined bolometer arrays achieve low-cost imaging. *Laser Focus World*, June, p. 101.

Wood, R. A., Cole, B. E., Han, C. J., Higashi, R. E., Nielsen, D., Weinstein, A., and Miller, J. E. (1991). HIDAD: A monolithic silicon uncooled infrared imaging focal plane. *Proc. GOMAC*.

Wood, R. A., Han, C. J., and Kruse, P. W. (1992). Integrated uncooled infrared detector imaging array. *Proc. IEEE Solid State Sens. Actuator Workshop*, Hilton Head Island, SC, pp. 132–135.

## Pyroelectric Arrays

Burgess, D. E. (1988). Pyroelectrics in a harsh environment. *Proc. SPIE* **930** (Infrared Detect. Arrays), 139.

Burgess, D. E., Manning, P. A., and Watton, R. (1985). The theoretical and experimental performance of a pyroelectric array imager. *Proc. SPIE* **572**, 2.

Butler, N. and Iwasa, S. (1992). Solid state pyroelectric imager. *Proc. SPIE* **1685** (Infrared Detect. Focal Plane Arrays II), 146.

Butler, N., McClelland, J., and Iwasa, S. (1988). Ambient temperature solid state pyroelectric IR imaging arrays. *SPIE Crit. Rev.* **930**, 151.

Manning, P. A., Burgess, D. E., and Watton, R. (1985). A linear pyroelectric array IR sensor. *Proc. SPIE* **590** (Infrared Technol. Appl.), 2.

Mansi, M. V., and Liddicoat, T. J. (1986). An uncooled linescan thermal imager for ground and airborne use. *SPIE Int. Tech. Symp. Opt. Optoelectron. Appl. Sci. Eng.*, 30th, San Diego, 1986.

Mueffelmann, W., and Iwasa, S. (1989). Second-generation technology enhances military imaging. *Laser Focus World*, August; p. 109.

Pham, L., Ye, C., and Polla, D. L. (1993). Integrated pyroelectric detectors based on solid-state micromachining and $PbTiO_3$ thin films. *Proc. IRIS Meet.*, Boston.

Polla, D. L., Ye, C., and Tamagawa, T. (1991). Surface-micromachined $PbTiO_3$ pyroelectric detectors. *Appl. Phys. Lett.* **59**, 3539.

Polla, D. L. et al. (1995). Micromachined infrared detectors based on pyroelectric thin films. *Proc. SPIE* **2552** (Infrared Technol. XXI), 602.

Porter, S. G., Watton, R., and McEwen, R. K. (1995). Ferroelectric arrays: The route to low cost thermal imaging. *Proc. SPIE* **2552** (Infrared Technol. XXI), 573.

Shorrocks, N. M. *et al.* (1990). Uncooled infrared thermal detector arrays. *Proc. SPIE* **1320**, 88.

Watton, R. (1989). Ferroelectric materials and devices in infrared detection and imaging. *Ferroelectrics* **92**, 87.

Watton, R. (1992). IR bolometers and thermal imaging: The role of ferroelectric materials. *Ferroelectrics* **133**, 5.

Watton, R., and Manning, P. A. (1987). The design of low-noise arrays of MOSFETs for pyroelectric array readout, LAMPAR. *Proc. SPIE* **807** (Passive Infrared Syst. Technol.).

Watton, R., and Mansi, M. V. (1987). Performance of a thermal imager employing a hybrid pyroelectric detector array with MOSFET readout. *Proc. SPIE* 965 (Focal Plane Arrays: Technol. Appl.), 79.

Whatmore, R. W. (1991). Pyroelectric ceramics and devices for thermal infrared detection and imaging. *Ferroelectrics* **118**, 241.

## Ferroelectric Bolometer Arrays

Flannery, R. E., and Miller, J. E. (1992). Status of uncooled IR imagers. *Proc. SPIE* **1689**, 379.

Hanson, C. M. (1993). Uncooled ferroelectric thermal imaging, *Proc. SPIE* 2020 (Infrared Technol. XIX).

Hanson, C. et al. (1992). Uncooled thermal imaging at Texas instruments. *Proc. SPIE* **1735** (Infrared Detect.: State-of-the-Art), 17.

## Thermoelectric Arrays

Il Hyun Choi and Wise, K. D. (1986). A silicon-thermopile-based infrared sensing array for use in automated manufacturing. *IEEE Trans. Electron Devices* **ED-33**, 72.

Kanno, T. et al. (1994). Uncooled infrared focal plane array having 128 × 128 thermopile detector elements. *Proc. SPIE* **2269** (Infrared Technol. XX).

Kruse, P. W. (1996). High-speed imaging radiometer employing uncooled thermoelectric linear array. *SPIE Thermosense Meeting*, Orlando, FL.

Listvan, M., Rhodes, M., and Wilson, M. L. (1991). On-line thermal profiling for industrial process control. *Proc. Soc. Instrum. Symp. Innovation Meas. Sci.*, Genevā, NY, 1991.

Sarro, P. M., and van Herwaarden, A. W. (1987). Infrared detector based on an integrated silicon thermopile. *Proc. SPIE* **807**, 113.

Wilson, M. L., Kubisiak, D., Wood, R. A., Ridley, J. A., and Listvan, M. (1991). An uncooled thermo-electric microthermopile camera developed using silicon microstructure sensor. *Proc. IRIS Spec. Group Infrared Detect.*, Boulder, CO, 1991.

Wood, R. A. et al. (1995). IR snapshot™ camera. *Proc. SPIE* **2552** (Infrared Technol. XXI), 654.

## Other Detection Mechanisms

Humphreys, R. G., and Tarry, H. A. (1988). An optically coupled thermal imager. *Infrared Phys.* **28**, 113.

Kenny, T. W. et al. (1991). Novel infrared detector based on a tunneling displacement transducer. *Appl. Phys. Lett.* **59**, 1820.

Vig, J. R., Filler, R. L. and Kim, Y. (1995). Microresonator sensor arrays. *Proc. IEEE Int. Freq. Control Symp., 1995.*

## Fundamental Limits

Kruse, P. W. (1993). A comparison of the limits to the performance of thermal and photon detector imaging arrays. *Proc. IRIS Detect. Spec. Group Meet.*, Boston, 1993.

Kruse, P. W. (1995). A comparison of the limits to the performance of thermal and photon detector imaging arrays. *Infrared Phys. Technol.* **36**, 869.

Kruse, P. W., McGlauchlin, L. D., and McQuistan, R. B. (1962). "Elements of Infrared Technology", Chapter 9. Wiley, New York.

CHAPTER 3

# Monolithic Silicon Microbolometer Arrays

*R. A. Wood*

Honeywell Inc.
Plymouth, Minnesota

| | | |
|---|---|---|
| I. | BACKGROUND | 45 |
| II. | RESPONSIVITY OF MICROBOLOMETERS | 47 |
| | 1. *Microbolometer Model* | 47 |
| | 2. *Resistance Changes in Microbolometer Materials* | 51 |
| | 3. *Microbolometer Heat Balance Equation* | 56 |
| | 4. *Solutions of the Heat Balance Equation* | 57 |
| | 5. *Heat Balance with No Applied Bias* | 57 |
| | 6. *Heat Balance with Applied Bias* | 59 |
| | 7. *Calculations of V-I Curves* | 61 |
| | 8. *Load Line* | 64 |
| | 9. *Low-Frequency Noise in Microbolometer with Applied Bias* | 68 |
| | 10. *Microbolometer Responsivity with Pulsed Bias or Large Radiation Signals* | 70 |
| | 11. *Numerical Calculation of Microbolometer Performance* | 71 |
| III. | NOISE IN BOLOMETERS | 75 |
| | 1. *Bolometer Resistance Noise* | 75 |
| | 2. *Noise from Bias Resistors* | 79 |
| | 3. *Thermal Conductance Noise* | 80 |
| | 4. *Radiation Noise* | 81 |
| | 5. *Total Electrical Noise* | 83 |
| | 6. *Preamplifier Noise* | 85 |
| IV. | MICROBOLOMETER SIGNAL-TO-NOISE | 86 |
| | 1. *Noise Equivalent Power (NEP)* | 86 |
| | 2. *Noise Equivalent Temperature Difference (NETD)* | 86 |
| | 3. *Detectivity* | 87 |
| | 4. *Comparison with the Ideal Bolometer* | 89 |
| | 5. *Johnson Noise Approximation* | 91 |
| V. | ELECTRONIC READ-OUT CIRCUITS FOR TWO-DIMENSIONAL MICROBOLOMETER ARRAYS | 91 |
| VI. | OFFSET COMPENSATION SCHEMES | 95 |
| VII. | GAIN CORRECTION | 97 |
| VIII. | MODULATION TRANSFER FUNCTION (MTF) | 98 |
| IX. | MICROBOLOMETER PHYSICAL DESIGN, FABRICATION, AND PACKAGING | 98 |
| | 1. *One-level Microbolometers* | 100 |
| | 2. *Two-level Microbolometers* | 102 |
| | 3. *Packaging* | 109 |

| | | 116 |
|---|---|---|
| X. Practical Camera Development | | 116 |
| References | | 119 |

## List of Symbols

| Name | Unit | Symbol |
|---|---|---|
| microbolometer front surface area | cm$^2$ | $A$ |
| unit cell area | cm$^2$ | $A_c$ |
| thermal coefficient of resistance | K$^{-1}$ | $\alpha$ |
| electrothermal parameter | | $\beta$ |
| velocity of light | m/sec | $c$ |
| thermal capacity | Joule/K (J/K) | $c$ |
| capacitance | Farad | $C$ |
| detectivity | cm$\sqrt{\text{Hz}}$/W | $D^*$ |
| activation energy | Joule | $\Delta E$ |
| energy gap | Joule | $E_g$ |
| emissivity | | $\varepsilon$ |
| array unit cell area fill factor | | $F_f$ |
| F-number of incident radiation cone | | $F_{no}$ |
| lower bandwidth frequency limit | Hz | $f_1$ |
| upper bandwidth frequency limit | Hz | $f_2$ |
| read-out system bandwidth | Hz | $\Delta f$ |
| thermal conductance of microbolometer legs | W/K | $g_{leg}$ |
| effective value of thermal conductance of microbolometer to substrate | W/K | $g_{eff}$ |
| radiative thermal conductance | W/K | $g_{rad}$ |
| Planck's constant | Joule.sec | $h$ |
| bias current through microbolometer | amperes | $I$ |
| 1/f noise parameter | | $k$ |
| Boltzmann's constant | Joule K$^{-1}$ | $K$ |
| radiance | Wcm$^{-2}$sr$^{-1}$ | $L$ |
| microbolometer resistance | Ohm | $R$ |
| load resistance | Ohm | $R_L$ |
| voltage responsivity | Volts/Watt (V/W) | $\Re_V$ |
| current responsivity | Amperes/Watt (A/W) | $\Re_I$ |
| temperature responsivity | K/Watt (K/W) | $\Re_T$ |
| effective value of thermal time constant | sec | $\tau_{eff}$ |
| bias pulse duration | sec | $\Delta t$ |
| time between shutter closures | sec | $T_{stare}$ |
| thermal time constant of microbolometer | sec | $\tau$ |
| temperature rise in bias pulse | K | $\Delta T$ |

| | | |
|---|---|---|
| target temperature | K | $T_t$ |
| substrate temperature | K | $T_s$ |
| microbolometer temperature | K | $T$ |
| semi–cone-angle of incident radiation | radians | $\theta$ |
| general thermodynamic quantity | | $U$ |
| radiation frequency | Hz | $v$ |
| bias voltage to load resistor | volts | $V_b$ |
| bias voltage across microbolometer | volts | $V$ |
| power dissipation in microbolometer | watts | $W$ |
| small signal V-I slope resistance | ohms | $Z$ |
| wavelength | m | $\lambda$ |
| numerical aperture | | NA |
| noise equivalent power | Watt | NEP |
| noise equivalent temperature difference | K | NETD |
| solid angle of incident radiation cone | steradian | $\Omega$ |
| joule power dissipated in microbolometer | Watt | $P$ |
| infrared power from surroundings | Watt | $P_s$ |
| power incident on microbolometer from target | Watt | $P_t$ |
| infrared radiation power | Watt | $Q$ |
| quality factor | | $Q$ |

The first part of this chapter summarizes the operating principles of two-dimensional arrays of microbolometers for infrared (IR) imaging applications. The second part describes how silicon (Si) micromachined microbolometers are fabricated and operated in practical IR imaging systems. The simplest possible derivations of the important equations are provided throughout, and the equations are cast in a form intended to be useful for practical numerical applications. To provide the reader with a feel for the magnitudes of microbolometer parameters, numerical examples are often provided in the text. Most information in this chapter is drawn from the work most familiar to the author, which is the work performed at the Honeywell Technology Center from 1982 to 1995, using silicon micromachined microbolometer arrays. Published work from other sources is included where appropriate for completeness.

## I. Background

A bolometer is a "thermal" IR sensor, which here we define as a sensor that detects incident IR by the induced increase in sensor temperature. A thermal sensor (a thermometer) was used by Herschel in 1800 (Barr, 1961) to first demonstrate the existence of invisible IR "heat" radiation. The temperature increase in a thermal sensor may be detected by many different means: the distinguishing feature of a "bolometer" sensor is that it uses an electrical conductivity change to measure the sensor temperature. Nature appears to have adopted this IR detection mechanism: the "pit organs" of a snake are depressions that focus IR radiation onto a thin membrane whose ionic conductance varies with temperature (Newman and Hartline, 1982).

A detailed account of the overall evolution of uncooled IR sensors and imaging systems is given in Chapter 1 of this book. The development of bolometers may be briefly summarized as follows: The first usefully sensitive bolometer was constructed by Langley in 1881, although the bolometer principle had been established earlier by others (Barr, 1963). Following Langley's work, bolometers were developed in improved forms, as reviewed by Richards (1994). However, until 1979, bolometers were usually large in size (e.g., 1 mm or larger), and hence impractical for use in two-dimensional arrays with a large number of sensors.

In 1978, Johnson (Higashi, 1996) proposed the use of silicon (Si) microminiature thermally isolated structures as room temperature thermal IR sensors, and in 1979, Johnson and Higashi constructed prototype sensors consisting of mechanical bridgelike silicon nitride ($Si_3N_4$) structures of about 100-$\mu$m lateral dimensions and 1 $\mu$m thickness, with metal thin-film temperature-sensitive resistors (Higashi, 1996). These devices were constructed using anisotropic etching of Si pits to produce thermally isolated $Si_3N_4$ microbridges, and they were found to be very sensitive to IR radiation, as Johnson had predicted. Kruse (1982) showed by calculation that Si micromachined microbolometers could have a performance approaching the ideal performance for a room temperature IR sensor, and proposed their construction as two-dimensional staring focal planes for low-cost uncooled IR imaging. Quantitative measurements on micromachined $Si_3N_4$ microbolometers were made by Arch and Heisler (1982), who measured a detectivity $D^*$ of 1E9 cm$\sqrt{Hz}$/W, in agreement with Kruse's predictions. Since a $D^*$ of this value is sufficient to yield an excellent noise equivalent temperature difference (NETD) in a staring focal plane, this measurement prompted a development program at Honeywell research laboratories, the goal of which was to produce two-dimensional arrays of microbolometers specifically for uncooled IR imaging. Infrared imaging with small arrays of micromachined microbolometers was demonstrated in 1983 (Wood, 1983). Array size and imaging performance of these microbolometer arrays improved rapidly, and by 1992, 240 × 336 microbolometer arrays with an NETD of 39 mK ($F_{no} = 1$, 30-Hz frame rate) had been demonstrated in portable uncooled cameras (Wood et al., 1992a,b). From 1981 to 1992, microbolometer development work in the United States was classified as either confidential or secret (Wood *et al.*, 1982 to 1991). The first public disclosure of the work in the United States was made in 1992 (Flannery and Miller, 1992; Wood *et al.*, 1992b; Horn and Buser, 1993).

In addition to the work at Honeywell's laboratories, independent development of micromachined microbolometers was performed by Kimura (1981), Downey *et al.* (1984), Liddiard (1984, 1986), and Neikirk *et al.* (1984). Since 1992, a large increase in reported work on microbolometers has occurred. References to much of this work are provided subsequently.

## II. Responsivity of Microbolometers

In this section, we use an electrical and thermal model of a microbolometer to derive useful practical equations and numerical estimates for the responsivity. The following sections will consider noise, NETD, noise equivalent power (NEP), and $D^*$ of microbolometer arrays. These equations will provide the insight necessary to understand the optimum design and practical operating requirements of microbolometer arrays. Microbolometer fabrication techniques, and practical application of these arrays in IR cameras, will be discussed in the last part of this chapter.

1. MICROBOLOMETER MODEL

The microbolometer array is assumed to consist of a rectangular array of microbolometers. Each microbolometer consists of a thin plate of thickness $t$ and front surface area $A$. Each microbolometer front surface area occupies a fraction $F_f$ of the total unit cell area $A_c$, where $F_f$ is called the *fill factor* of the unit cell, that is, $A = F_f A_c$ (Fig. 1). Each microbolometer thus has a total surface area of $2A$, with front surface $A$ receiving radiation from an optical system. We assume that the emissivity of the microbolometer surface is $\varepsilon$, that the microbolometer has thermal capacity $c$, is at absolute temperature $T$, and that each microbolometer is suspended from a thermal reservoir (the Si substrate) at temperature $T_s$ by two electrically conducting supporting "legs" (Fig. 1). We also assume that there is a thermal conductance

$$g = \frac{dW}{dT} \qquad (1)$$

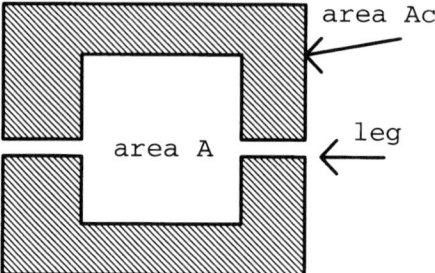

FIG. 1. Microbolometer model. $A_c$, total unit cell area; A, front plate area.

between the bolometer and its supporting structure, where $W$ is the power dissipated inside the microbolometer at temperature $T$. Thermal conductance $g$ is due to thermal conductivity of the supporting legs and any surrounding gas, and to radiation transfer. We assume that the only electrical resistance present is within each microbolometer plate, presenting a temperature-dependent electrical resistance between the two supporting legs, of value $R(T)$.

The microbolometer is mounted within a package (Fig. 2) with blackbody walls at temperature $T_s$. The IR radiation from a distant target at temperature $T_t$ is imaged onto the microbolometer with a lens, as shown. The package possesses a perfectly transparent circular optical window subtending a solid angle $\Omega$ with semicone angle $\theta$ at the microbolometer. The front surface of the microbolometer thus receives radiation from outside the package over a solid angle $\Omega$ given by (Fig. 3)

$$\Omega = \int_{\theta=0}^{\theta} \int_{\phi=0}^{2\pi} \sin\theta.d\theta.d\phi = 2\pi(1 - \cos\theta) \qquad (2)$$

The F-number of this optical arrangement of Fig. 2 may be defined in several ways (Slater, 1980), one of which is

$$F_{no} = \frac{1}{2\sin\theta} \qquad (3)$$

where $\sin\theta$ is called the numerical aperture (NA) of the arrangement.

Referring to Fig. 2, the IR radiation power $P_t$ incident on the front surface area $A$ of a sensor in semicone angle $\theta$ from a distant target with radiance $L$ (emitted power per unit area per solid angle) is (Slater, 1980)

$$P_t = \pi L A \sin^2\theta \qquad (4)$$

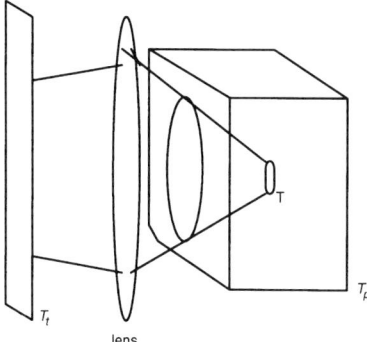

FIG. 2. Collection optics. $T_t$, target temperature; $T_p$, package temperature; $T$, sensor temperature.

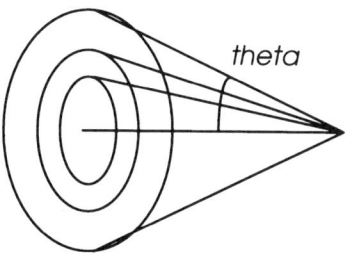

FIG. 3. Light cone geometry. $\theta$, semicone angle of incident radiation

which can also be expressed in the following useful alternative forms:

$$P_t = LA\pi \left( \frac{\Omega}{\pi} - \left( \frac{\Omega}{2\pi} \right)^2 \right) = \frac{LA\pi}{4F_{no}^2} \quad (5)$$

Thus the IR power $Q$ absorbed within a microbolometer with front area $A$ and emissivity $\varepsilon$ due to a distant blackbody target with radiance $L$ is

$$Q = \frac{A\varepsilon\pi}{4F_{no}^2} L \quad (6)$$

(Note. The commonly used approximation $F_{no} \approx 1/2 \tan \theta$ gives significant numerical errors if $F_{no}$ is less than about 2.0 (Slater, 1980).)

In Eq. (6), the radiance $L$ of the target is given by Planck's law

$$L = 2hc^2\varepsilon_t \int_{\lambda_1}^{\lambda_2} \frac{d\lambda}{\lambda^5(\exp(hc/KT_t\lambda) - 1)} \qquad (7)$$

where the transmitted IR wavelengths lie between $\lambda_1$ and $\lambda_2$, $T_t$ is the target temperature, and $\varepsilon_t$ is the target emittance.

In Fig. 2, we can calculate the temperature change induced in a microbolometer by a change in the temperature of the distant blackbody target at temperature $T_t$:

$$\frac{\delta T}{\delta T_t} = \frac{\delta T}{\delta P_t} \cdot \frac{\delta P_t}{\delta T_t} = \frac{1}{g} \cdot \frac{A\varepsilon\pi}{4F_{no}^2} \frac{dL}{dT_t} \qquad (8)$$

Numerical values of $dL/dT_t$ are especially useful in imaging calculations (Klein, 1976) and are listed in Table I.

As a numerical example assume that

$$g = 1\text{E-}7\,\text{W/K}$$
$$A = 50\,\mu\text{m square}$$
$$\varepsilon = 0.8$$
$$F_{no} = 1$$
$$\lambda_1 = 8\,\mu\text{m}$$
$$\lambda_2 = 12\,\mu\text{m}$$
$$T_t = 300\,\text{K}$$

TABLE I

$dL/dT_t$ Values (W.cm$^{-2}$.sr$^{-1}$.K$^{-1}$) for Target Temperatures ($T_t$ = 290, 300, and 310 K)

| Wavelength Interval $\lambda_1$ to $\lambda_2$ ($\mu$m) | 290 K | 300 K | 310 K |
|---|---|---|---|
| 3–4 | 8.9E-7 | 1.3E-6 | 1.9E-6 |
| 4–5 | 4.0E-6 | 5.5E-6 | 7.2E-6 |
| 8–10 | 3.1E-5 | 3.5E-5 | 3.9E-5 |
| 10–12 | 2.6E-5 | 2.8E-5 | 3.1E-5 |
| 12–14 | 1.9E-5 | 2.1E-5 | 2.2E-5 |

Then from Table I, $dL/dT_t = 6E - 5$, and for $dT_t = 1°C$, Eq. (8) shows that the temperature change in the microbolometer is $dT = 10$ mK.

## 2. Resistance Changes in Microbolometer Materials

A bolometer at temperature $T$ is a thermal sensor that uses a material with a temperature-dependent resistance $R(T)$ to measure the heating effect of absorbed IR radiation. The parameter used to quantify the temperature dependence of $R$ is the *temperature coefficient of resistance* (TCR), defined as

$$\alpha = \frac{dR}{RdT} \qquad (9)$$

If $\alpha$ is independent of temperature, we can integrate Eq. (8) to show that

$$R(T) = R(T_s)(1 + \alpha(T - T_s)) \qquad (10)$$

Metals show little change in free-carrier density with temperature, but the mobility of the free carriers decreases with temperature, producing a positive $\alpha$, typically about $+0.002$ K$^{-1}$. Thin metal films usually have a TCR less than the bulk value (Brunetti and Monticone, 1993). The TCR of metal films usually varies little with temperature, so that Eq. (10) well-describes metals $R(T)$.

The first microbolometer arrays (Wood et al., 1983) employed nickel–iron (Ni–Fe) metal films with a TCR of $+0.0023$ K$^{-1}$, and achieved NETDs of 0.250°C (75 µm pixels, $F_{no} = 1$, 30-Hz frame rate (Wood et al., 1983)). Because the resistivity of metal is low, serpentine resistors about 100 squares long were used to form the bolometer resistors. Titanium (Ti) films ($\alpha = +0.0042$ K$^{-1}$) also have been successfully used in 128 × 128 microbolometer arrays to demonstrate an NETD of 0.7°C (50 µm pixels, $F_{no} = 1$, 30-Hz frame rate; Tanaka et al., 1995). The principal attractions of metal TCR materials are the ease of achieving controllable deposition and the low 1/f noise in metal films and contacts (Voss and Clarke, 1976; Dagge et al., 1996). Platinum (Pt) ($\alpha = +0.0018$ K$^{-1}$) has been employed by Liddiard (1984) and Shie (1992), Ni by Brunetti and Monticone (1993), Lang et al. (1990), and Katsan et al. (1994), niobium (Nb) by MacDonald and Grossman (1995), and gold (Au) by Lang et al. (1994).

Semiconductor materials have mobile carrier densities that increase with increasing temperature, as well as carrier mobilities that change with temperature (Fisher, 1982), producing a higher negative $\alpha$ that is generally temperature-dependent. A typical $R(T)$ behavior for a semiconductor whose mobile carrier density is controlled by thermal excitation across a bandgap is

$$R(T) = R_0 \exp\left(\frac{\Delta E}{KT}\right) \tag{11}$$

where $\Delta E$ is the activation energy, equal to half the bandgap $E_g$, and $R_0$ is a constant.

From Eqs. 9 and 11, we have for a semiconductor material

$$\alpha = \frac{dR}{RdT} = -\frac{\Delta E}{kT^2} \tag{12}$$

For $E_g = 0.6\,\text{eV}$ and $T = 300\,\text{K}$, Eq. (12) gives $\alpha = -0.04\,\text{K}^{-1}$.

Since the number of mobile carriers decreases as $\Delta E$ increases, the general trend is for higher resistance materials to have higher $\alpha$. Unfortunately, higher resistivity materials also tend to have greater "excess" noise (1/f noise).

Many semiconductor materials are candidates for microbolometer TCR material. Unewisse et al. (1995a), and Mori et al. (1994) report use of films of Si, germanium (Ge) and silicon–germanium (Si–Ge) as TCR materials for microbolometers. Films were grown by sputtering and plasma-enhanced chemical vapor deposition (PECVD). PECVD allowed well-controlled films with low stress and high uniformity, and phosphorus doping allowed control of resistivity. PECVD amorphous Si films had about twice the TCR of sputtered films. Excess noise levels in PECVD films were found to render the material unsuitable for microbolometer use. Resistivity of sputtered films were controlled by allowing small amounts of hydrogen in the sputtering gas: Control and uniformity were reported to be difficult. In sputtered Si, 1/f noise was found to be lower than in PECVD Si. Si–Ge films are reported to show encouraging combinations of TCR, resistivity, and excess noise by Unewisse et al. (1995b), who also propose other new TCT materials. The temperature-dependent conductivity of a silicon–molybdenum silicide (Si/MoSi$_2$) Schottky barrier also has been employed as a bolometer resistor (Mori et al., 1994). Tellurium (Te) and bismuth (Bi) thin-film resistors have been used (Kimura, 1981; Neikirk et al., 1984; Wentworth and Neikirk, 1992; Ling et al., 1994) for microbolometers coupled to thin-film microantennas sensitive to long wavelengths (100 $\mu$m to 3 mm). The TCR values reported are $-0.003$ to $-0.01\,\text{K}^{-1}$. The 1/f noise in Bi films is reported to be low compared with that of metals (Schnelle and Dillner, 1989). Thin-film sputtered mixed nickel–cobalt–manganese (Ni)–Co–Mn) "thermistor" oxides have been reported to show TCR values of about $-0.04\,\text{K}^{-1}$ (Baliga et al., 1994). Gallium arsenide (GaAs) has been proposed as a possible TCR material up to $-0.09\,\text{K}^{-1}$ (Estill and Brozel, 1994).

High-temperature superconductor materials have also been demonstrated to provide very high TCR values over the normal-superconductor transition interval and bolometer action demonstrated (Stratton, 1990; Johnson et al., 1993). To date, however, the transition temperature of these materials is well below room temperature.

Thin films of mixed vanadium oxides ($VO_x$) were selected for microbolometer development (Wood et al., 1988), because the technology for their deposition in thin films had been fairly well developed (for other applications). In addition, early tests of TCR, resistivity, and 1/f noise indicated that this material would perform better than a metal TCR resistor. Thin films (500 to 1000 Å) of polycrystalline mixed oxides ($VO_2$, $V_2O_3$, and $V_2O_5$)

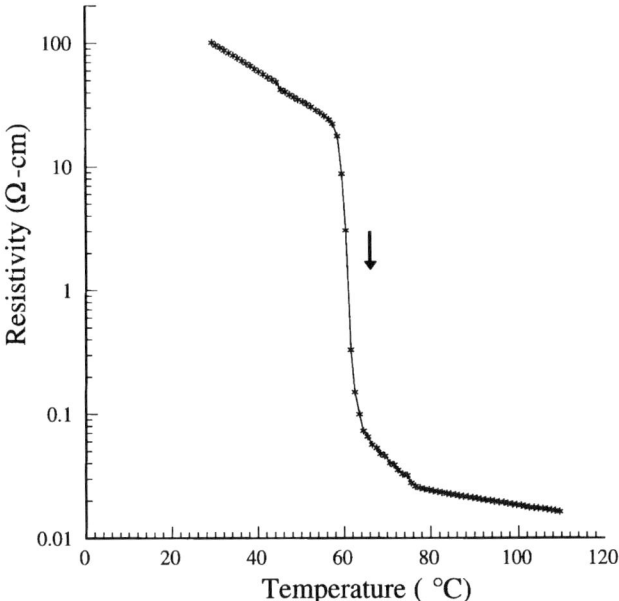

FIG. 4. Resistance versus temperature for vanadium dioxide. (Reprinted from Jerominek et al. (1993). Vanadium oxide films for optical switching and detection. Opt. Eng. 32(9), 2092–2099; with permission from the National Optics Institute, Quebec.)

were found to be readily deposited by sputtering at deposition temperatures compatible with microbolometer materials (Jerominek et al., 1993). A material commonly used for constructing microbolometers ($Si_3N_4$) was found to be good substrate and passivant for vanadium oxides. Vanadium oxide films were found to be conveniently deposited with a suitable resistance for microbolometer read-out circuits (about 20 KΩ per square at 25°C), 1/f-noise was found to be moderately low when properly applied metal contacts were applied, optical properties were found to allow high bolometer IR absorption to be obtained, and $\alpha$ was about $-0.02\,K^{-1}$ at 25°C (which is 5 to 10 times better than the $\alpha$ of most metals). Although sputtered vandium oxides have been demonstrated to be useful microbolometer TCR materials, maximum TCRs are limited to about $-0.06\,K^{-1}$, which is lower than that of some other known materials. In addition, deposition conditions must be carefully controlled to obtain reliable properties, and some hysteresis is present in the dioxide form (Umadevi et al., 1991; Jerominek et al., 1993). Figure 4 shows the overall variation of resistance with temperature for $VO_2$ thin films: these undergo a semiconductor–metal phase transition at 68°C, with a latent heat of 750 cal/mole

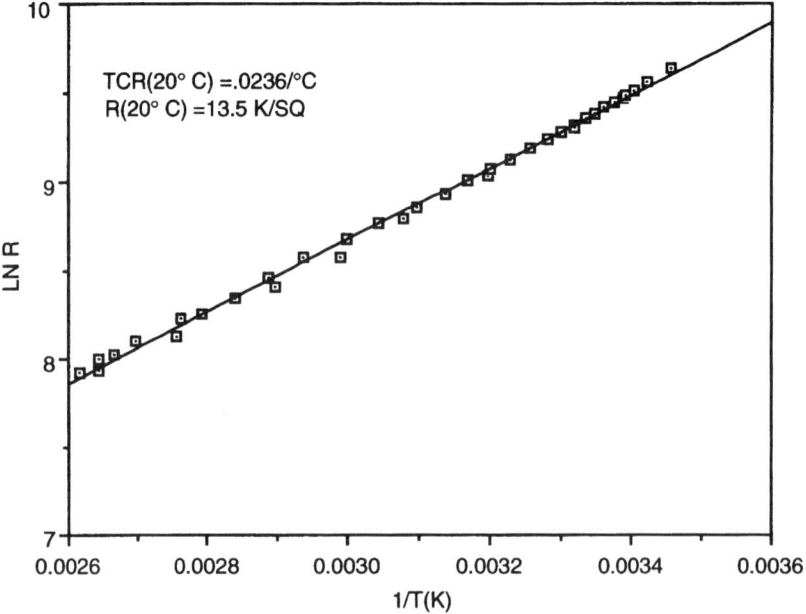

FIG. 5. Resistance versus temperature for thin films of mixed vanadium oxides. (Reprinted from Wood and Stelzer (1992) with permission from the authors.)

(Kawakuba and Kakagawa, 1964; Andreev *et al.*, 1978), which, with other materials that show semiconductor-metal phase transitions, may be suitable for microbolometer use (Scott and Fredericks, 1976). Mixed vanadium oxide ($VO_x$) films of 500 Å show no phase transitions near room temperature but have TCRs as large as $-0.03 \text{ K}^{-1}$ (Figs. 5 and 6; Wood and Stelzer, 1992).

At the time of this writing, comparison of the practical merits of microbolometer resistor materials is made difficult by the different designs of bolometers employed by different workers and by sparse quantitative information, particularly on 1/f noise. Data on material controllability (which is vital for commercial production) is also sparse, because this is usually held as proprietary information. At the time of this writing, the best practical IR imaging results have been achieved with sputtered vanadium oxide films about 500-Å thick. However, because of the wide range of materials with $\alpha$ values much greater than vanadium oxides, development of other TCR materials is likely to produce improved results.

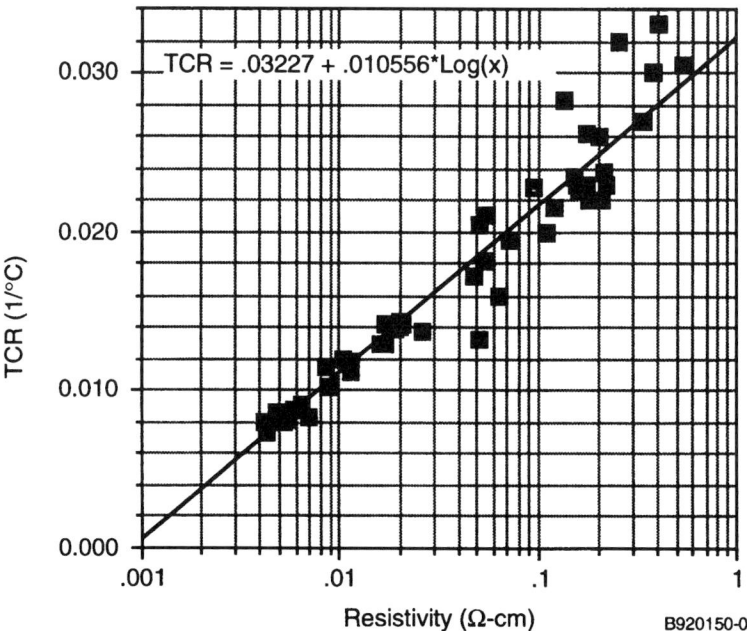

FIG. 6. Temperature coefficient of resistance (TCR) versus resistivity for thin films of mixed vanadium oxides. (Reprinted from Wood and Stelzer (1992) with permission from the authors.)

## 3. Microbolometer Heat Balance Equation

In this section we derive useful expressions describing how a microbolometer's temperature $T$ depends on the energy exchange with its heat reservoir at temperature $T_s$; IR radiation power from the target and the reservoir, $P_t$ and $P_s$, respectively; and joule power $P$ produced by an applied bias voltage or current.

Electronic read-out of two-dimensional arrays of microbolometers requires that the resistance changes of a microbolometer be converted to a voltage or current signal that can be accepted by IR camera circuitry. This is commonly achieved by applying a "bias" voltage $V$ or current $I$, and measuring the resulting current or voltage signal. The applied bias produces a temperature increase in the microbolometer due to the joule power dissipation $P = IV$ in the microbolometer. We shall see that in typical microbolometer operation, the joule heating from the applied bias is enormously greater than is the heating due to IR signals from typical targets. The joule power $P$ therefore must be managed carefully, and means must be employed to distinguish joule heating from that caused by incident IR radiation.

The *heat balance equation* for a microbolometer at temperature $T$ is

$$c\frac{dT}{dt} = IV + \varepsilon P_t + \varepsilon P_s - g(T - T_s) - (2A)\varepsilon\sigma T^4 \tag{13}$$

The last term is Stefan's law. For small changes in microbolometer temperatures $dT$, the corresponding increments in the radiation power emitted from the microbolometer $P_{rad} = (2A)\varepsilon\sigma T^4$ can be written in terms of a *radiation thermal conductance* $(g_{rad})$:

$$g_{rad} = \frac{d\{(2A)\varepsilon\sigma T^4\}}{dT} = 4(2A)\varepsilon\sigma T^3 \ [\text{W/K}] \tag{14}$$

The total thermal conductance $g$ of a microbolometer to its surroundings is therefore the sum of $g_{rad}$ and $g_{leg}$, where $g_{leg}$ is the thermal conductance of the supporting legs (Fig. 7).

As will be seen in a subsequent section of this chapter, a typical value of $g_{leg}$ for a 50-$\mu$m square microbolometer is 2E-7 W/K, while numerical evaluation of Eq. (14) gives a value of about 2E-8 W/K for $g_{rad}$, assuming typical values ($A = 50 \times 50\ \mu$m, $\varepsilon = 0.8$, $T = 25°$C). Since $g_{rad} \ll g_{leg}$ in present-day microbolometer, we will often omit the $g_{rad}$ term in the subsequent calculations where doing so simplifies the equations.

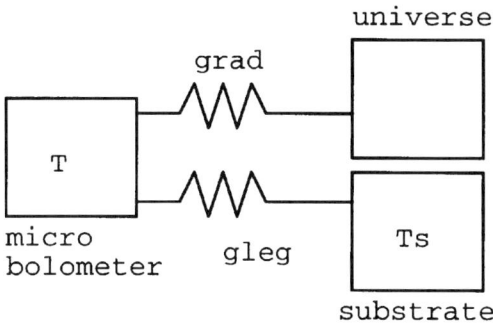

FIG. 7. Thermal conductances between microbolometer and surroundings. $g_{rad}$, radiative thermal conductance; $g_{leg}$, thermal conductance of the microbolometer legs; $T_s$, substrate temperature.

## 4. Solutions of the Heat Balance Equation

Useful practical solutions of the heat balance equation can best be obtained by solving it under different simplifying assumptions, as appropriate to different topics of interest. Here we solve the heat balance equation to illustrate how a microbolometer interacts with (1) its heat reservoir (the substrate) and (2) its bias supply. The first interaction is simple to understand; however, as will be seen, the interaction of a microbolometer with its bias supply is surprisingly complicated.

## 5. Heat Balance with No Applied Bias

The zero-bias solution to the heat equation illustrates how the microbolometer interacts with the heat reservoir (the substrate):

$$c\frac{dT}{dt} = Q - g(T - T_s) \qquad (15)$$

where all the radiation terms have been written as a net absorbed radiation power $Q$

$$Q = \varepsilon P_t + \varepsilon P_s - (2A)\varepsilon\sigma T^4 \qquad (16)$$

If $Q$ is zero for $t < 0$ and $Q$ for $t > = 0$, we can solve this in the usual way to obtain

$$T(t) = T_s + \frac{Q}{g}\{1 - e^{-t/\tau}\} \qquad (17)$$

showing that the microbolometer temperature responds to IR power $Q$ with an exponential *thermal time constant* $\tau = c/g$. After a period of many time constants, the microbolometer temperature approaches the steady-state value

$$T(t \gg \tau) = T_s + \frac{Q}{g} \qquad (18)$$

The responsivity of the bolometer temperature to absorbed radiation power is

$$\Re_T = \frac{\delta T}{\delta Q} = \frac{1}{g} \qquad (19)$$

If the net absorbed radiation power $Q$ is reduced to zero, the bolometer temperature falls back toward the substrate temperature $T_s$, with the same exponential time constant $\tau$ (Fig. 8). If the input heating power $Q$ is a sine wave $Q = Q_0 e^{j\omega t}$, then the solution to the heat balance equation is

$$\Delta T(\omega) = \frac{Q(\omega)}{g(\omega)} \qquad (20)$$

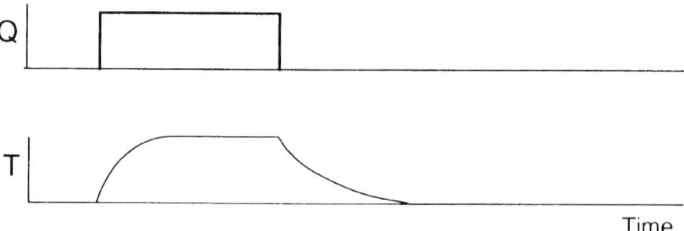

FIG. 8. Response of microbolometer to pulse of radiation. $Q$, net absorbed infrared radiation power; T, microbolometer temperature.

where (Jones, 1953)

$$g(\omega) = \frac{c}{\tau}(1 + j\omega\tau) \qquad (21)$$

The amplitude of the resultant temperature signal therefore varies as $1/\sqrt{1 + \omega^2\tau^2}$, that is, has a 3 dB "roll-off" frequency of $f_c = 1/2\pi\tau$.

6. HEAT BALANCE WITH APPLIED BIAS

With an applied bias, the microbolometer is heated by a total power $W = Q + P$, where $Q$ is the net absorbed IR power and $P = IV$ is the joule power from the applied bias. In all normal situations, $P \gg Q$, so that $W$ may be replaced by $P$ whenever no differentiation of $W$ is involved. Because an applied bias heats the bolometer and causes the microbolometer resistance $R = V/I$ to change, the V-I curve of a bolometer is not a straight line. To calculate the shape of the V-I curve of a bolometer, we assume the radiation power $Q = 0$ and that $P$ is slowly increased (zero-frequency approximation). The bolometer temperature increases, which causes either an increase or a decrease in the bolometer resistance $V/I$, depending on whether the TCR is positive (*positive microbolometer*) or negative (*negative microbolometer*). The V-I curves of positive and negative microbolometers are therefore as sketched in Fig. 9.

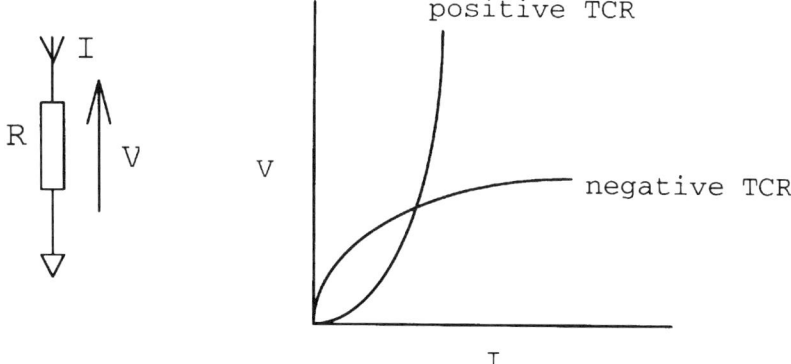

FIG. 9. V-I curves of microbolometers. TCR, temperature coefficient of resistance; R, microbolometer resistance; V, I, microbolometer voltage and current.

To describe the curvature of the V-I curve, it is convenient to define an *electrothermal parameter* $\beta$ as the ratio of the fractional change in resistance to the fractional change in power dissipation within the microbolometer:

$$\beta = \frac{\delta R}{R} \bigg/ \frac{\delta W}{W} = \frac{W}{R} \frac{\delta R}{\delta W} = \frac{\delta \ln W}{\delta \ln R} \quad (22)$$

If we assume that the V-I curve is measured with a fixed small value of $Q$, then $W \Rightarrow P$ and $\delta W \Rightarrow \delta P$, and we can expand $\beta$ to find the small signal slope of the V-I curve $Z = dV/dI$. Thus (Jones, 1953)

$$\beta = \frac{P}{R} \frac{\delta R}{\delta P} = \frac{P}{R} \frac{\delta(V/I)}{\delta(IV)} = \frac{I\delta V - V\delta I}{I\delta V + V\delta I} = R \frac{Z - 1}{Z + 1} \quad (23)$$

that is,

$$Z = \frac{\delta V}{\delta I} = R \frac{1 + \beta}{1 - \beta} \quad (24)$$

The parameter $\beta$ changes in value along the V-I curve (Fig. 10) and has the same sign as does $\alpha$. At any point on the V-I curve, $\beta$ has the numerical value

$$\beta = \frac{W}{R} \cdot \frac{\delta R}{\delta T} \cdot \frac{\delta T}{\delta W} = \frac{(IV + Q)\alpha}{g} = \alpha \Delta T \quad (25)$$

where $\Delta T$ is the temperature elevation of the microbolometer, $\alpha$ is evaluated at the operating point on the IV curve, and $g = dW/dT$. We see that as bias

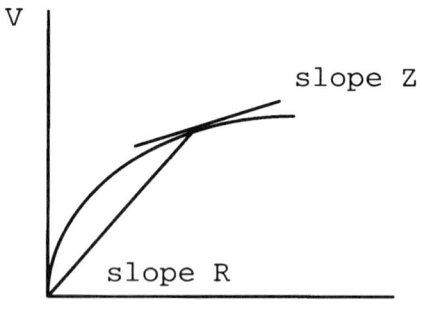

FIG. 10. Resistance parameters $R$ and $Z$ of microbolometer.

is increased, the slope of the V-I curve of a positive or a negative microbolometer reaches infinity or zero when $\beta = 1$ or $-1$, respectively, when the bolometer temperature elevation $\Delta T = |1/\alpha|$, under which condition the microbolometer current or voltage reach maximum values for positive or negative bolometers, respectively.

The previous equations give the zero-frequency expressions for $\beta$ and $Z$. As we have seen earlier, the frequency response of a microbolometer can be expressed by using a frequency-dependent $g$:

$$g(\omega) = \frac{c}{\tau}(1 + j\omega\tau) \tag{26}$$

Therefore we can write corresponding frequency-dependent forms of the previous equations for $\beta$ and $Z$ (Clark Jones, 1953):

$$\beta(\omega) = \frac{IV\alpha}{g(\omega)} = \frac{\beta(\omega = 0)}{1 + j\omega\tau} \tag{27}$$

$$Z(\omega) = R\frac{1 + \beta(\omega)}{1 - \beta(\omega)} \tag{28}$$

As a numerical example, for a metal-like $\alpha$ of $+0.002$ K$^{-1}$, the V-I slope becomes infinite at a temperature increase of $1/\alpha = 500°$C. For a semiconductor-like $\alpha$ of $-0.02$ K$^{-1}$, the V-I slope becomes zero at a temperature increase of about $1/\alpha = 50°$C. For $g = 1$E-7, this requires joule power of about 500E-7 and 50E-7 W, respectively; that is, for $R = 10$ K, the microbolometer currents required are about 70 and 22 $\mu$A, respectively.

7. CALCULATIONS OF V-I CURVES

The heat balance equation Eq. (13), in thermal equilibrium is

$$0 = Q + I^2 R(T) - g(T - T_s) \tag{29}$$

where $Q$ represents all the radiation input power and $T$ is the microbolometer temperature. That is,

$$T - T_s = \frac{IV + Q}{g} \tag{30}$$

Also, from Ohm's law

$$V = IR(T) \tag{31}$$

where $R(T)$ is the equation for the temperature-dependent microbolometer resistance. The simultaneous equations, Eqs. (30) and (31), can be used to numerically calculate the V-I curve of a microbolometer. We will solve these equations for the case of (1) a metal-like TCR and (2) a semiconductor-like TCR.

a. *Metal-like Temperature Coefficient of Resistance*

For a microbolometer with a metal sensing resistor, $\alpha$ is typically about $+0.002\ \mathrm{K}^{-1}$, and is almost independent of temperature. In this case the simultaneous equations, Eqs. (30) and (31), become

$$T - T_s = \frac{IV + Q}{g} \tag{32}$$

$$V = IR(T_s)(1 + \alpha(T - T_s)) \tag{33}$$

and can be analytically solved by eliminating $T$ to yield the I-V relation of a metal-like TCR:

$$V = IR(T_s)\left(1 + \alpha \frac{IV + Q}{g}\right) \tag{34}$$

The I-V curve for a microbolometer with a metal-like TCR can be computed using personal computer tools (Fig. 11) ($\alpha = +0.002\ \mathrm{K}^{-1}$, $R(T_s) = 10\ \mathrm{K}\Omega$, $g = 1\mathrm{E}{-}7\ \mathrm{W/K}$, $Q = 0$).

b. *Semiconductor-like Temperature Coefficient of Resistance*

For a semiconductor-like TCR, the steady-state V-I curve of the microbolometer at temperature $T$ is given by the coupled equations

$$T - T_s = \frac{IV + Q}{g} \tag{35}$$

$$V = IR(T_s) \exp\left\{\frac{\Delta E}{KT}\right\} \tag{36}$$

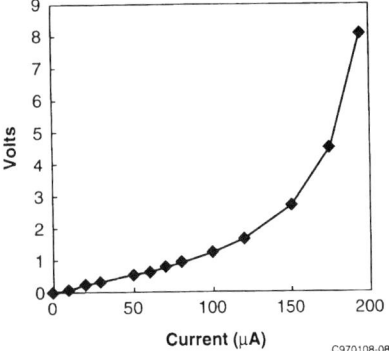

FIG. 11. Computed V-I curves of the metal-like temperature coefficient of resistance (TCR) microbolometer (eq. 34), with $Q = 0$, $\alpha = 0.002 \, K^{-1}$, $R = 10 \, K\Omega$ at 300 K, $g = 1E - 7 \, W/K$.

Eliminating $T$ between these equations gives the V-I relation

$$V = IR(T_s) \exp\left\{\frac{\Delta E}{K\left(T_s + \dfrac{IV + Q}{g}\right)}\right\} \quad (37)$$

Numerical results for equation 37 can be readily calculated using personal computer tools, as shown in Fig. 12.

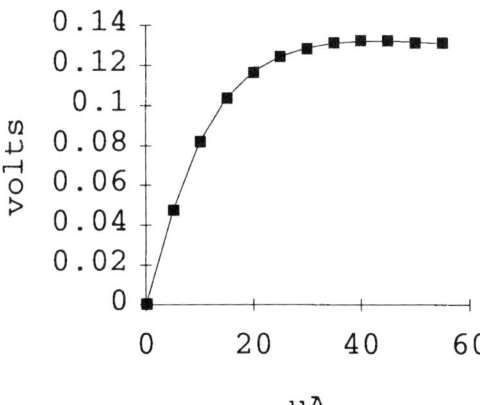

FIG. 12. Numerical values of semiconductor-like temperature coefficient of resistance (TCR) microbolometer steady-state V-I curve (Eq. 37), with $Q = 0$, $\alpha = -0.027 \, K^{-1}$ at 300 K, $R = 10 \, K\Omega$ at 300 K, $g = 1E\text{-}7 \, W/K$).

## 8. LOAD LINE

A microbolometer interacts with the circuit that applies the bias in a manner that significantly changes the operation of a bolometer. This can be understood by the following argument: An absorbed radiation power change $dQ$ heats the bolometer, which changes the resistance of the microbolometer, which changes the joule power $P = IV$ applied by the bias circuit. This results in a different net temperature change from the value $dQ/g$ expected without bias (i.e., the effective microbolometer $g_{eff}$ for changes in $Q$ is different from the physical conductance $g$). Because the temperature change is modified, the bolometer also takes a different time to reach equilibrium (i.e., the effective microbolometer $\tau_{eff}$ is different from the value $c/g$ expected without bias). This is termed the *electrothermal effect*.

To quantify the electrothermal effect, we use the circuit of Fig. 13, where a bias generator is represented by its Thevenin equivalent circuit: an ideal (zero-impedance) bias voltage source $V_b$ in series with a "load" resistor $R_L$. The operating point of the microbolometer $(V, I)$ is given by the intersection of the load line $V_b = V + IR_L$ with the microbolometer V-I curve.

The zero-frequency voltage responsivity $\Re_V = dV/dQ$ when the microbolometer is coupled to the bias generator circuit can be calculated by expanding the slope parameter $\beta$, assuming that $P \gg Q$ so that we can write $W \Rightarrow P$ (Jones, 1953):

$$\beta = \frac{W\delta(V/I)}{R\delta(IV+Q)} = \frac{\left(I^2\left(\frac{\delta V}{I} - \frac{V\delta I}{I^2}\right)\right)}{(I\delta V + V\delta I + \delta Q)} = \frac{1 + \frac{V}{I}\frac{1}{R_L}}{1 - \frac{V}{I}\frac{1}{R_L} + \frac{1}{I}\frac{1}{\Re_V}} \quad (38)$$

hence

$$\Re_V = \frac{\delta V}{\delta Q} = V_b \frac{RR_L\alpha}{(R+R_L)^2} \frac{1}{g} \frac{1}{1 + \beta\frac{R-R_L}{R+R_L}} \quad (39)$$

at zero frequency. Or, in terms of sensitivity to target temperature $T_t$,

$$\frac{\delta V}{\delta T_t} = \frac{\delta V}{\delta Q} \cdot \frac{\delta Q}{\delta T_t} = V_b \frac{RR_L\alpha}{(R+R_L)^2} \frac{1}{1 + \beta\frac{R-R_L}{R+R_L}} \frac{1}{g} \frac{A\pi}{F_{no}^2} \frac{\delta L}{\delta T_t} \quad (40)$$

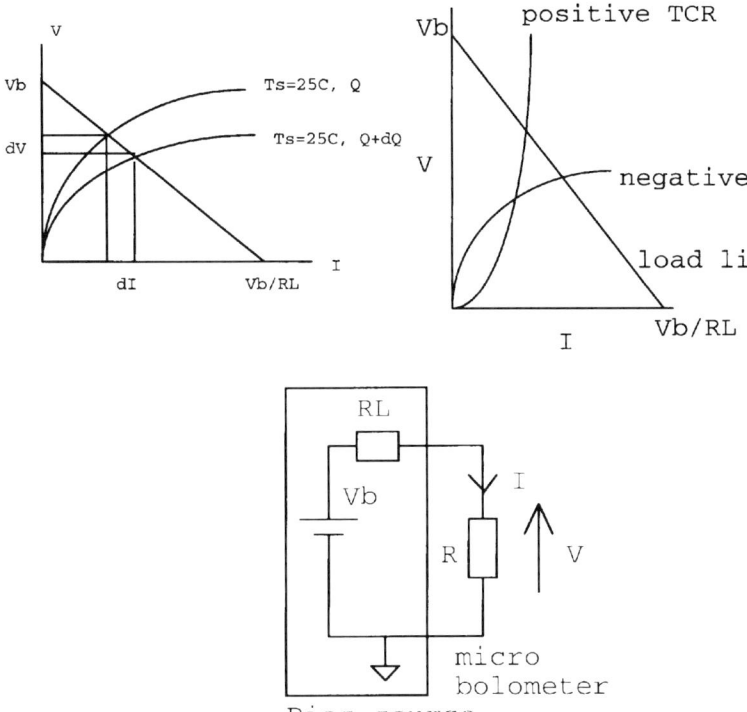

FIG. 13. Load lines of a bias circuit. $R_L$, load resistance; $V_b$, bias voltage to load resistor; $T_s$, substrate temperature; $Q$, infrared radiation input power; $dQ$, absorbed radiation power change; $dV$, $dI$, increments of microbolometer voltage and current $V$, $I$; $R$, microbolometer resistance.

$\Re_I$ can be found from

$$\Re_I = \frac{\delta I}{\delta Q} = \frac{\delta I}{\delta V}\frac{\delta V}{\delta Q} = -\frac{1}{R_L} \cdot \Re_V \qquad (41)$$

A similar expansion of $\beta$ can be used to find the low-frequency temperature responsivity $\Re_T = dT/dQ$:

$$\beta = \frac{W\delta(V/I)}{R\delta(VI+Q)} = \frac{I^2\left(\dfrac{\delta V}{I} - \dfrac{V\delta I}{I^2}\right)}{(I\delta V + V\delta I + \delta Q)} = \frac{\left(I\dfrac{\delta V}{\delta I}\dfrac{\delta I}{\delta T} - V\dfrac{\delta I}{\delta T}\right)}{\left(I\dfrac{\delta V}{\delta I}\dfrac{\delta I}{\delta T} + V\dfrac{\delta I}{\delta T} + \dfrac{\delta Q}{\delta T}\right)} \qquad (42)$$

which can be manipulated to the result

$$\mathfrak{R}_T = \frac{\delta T}{\delta Q} = \frac{1}{g} \cdot \frac{1}{1 + \beta \frac{R - R_L}{R + R_L}} \quad (43)$$

The temperature responsivity at zero applied bias earlier was shown to be $1/g$. Thus from Eq. (43), the "effective" $g$ value for changes in $Q$ with bias applied is

$$g_{eff} = g \left\{ 1 + \beta \frac{R - R_L}{R + R_L} \right\} \quad (44)$$

where $g$ is the physical thermal conductance between the bolometer and its surroundings. A bolometer will be stable provided that $g_{eff} > 0$, which is satisfied for any value of $R_L$ provided the bias point is below the maximum on the V-I curve.

Since the thermal capacity of a microbolometer is unchanged by applied bias, the effective $g$ value results in an "effective" thermal time constant for changes in $Q$, given by

$$\tau_{eff} = \tau \frac{1}{1 + \beta \frac{R - R_L}{R + R_L}} \quad (45)$$

where $\tau = c/g$ is the time constant with no applied bias, and

$$\beta = \alpha \Delta T \quad (46)$$

The previous responsivity equations assume that the radiation power varies at a slow rate compared with $1/2\pi\tau_{eff}$. As the frequency of the IR power signals increases, the responsivity decreases, with a 3 dB decrease (Eq. (21)) in responsivity at the condition $2\pi\tau_{eff} = 1$.

From these equations we see that the responsivity increases with bias voltage $V_b$. Bias voltage is usually limited by one of a number of practical limits: monolithic read-out electronics may reach a maximum voltage (e.g., about 8 V for typical CMOS), the temperature rise in the microbolometers may increase the signal dynamic range to an unacceptable level, or IR radiation from the focal plane may become unacceptable for military purposes. In addition, at higher bias levels, we will subsequently see that the microbolometer noise begins to increase, negating the advantage of higher bias. It also can be seen that responsivity increases with $1/g$. For a given microbolometer cell area (which usually is set by imaging system requirements) the thermal capacity $c$ should be made as small as possible, so that $g$ may be decreased until $\tau = c/g$ becomes close to the frame time. For 30-Hz frame rate operation, a typical $\tau$ might be as long as 20 msec, so that for a microbolometer with $c = $ 3E-9 J/K, $g$ can be made as amall as 1.5E-7 W/K. For best responsivity at any given microbolometer joule power dissipation, the bias load resistor $R_L$ should be made very large or very small compared with $R$; that is, the microbolometer should be operated with constant current or constant voltage bias.

a. *Example of the Constant Current Bias of Metal–Temperature Coefficient of Resistance Microbolometer*

For constant current bias ($R_L \gg R$) and a positive $\alpha$, the previous zero-frequency voltage responsivity becomes

$$\Re_V = \frac{\delta V}{\delta Q} = \frac{\beta}{I(1-\beta)} = \frac{IR\alpha}{(g - I^2 R\alpha)} = \frac{IR\alpha}{g_{eff}} \quad (47)$$

The responsivity can be attributed to an effective $g$ value

$$g_{eff} = g - I^2 R\alpha \approx g(1 - \alpha \Delta T) \quad (48)$$

Provided $\alpha \Delta T \ll 1$, $g_{eff} \approx g$ but becomes zero at a critical current

$$I_c = \sqrt{\frac{g}{R\alpha}} \quad (49)$$

The effective time constant $\tau_{eff}$ of the bolometer is

$$\tau_{eff} = \tau \frac{1}{1-\beta} \approx \tau \frac{1}{1 - \alpha \Delta T} \quad (50)$$

The changed $g$ and $\tau$ values are due to the "electrothermal" interaction between the joule heating of the bias circuit and the incident radiation. These effects are small, provided the heating of the bias is such that $\alpha \Delta T \ll 1$.

### b. *Example of the Constant Voltage Bias with Semiconductor-like Temperature Coefficient of Resistance*

For a constant-voltage bias source ($R_L \ll R$) the current responsivity is

$$\mathfrak{R}_I = -\frac{\beta}{V(1+\beta)} = \frac{I\alpha}{g + IV\alpha} = \frac{I\alpha}{g_{eff}} \tag{51}$$

where

$$g_{eff} = g + IV\alpha \approx g(1 + \alpha \Delta T) \tag{52}$$

and

$$\tau_{eff} = \tau \frac{1}{1+\beta} = \tau \frac{1}{1 + \alpha \Delta T} \tag{53}$$

Since $\alpha$ is a negative value, $\tau_{eff} > \tau$ and $g_{eff} < g$.

## 9. Low-Frequency Noise in Microbolometer with Applied Bias

Assume that a microbolometer contains an internal low-frequency ($\omega = 0$) voltage noise disturbance $dV_n$, as shown in Fig. 14. We calculate the resulting voltage change $dV$ that appears across the bolometer to determine if the full internal noise $dV_n$ appears across the bolometer:
Differentiating

$$V = \frac{V_b - V_n}{R + R_L} R + V_n \tag{54}$$

we obtain

$$\frac{dV}{dV_n} = \frac{R_L}{R + R_L} \left\{ 1 + I \frac{dR}{dV_n} \right\} \tag{55}$$

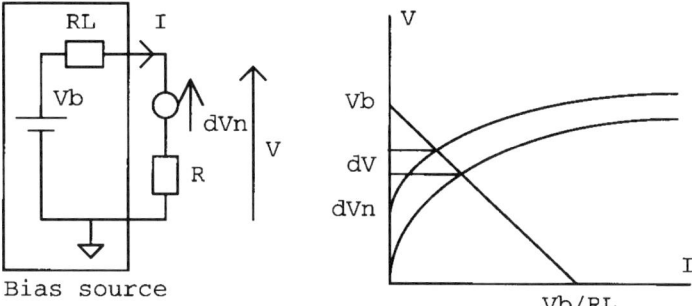

FIG. 14. Load line of a microbolometer with internal low-frequency noise. $R_L$, load resistance; $V_b$, bias voltage to load resistor; $dV_n$, voltage noise disturbance; $dV$, increment of voltage; $V$, $I$, microbolometer voltage and current; $R$, microbolometer resistance.

where $dR/dV_n$ can be found from

$$\frac{dR}{dV_n} = \frac{R\beta}{W}\frac{dW}{dV_n} \tag{56}$$

where

$$W = I^2R + IV_n = \left\{\frac{V_b - V_n}{R + R_L}\right\}^2 R + \frac{V_b - V_n}{R + R_L}V_n \tag{57}$$

This leads to the result ($\omega = 0$):

$$\frac{dV}{dV_n} = \frac{R_L}{R + R_L}\frac{1}{1 + \beta\dfrac{R - R_L}{R + R_L}} \tag{58}$$

This interesting result (Mather, 1982) shows that the electrothermal effect can affect how much of the noise from an internal noise voltage source actually appears across the microbolometer terminals. This is because the applied bias current dissipates power $IdV_n$ in the microbolometer, thereby changing the microbolometer temperature, thereby changing the bias point, and thereby changing the net voltage disturbance appearing across the microbolometer terminals. The effect only occurs at frequencies below the thermal roll-off frequency.

A similar result can be shown to occur for noise sources that may be modeled as low-frequency resistance fluctuations in the microbolometer resistance. A 1/f noise voltage is proportional to the applied current bias, indicating that it can be considered to be due to a fluctuating internal resistance $dR$ internal to the microbolometer resistance $R$. This fluctuating resistance causes a change in the bias heating power in the bolometer, which modifies the total voltage disturbance appearing across the bolometer. Manipulation of the circuit equations in a manner similar to that used previously shows that the voltage disturbance appearing across the bolometer is given by

$$\frac{dV}{dR} = I \frac{R_L}{R + R_L} \frac{1}{1 + \beta \frac{R - R_L}{R + R_L}} \quad (59)$$

which may be compared with the value $IR_L/(R + R_L)$ expected without electrothermal feedback ($\beta = 0$).

By a similar method, or by using the equivalent circuit of Fig. 15, we can show that a low-frequency noise voltage $dV_n$ in series with the load resistor $R_L$ produces a voltage noise $dV$ across the bolometer given by (Mather, 1982)

$$\frac{dV}{dV_n} = \frac{Z}{Z + R_L} = \frac{R}{R + R_L} \frac{1 + \beta}{1 + \beta \frac{R - R_L}{R + R_L}} \quad (60)$$

## 10. MICROBOLOMETER RESPONSIVITY WITH PULSED BIAS OR LARGE RADIATION SIGNALS

The previous equations have been derived assuming that the applied bias does not vary with time and that the signals are small. These equations are

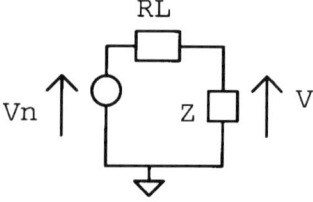

FIG. 15. Small signal equivalent circuit of a bias circuit. $R_L$, load resistance; $V_n$, voltage noise. Z, microbolometer slope impedance; $V$, $I$, microbolometer voltage and current.

very useful in understanding the way in which the microbolometer parameters trade-off against each other, and in providing insight into the physics of microbolometer operation. However, as will be discussed later in this chapter, read-out of two-dimensional arrays of microbolometers is usually performed using large pulsed applied bias, with any individual microbolometer being biased only during the instants when that particular microbolometer is being interrogated by the external camera circuits. With pulsed bias, the microbolometer never reaches a steady-state temperature, and performance parameters change rapidly during the bias pulse. The external camera circuits usually integrate the signal voltage or current over each bias pulse so that a performance-average is measured over the bias pulse. Radiation-induced temperature signals also can be quite large in some circumstances—for example, in industrial radiometry of very hot targets—so that the small-signal equations do not strictly apply. Microbolometer performance in these non–steady-state or large-signal conditions can be best calculated by numerical techniques, which easily can be performed on personal computers, as shown in the following examples. If the numerical results are compared with the results obtained from the steady-state small-signal equations, for many applications it will be found that the steady-state small-signal equations do produce results that are accurate enough for practical use.

## 11. Numerical Calculation of Microbolometer Performance

Figure 16 shows the load-line for a microbolometer under pulsed bias. For rapidly changing applied bias, the microbolometer V-I curve has a slope

$$Z(\omega) = R\frac{1 + \beta(\omega)}{1 - \beta(\omega)} \Rightarrow R \quad \text{as } \omega \to \infty$$

where $R = V/I$ at the bias point. Thus at any instant during the bias pulse the "instantaneous" microbolometer V-I curve is a straight line. During the duration of an applied bias pulse, the microbolometer temperature increases due to joule heating, and therefore the slope of the V-I line changes progressively during the bias pulse. The operating point moves as shown in Fig. 16 for a negative TCR. Changes in the net absorbed radiation power $Q$ also shift the overall operating point of the bolometer, and the resulting change in microbolometer voltage or current comprises the signal response to $Q$.

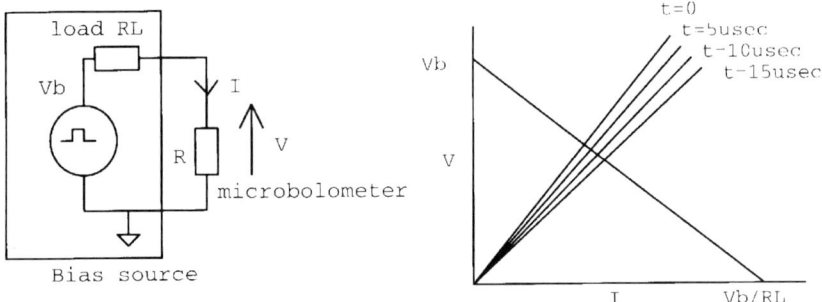

FIG. 16. Microbolometer load line with pulsed bias. $R_L$, load resistance; $V_b$, bias voltage to load resistor; $V$, $I$, microbolometer voltage and current; $R$, microbolometer resistance.

The quantitative time-dependent change in microbolometer bias point under pulsed bias and net absorbed radiation power $Q$ can be calculated using the heat balance equation and the load line equation for a microbolometer temperature $T$:

$$c\frac{\delta T}{\delta t} = IV + Q - g(T - T_s) \quad (61)$$

where

$$V = V_b \frac{R(T)}{R_L + R(T)} \quad (62)$$

and

$$I = \frac{V_b}{R_L + R(T)} \quad (63)$$

$Q$ can be expressed in terms of Stefan's law, if required.

Combining these equations, Eqs. (61) to (63), we have

$$\delta T = \frac{1}{c}\left\{\frac{V_b}{R_L + R(T)} V_b \frac{R(T)}{R_L + R(T)} + Q - g(T - T_s)\right\} \delta t \quad (64)$$

where, for a semiconductor resistor material,

$$R(T) = R(T_s) \exp\left(\frac{\Delta E}{KT}\right) \quad (65)$$

Equations (64) and (65) can be used to numerically compute curves of the time-dependent microbolometer temperature under varying bias and radiation power, for example, by summing the effects of small increments of time $\delta t$ using a personal computer. This allows computation of the time-dependent microbolometer temperature under bias pulses with large radiation signal power. Once the time-dependent microbolometer temperature is calculated, microbolometer responsivity can be derived. If desired, radiation heat transfer can also be included in the previous equations.

If short bias pulses $\Delta t$ are used, separated by relatively long intervals, large pulsed bias voltages can be applied without producing excessive microbolometer heating. Bias currents are commonly used in microbolometer read-out that, if applied continuously, would produce extremely high microbolometer temperatures. Microbolometer temperature elevations in a normal read-out are typically a few degrees (°C) above ambient temperature. The approximate microbolometer heating as a function of applied bias can be calculated as follows. If the applied bias pulses each dissipate joule energy $P\Delta t$, then the temperature increase in each pulse is $P\Delta t/c$. If the bias pulses are applied at frame rate $F$ and the thermal time constant of the microbolometer is $\tau$, then the mean power dissipation is $P\Delta tF$ and the mean temperature increase is $\overline{\Delta T} = P\Delta tF/g$.

Provided $\beta \ll 1$, we find that use of dc-bias equations and room temperature parameter values produces numerical results that agree acceptably well — for many applications — with numerical calculations that take into account the full time dependence of the microbolometer parameters. The time-dependent numerical approach can be very powerful and easy to use if a few hours are invested in preparing personal computer programs and spreadsheets. Many commercial computer programs suitable for this purpose are available.

As an example, Fig. 17 shows computed values of some microbolometer parameters, assuming the parameters summarized in Table II and assuming a net radiation power of 1 nW is absorbed within the microbolometer. The numerical calculations of Fig. 17 were performed using a BASIC program and an EXCEL™ spreadsheet, with 10 $\mu$sec time increments over an elapsed time of 30 msec, assuming a microbolometer initial temperature $T_s = 300$ K: the time interval plotted in Fig. 17 covers only the last 200 $\mu$sec of the 30 msec period, at the end of which the second of two 100-$\mu$sec applied bias pulses $\Delta t$ is assumed to occur. The microbolometer temperature changes very rapidly during the bias pulse — at a rate of $VI/c \approx 50,000°C$ per second. However, the 100-$\mu$sec duration of the bias pulse limits the total temperature increase to about 5°C. During the interval between bias pulses (30 msec) the microbolometer cools, and it reaches about 300.2 K — close to the substrate temperature (300 K) — by the time the next bias pulse begins. At any instant

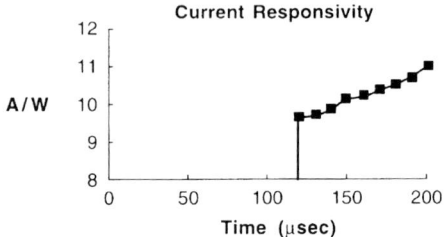

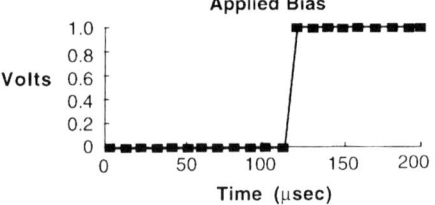

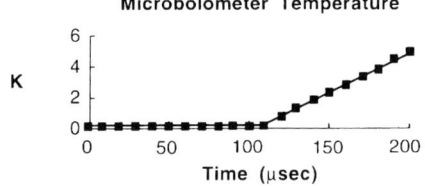

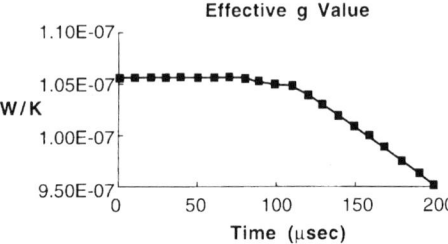

FIG. 17. Numerical calculations made assuming the parameters in Table II, as discussed in the text. $g_{eff}$, effective value of thermal conductance of microbolometer to substrate.

the effective $g$ value is $g_{eff} = Q/\Delta T$, where $\Delta T$ is the temperature perturbation caused by the presence of $Q$. This is computed and shown in Fig. 17. In the initial part of the time period plotted in Fig. 17, $g_{eff}$ is slightly greater than its assumed physical value ($g$) of 1E-7 W/K (Table II), because insufficient time has elapsed (30 msec) for the microbolometer to reach the final equilibrium temperature expected from $Q$. The rapid decrease in $g_{eff}$ to

## 3 MONOLITHIC SILICON MICROBOLOMETER ARRAYS

TABLE II

ASSUMED PARAMETER VALUES FOR FIG. 17

| Parameter | Value |
|---|---|
| $R(T_s)$ | 20 KΩ |
| $R_L$ | 0 |
| $V_b$ | 1 V |
| $\Delta t$ | 100 μsec |
| $1/F$ | 30 msec |
| $T_s$ | 300 K |
| $\alpha$ | $-0.02 \text{ K}^{-1}$ at 300 K |
| $c$ | 1E-9 J/K |
| $g$ | 1E-7 W/K |

values less than $g$ during the bias pulse is due to the electrothermal effect (Eqs. (44)). The time-dependent microbolometer responsivity is computed and shown in Fig. 17: The current responsivity $\Re_I = dI/dQ$ is plotted rather than the voltage responsivity $\Re_V$ because, in this example, the microbolometer is assumed to be operated at a constant voltage ($R_L = 0$, Table II) during a bias pulse.

### III. Noise in Bolometers

The noise effects that influence microbolometer systems are

1. Johnson noise in the bolometer resistance
2. variations in the bolometer resistance
3. variations in the bolometer temperature
4. noise present on the incident IR radiation stream

We shall not consider electronic noise in read-out circuits in this chapter, because this is a matter of proper electrical circuit design using established electronic engineering principles.

1. BOLOMETER RESISTANCE NOISE

A resistance $R$ at temperature $T$ develops a fluctuating voltage across its terminals (Johnson noise) due to thermal agitation of charge carriers. The electrical noise power is "white" for frequencies $f \ll KT/h$, that is, the electrical noise power per unit electronic bandwidth is independent of frequency. The Johnson noise of a microbolometer is usually represented by an equivalent circuit consisting of a noiseless resistor $R$ in series with a Johnson voltage noise source (Fig. 18), together with its noiseless self-capacitance $C$. The voltage noise fluctuates in a manner such that

$$\sqrt{\overline{(V - \bar{V})^2}} = \sqrt{\overline{\Delta V^2}} = \sqrt{4KTR\Delta f}$$

where $\overline{\Delta V^2}$ is the *variance* of $V$ from its average value $\bar{V}$, that is, the mean square value of the fluctuating noise voltage. A convenient benchmark is that a 1-K$\Omega$ resistor at room temperature has an root mean square (rms) Johnson noise voltage of $4\,\text{nV}/\sqrt{\text{Hz}}$. The total voltage variance $\overline{\Delta V^2}$ across the resistor terminals is

$$\overline{\Delta V^2} = \int_0^\infty 4KTR \left| \frac{1/j\omega C}{1/j\omega C + R} \right|^2 d\omega = \frac{KT}{C} \qquad (66)$$

For convenient numerical estimates, the voltage noise can be considered to have an rms value of $\sqrt{4KTR}\,V/\sqrt{\text{Hz}}$ extending uniformly over an *effective noise bandwidth* of $1/4CR$ (Motchenbacher and Fitchen, 1973).

Microbolometer resistor materials usually show some degree of fluctuation in resistance (Voss and Clarke, 1976), with a variance $\overline{\Delta R^2}$. If a bias $I$ flows through the resistor, this resistance fluctuation produces a bias-dependent "excess" voltage noise $I\sqrt{\overline{\Delta R^2}}$, over and above the Johnson noise that exists even at zero bias.

Two types of excess noise are reported: 1/f noise and telegraph noise. The first type has a power per unit bandwidth that varies approximately as 1/f, and is therefore termed *1/f noise*. This noise source is represented as an additional voltage source, uncorrelated with the Johnson noise source, of mean square magnitude in a 1 Hz bandwidth.

$$\overline{\Delta V^2} = V^2 \frac{k}{f} \qquad \text{at frequency } f \qquad (67)$$

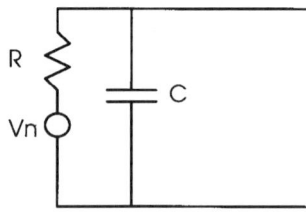

FIG. 18. Johnson noise equivalent circuit of a resistor. $V_n$, voltage noise; $R$, resistance; $C$, capacitance.

where the parameter $k$ is called the *1/f noise parameter* (Motchenbacher and Fitchen, 1973). The parameter $k$ is strongly dependent on the particular resistor—on its material, deposition technique, dimensions, and electrical contacts, etc. Theoretic work indicates that an (extremely low) level of 1/f noise is inescapable for quantum mechanical reasons. However, the 1/f noise level measured in all real resistors is many orders of magnitude above this level, and appears to result from imperfections in resistor material structure and electrical contacts. Unewisse *et al.* (1995a), for example, report that 1/f noise in amorphous Si varies by three orders of magnitude, depending on the growth technique. The parameter $k$ is related to the *Hooge parameter* $\alpha_H$ (Voss and Clarke, 1976) by the relation $k = \alpha_H/nV$, where $n$ is the mobile charge carrier density and $V$ is the volume of the resistor material. Typical $k$ values in different resistor materials are listed by Motchenbacher and Fitchen (1973). Measurements in $VO_x$ microbolometer resistors have shown $k$ values typically of 1E-13 for $\alpha = -0.02 \, K^{-1}$ (Wood and Stelzer, 1992).

Telegraph noise (so-called popcorn noise) is the name given to sudden voltage steps (Unewisse *et al.*, 1995b) caused by unknown imperfections in material structure and contacts.

The excess noise (equation 67) increases with increasing bias across the resistor, and is always zero if no bias exists across the resistor. At zero applied bias, only Johnson noise is present. No shot noise is associated with the bias current flow in a resistor because shot noise only occurs when current flows across a potential barrier, such as a p-n junction (Smith *et al.*, 1958).

The noise equivalent circuit of a microbolometer therefore is a noiseless resistor $R$ in series with two uncorrelated rms voltage noise sources, and the resistor's noiseless self-capacitance (Fig. 19).

For microbolometers, we can usually assume that $C$ is low enough that signals are confined to the range $f \ll 1/2\pi CR$, where the rms noise per unit

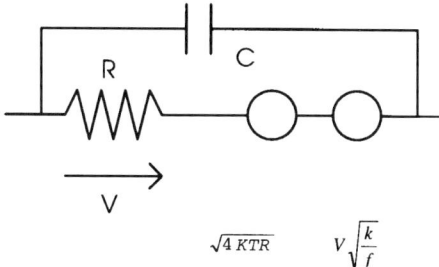

FIG. 19. Noise equivalent circuit of a resistor for 1-Hz bandwidth at frequency $f$. $R$, resistance; $C$, capacitance.

bandwidth is

$$\sqrt{4KTR + V^2 \frac{k}{f}}$$

The electrical noise increases at lower frequencies due to the increasing contribution of 1/f noise, and at higher frequencies becomes equal to the Johnson noise. The *knee frequency* is defined as the frequency at which the 1/f noise power equals the Johnson noise power in a 1-Hz interval, that is,

$$4KTR = V^2 \frac{k}{f_{knee}} \quad \text{that is,} \quad f_{knee} = \frac{V^2 k}{4KTR} \qquad (68)$$

In a measurement bandwidth extending from frequency $f_1$ to $f_2$, these two noise powers combine to produce a total mean square microbolometer voltage noise ($f \ll 1/2\pi CR$):

$$\overline{\Delta V^2} = 4KTR(f_2 - f_1) + V^2 \int_{f_1}^{f_2} \frac{1}{f} df = 4KTR(f_2 - f_1) + V^2 \ln\left(\frac{f_2}{f_1}\right) \qquad (69)$$

Microbolometer signals (currents or voltages) are usually integrated electronically over the duration of each bias pulse $\Delta t$. The upper noise bandwidth limit $f_2$ of an ideal integrator with integration time $\Delta t$ is (Boyd, 1983)

$$f_2 = \frac{1}{2\Delta t} \qquad (70)$$

The lower bandwidth limit $f_2$ is determined by the *staring time* $T_{stare}$ by the relation

$$f_1 \approx \frac{1}{4T_{stare}} \qquad (71)$$

As a numerical example, for a microbolometer with resistance 10 KΩ, applied bias 1V, $k = $ 1E-13, $T = $ 300 K, the 1/f knee is at 600 Hz (Fig. 20). The white (Johnson) noise level is 13 nV per root Hz. In a bandwidth from 0.001 to 10 KHz, the total Johnson noise is 1.3 μV rms, the total 1/f noise is 0.7 μV rms, and the total noise from both noise contributions is 1.5 μV rms. Most of the microbolometer noise therefore is Johnson noise with these assumed parameters. Increasing $V$ from 1V to 2V reverses this conclusion, that is, 1/f noise dominates the total noise with 2V applied bias.

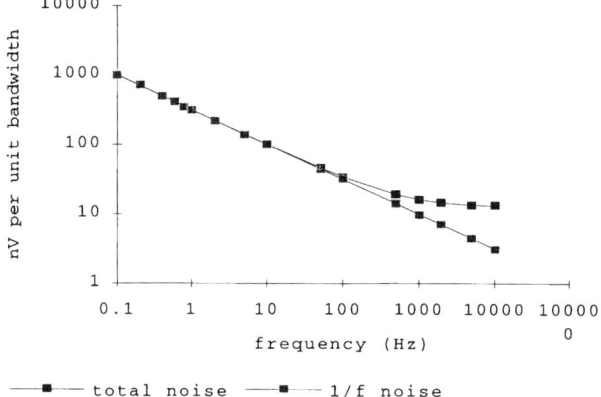

FIG. 20. Calculated resistor noise for assumed parameters listed in the text and $V = 1$ Volt.

## 2. Noise from Bias Resistors

For the bias circuit of Fig. 21, the voltage noise appearing across the microbolometer is increased by the noise from the additional resistor $R_L$. Therefore, the mean square voltage noise across the microbolometer is (assuming $R_L$ has negligible 1/f noise)

$$\overline{\Delta V^2} = 4KT\frac{RR_L}{R + R_L}(f_2 - f_1) + V_b^2 k \ln\left(\frac{f_2}{f_1}\right)\left\{\frac{RR_L}{(R + R_L)^2}\right\}^2 \quad (72)$$

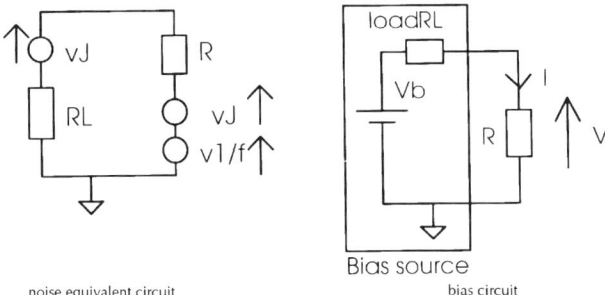

FIG. 21. Bias circuit and noise equivalent circuit for microbolometer $R$ and load resistor $R_L$. $V_b$, bias voltage to load resistor; $vJ$, Johnson noise voltage source; $v1/f$, 1/f noise voltage source.

and the current noise flowing through the microbolometer is

$$\overline{\Delta I^2} = 4KT\frac{1}{R + R_L}(f_2 - f_1) + V_b^2 k \ln\left(\frac{f_2}{f_1}\right)\left\{\frac{R}{(R + R_L)^2}\right\}^2 \quad (73)$$

As noted previously, these noise terms are modified at frequencies less than $1/2\pi\tau$ because of the electrothermal interaction. Because the system bandwidth of a microbolometer read-out is typically much greater than $1/2\pi\tau$, the electrothermal interaction produces a negligible effect on the total (rms) white noise. The electrothermal effect, however, may produce a noticeable effect on the total noise from effects whose major contribution lies at frequencies below $1/2\pi\tau$.

3. THERMAL CONDUCTANCE NOISE

The variance $\overline{(U - \bar{U})^2}$ of a quantity $U$ in thermal equilibrium with a large thermal reservoir at temperature $T$ is given by

$$\overline{(U - \bar{U})^2} = \overline{\Delta U^2} = \frac{-k}{\partial^2 S/\partial U^2} \quad (74)$$

Therefore, putting $U = T$ and using

$$\frac{\partial S}{\partial T} = \frac{1}{T}\frac{dQ}{dT} = \frac{c}{T}$$

we obtain, for a microbolometer at zero bias,

$$\overline{\Delta T^2} = kT^2/c \quad (75)$$

If thermal noise is the only noise present (the ideal case), then the noise equivalent power of the bolometer is

$$\text{NEP (ideal)} = g\sqrt{KT^2/c} \quad (76)$$

This temperature noise has a power density of (Marasco and Dereniak, 1993; Knise, 1995; Jensen, 1993)

$$\Delta T^2(f) = 4KT^2 g \cdot \frac{1}{g^2(f)} = \frac{4KT^2}{g} \cdot \frac{1}{1 + \omega^2\tau^2} \quad (77)$$

that is, the zero-frequency temperature noise density is $\sqrt{4KT^2/g}$, and the equivalent noise bandwidth $\Delta f$ (Motchenbacher and Fitchen, 1973) is the same as an electrical RC filter, that is, $\Delta f = 1/4\tau$ Hz.

As a numerical example, a microbolometer with $c =$ 3E-9 J/K at 300 K has a thermal noise of 20 $\mu$K rms.

The previous results are strictly correct only for a microbolometer in thermal equilibrium with its heat reservoir, that is, the substrate. If the microbolometer is heated above the substrate temperature $T_s$ by an applied bias, Mather (1982) has shown that the previous thermal noise equations are modified due to the thermal gradient across the supporting legs of the microbolometer. This is not a significant effect with typical microbolometer conditions.

### 4. Radiation Noise

The variance $\overline{\Delta p^2} = \overline{(p - \bar{p})^2}$ in the radiation power $p$ incident in a solid angle $\Omega$ on an area $A$ of the wall of a backbody at temperature $T$ is (Dereniak and Crowe, 1984)

$$\overline{\Delta p^2} = \left(\frac{cA}{4}\right)\left\{\left[\frac{\Omega}{\pi}\right] - \left[\frac{\Omega}{2\pi}\right]^2\right\}\left(\frac{8\pi}{c^3}\right)\left(\frac{KT}{h}\right)^5 h^2 \int_{x_1}^{x_2} x^4 \frac{e^x}{(e^x - 1)^2} dx \quad (78)$$

where $x = h\nu/KT$, where $\nu$ is the radiation frequency. In thermal equilibrium, there is an equal noise on the outgoing (re-emitted) radiation from the surface $A$. For a microbolometer in thermal equilibrium with its surroundings and having two surfaces, each of area $A$, the total radiation noise power affecting the bolometer is therefore

$$\overline{\Delta p^2} = 2\left(\frac{c(2A)}{4}\right)\left\{\left[\frac{\Omega}{\pi}\right] - \left[\frac{\Omega}{2\pi}\right]^2\right\}\left(\frac{8\pi}{c^3}\right)\left(\frac{KT}{h}\right)^5 h^2 \int_{x_1}^{x_2} x^4 \frac{e^x}{(e^x - 1)^2} dx \quad (79)$$

Integrating over all wavelengths, putting $\Omega = 2\pi$, and using Stefan's constant $\sigma = 2\pi^5 K^4/15c^2h^3$, this reduces to

$$\overline{\Delta p^2} = 8(2A)\varepsilon\sigma KT^5 = 2KT^2 g_{rad} \quad \text{where} \quad g_{rad} = 4(2A)\varepsilon\sigma T^3 \quad (80)$$

Equation (80) gives the total radiation noise power, which is spread over a very wide frequency range, extending up to much higher than normal signal frequencies. The radiation noise power variance per unit bandwidth at low frequencies $S(f)_{f\to 0}$ (here low means compared with the photon arrival rate) can be shown to be (Lewis, 1947; Van der Ziel, 1986)

$$S(f)_{f \to 0} = 2\overline{\Delta p^2} = 4KT^2 g_{rad} \qquad (81)$$

The temperature fluctuation in a bolometer induced by this radiation noise is

$$\overline{\Delta T^2} = \int_0^\infty \frac{S(f)}{g^2} \cdot \frac{1}{1 + (2\pi f \tau)^2} df$$
$$= \frac{4(2A)\varepsilon\sigma KT^5}{cg} = \frac{KT^2}{c} \cdot \frac{g_{rad}}{g_{leg} + g_{rad}} \qquad (82)$$

The difference between the total thermodynamic temperature noise $kT^2/c$ and the temperature noise induced by radiation noise can be considered to be due to thermal noise from power fluctuations in the leg thermal conductance $g_{leg}$. The temperature variance in a microbolometer due to the leg thermal conductance $g_{leg}$ is therefore

$$\overline{\Delta T^2} = \frac{KT^2}{c} - \frac{KT^2}{c} \frac{g_{rad}}{g_{leg} + g_{rad}} = \frac{KT^2}{c} \left\{ \frac{g_{leg}}{g_{leg} + g_{rad}} \right\} \qquad (83)$$

The total temperature noise $\overline{\Delta T^2} = kT^2/c$ induced by radiation and leg thermal conductance is called *thermal noise*, and it constitutes the lowest possible noise that a microbolometer can have.

As shown by Eqs. (77) and (81), the noise power variance density $S(f)_{f \to 0}$ due to random energy flow fluctuations in any thermal conductance $g$ can be considered (Mather, 1982) to be $4KT^2g$, where, for a microbolometer, $g$ is $g_{leg}$ and $g_{rad}$, and the equivalent noise bandwidth (Motchenbacher and Fitchen, 1973) for a microbolometer is $1/4\tau$.

When a bias is applied to a microbolometer, the voltage variance $\overline{\Delta V^2}$ across the microbolometer due to the total thermal noise power $\overline{\Delta Q^2} = (4KT^2g)(1/4\tau) = (KT^2/c)g^2$ is therefore (Eq. (39)):

$$\overline{\Delta V^2} = \left\{ V_b \frac{RR_L\alpha}{(R + R_L)^2} \right\}^2 \left\{ \frac{1}{1 + \beta \frac{R - R_L}{R + R_L}} \right\}^2 \left\{ \frac{KT^2}{c} \right\} \qquad (84)$$

Equation (84) is written using the assumption that the electrothermal effect acts with full force at all thermal noise frequencies: This is only strictly true for frequencies $f \ll 1/2\pi\tau$. To allow for this, the electrothermal factor can be replaced with its frequency-averaged form (Eq. (27)).

## 5. TOTAL ELECTRICAL NOISE

Now, summing the thermal, Johnson, and 1/f voltage noise powers, we can write the total mean square voltage noise $\overline{\Delta V^2}$ across the microbolometer in bandwidth $f_1$ to $f_2 > 1/2\pi\tau$:

$$\overline{\Delta V^2} = \left\{ V_b \frac{RR_L \alpha}{(R+R_L)^2} \right\}^2 \left\{ \frac{1}{1+\beta \frac{R-R_L}{R+R_L}} \right\}^2 \left\{ \frac{KT^2}{c} \right\}$$

$$+ 4KT \frac{RR_L}{R+R_L}(f_2 - f_1) \qquad (85)$$

$$+ V_b^2 k \ln\left(\frac{f_2}{f_1}\right) \left\{ \frac{RR_L}{(R+R_L)^2} \right\}^2 \left\{ \frac{1}{1+\beta \frac{R-R_L}{R+R_L}} \right\}^2$$

Similarly, the current noise through the microbolometer is

$$\overline{\Delta I^2} = \left\{ V_b \frac{R\alpha}{(R+R_L)^2} \right\}^2 \left\{ \frac{1}{1+\beta \frac{R-R_L}{R+R_L}} \right\}^2 \left\{ \frac{KT^2}{c} \right\}$$

$$+ 4KT \frac{1}{R+R_L}(f_2 - f_1) \qquad (86)$$

$$+ V_b^2 k \ln\left(\frac{f_2}{f_1}\right) \left\{ \frac{R}{(R+R_L)^2} \right\}^2 \left\{ \frac{1}{1+\beta \frac{R-R_L}{R+R_L}} \right\}^2$$

To minimize complexity, we write Eqs. (85) and (86) using the simplifying assumption that the electrothermal effect acts with full force on the thermal noise and 1/f noise components, because these are strongest at low frequencies, and can be ignored for Johnson noise, which extends to very high frequencies. The electrothermal terms can be integrated over the system bandwidth using their frequency-dependent form (Eq. (27)), to obtain an effective value of the electrothermal term $\overline{f(\beta)}$:

$$\overline{f(\beta)} = \overline{\left\{ 1 + \beta \frac{R-R_L}{R+R_L} \right\}^2} \qquad (87)$$

Here we also note that the *average voltage noise per unit bandwidth* is

$$\sqrt{\overline{V^2}}/\sqrt{f_2 - f_1} \qquad (88)$$

This parameter is of interest because this term appears in the classic expression for $D^*$. From the above equations, we can see that, with pulsed bias, the average voltage noise per unit bandwidth approaches the Johnson noise value $\sqrt{4KTR//R_L}$ when $f_2$ becomes large, that is, when short bias pulses are used. We will draw on this result in a subsequent section that deals with the calculation of the detectivity $D^*$.

As a numerical example from Eqs. (85) and (86), and Table III we calculate the following at $4V$ bias:

| | |
|---|---|
| Thermal voltage noise: | 0.4 $\mu$V rms |
| Johnson voltage noise: | 1.6 $\mu$V rms |
| 1/f voltage noise: | 2.0 $\mu$V rms |
| Total voltage noise: | 2.6 $\mu$V rms |
| Total current noise: | 65 pA rms |
| Mean noise per unit bandwidth: | 22 nV/$\sqrt{Hz}$ |
| Johnson noise: | 13 nV/$\sqrt{Hz}$ |
| Thermal noise at low frequencies: | 138 nV/$\sqrt{Hz}$ |

The dependence of the microbolometer noise on bias for the parameters of Table III is plotted in Fig. 22. We see that for bias less than 100 $\mu$A, Johnson noise is the dominant contributor to the microbolometer noise. Above a bias of 100 $\mu$A, 1/f noise becomes the dominant noise contributor.

TABLE III

ASSUMED PARAMETERS FOR NUMERICAL EXAMPLES

| Parameter | Value | Parameter | Value |
|---|---|---|---|
| $R(T_s)$ | 20 K$\Omega$ | $A_{cell}$ | 50 $\mu$m square |
| $R_L$ | 20 K$\Omega$ | $F_f$ | 0.75 |
| $V_b$ | <4 V | $\varepsilon$ | 0.8 |
| $\Delta t$ | 35 $\mu$sec | $f_2$ | 15 KHz[b] |
| Bias pulse period | 33 msec[a] | $f_1$ | 0.0001 Hz[c] |
| $T_s$ | 300 K | $k$ | 1E-13 |
| $\alpha$ | $-0.02$ K$^{-1}$ at 300 K | $F_{no}$ | 1.0 |
| $c$ | 3E-9 J/K | $dL/dT_t$ | 8E-5 W cm$^{-2}$ sr$^{-1}$ K$^{-1}$ [d] |
| $g$ | 1E-7 W/K | | |

[a]30-Hz frame rate.
[b]1/(2∗35 $\mu$sec), equation 70.
[c]$T_{stare} = 1/4f_1 = 40$ min.
[d]300 K blackbody target, 8 to 12 $\mu$m, Table I.

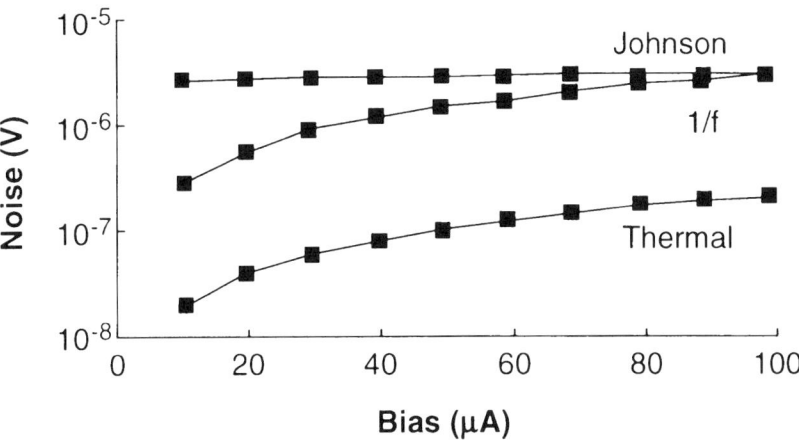

FIG. 22. Calculated root mean square microbolometer voltage noise versus bias current for the assumed parameters in Table III.

6. PREAMPLIFIER NOISE

As seen in the previous numerical example, the noise level for a microbolometer and load resistor is a typically few $\mu$V rms, or about 100 pA rms, over a bandwidth of a few tens of KHz. As discussed subsequently, the low bandwidth limit may extend down to very low frequencies in some modes of operation of microbolometer IR imagers (chopperless operation). Ideally, the amplifier used to amplify the microbolometer (voltage or current) signals will have a noise level significantly less than the microbolometer noise. This is more readily achieved with bipolar circuits, and is more difficult to achieve with complementary metal-oxide semiconductor (CMOS) circuits because of the 1/f noise associated with field-effect transistor (FET) gate insulator materials. The preamplifier noise contribution can be calculated using the frequency-dependent small-signal microbolometer impedance (Zwerdling et al., 1968; Mather, 1982):

$$Z(\omega) = R\frac{1 + \beta(\omega)}{1 - \beta(\omega)} \quad \text{where} \quad \beta(\omega) = \frac{\beta}{1 + j\omega\tau} \tag{89}$$

Because of the low noise levels of microbolometers, the design of suitable preamplifiers is best done with care, using established electrical engineering techniques.

## IV. Microbolometer Signal-to-Noise

Using the previous expressions for signal and noise, we can write expressions for the signal-to-noise ratio in several different forms.

### 1. Noise Equivalent Power (NEP)

We define the NEP as the absorbed (i.e., internal) power change that produces a signal equal to the total (rms) noise. In terms of voltage noise and voltage signals, then

$$\text{NEP} = \frac{\sqrt{\overline{V^2}}}{\Re_V} \tag{90}$$

$$\text{NEP} = \sqrt{g^2 \left\{\frac{KT^2}{c}\right\} + \frac{4KT\dfrac{RR_L}{R+R_L}(f_2 - f_1)\overline{f(\beta)} + V_b^2 k \ln\left(\dfrac{f_2}{f_1}\right)\left\{\dfrac{RR_L}{(R+R_L)^2}\right\}^2}{\left\{V_b \dfrac{RR_L \alpha}{(R+R_L)^2} \dfrac{1}{g}\right\}^2}} \tag{91}$$

In terms of current noise and current signals, then

$$\text{NEP} = \frac{\sqrt{\overline{I^2}}}{\Re_I}$$

$$= \sqrt{g^2 \left\{\frac{KT^2}{c}\right\} + \frac{4KT\dfrac{1}{R+R_L}(f_2 - f_1)\overline{f(\beta)} + V_b^2 k \ln\left(\dfrac{f_2}{f_1}\right)\left\{\dfrac{R}{(R+R_L)^2}\right\}^2}{\left\{V_b \dfrac{R\alpha}{(R+R_L)^2} \dfrac{1}{g}\right\}^2}} \tag{92}$$

In both cases, the best possible (smallest) NEP is $g\sqrt{KT^2/c}$.

### 2. Noise Equivalent Temperature Difference (NETD)

We define the NETD as the temperature change at the target that produces a signal in the bolometer equal to the total (rms) noise. The NETD can be calculated directly from NEP using the relation

$$\text{NEP} = \frac{dQ}{dT_t}\text{NETD} = \frac{A\varepsilon\pi}{4F_{no}^2}\frac{dL}{dT_t}\text{NETD} \quad (93)$$

Therefore, the NETD is

$$\text{NETD} = \frac{4F_{no}^2}{A\varepsilon\pi(dL/dT_t)}\text{NEP} \quad (94)$$

where NEP is given by Eq. (91) or (92).

3. DETECTIVITY

Assuming voltage signals are measured, $D^*$ can be defined as

$$D^* = \frac{\mathfrak{R}_V\sqrt{A}\sqrt{f_2 - f_1}}{\sqrt{\overline{V^2}}} \quad (95)$$

Microbolometer performance varies with applied bias, so to be a meaningful figure of merit, microbolometer $D^*$ should be calculated using the same bias as in the intended applications. For two-dimensional microbolometer arrays, this usually means a high bias applied in short pulses. We have derived expressions for $\mathfrak{R}_V$ and $\sqrt{\overline{V_n^2}}/\sqrt{f_2 - f_1}$ in the preceding sections, and these terms can be inserted in the expression for $D^*$. This produces an unexpected result, however. As noted previously, the term $\sqrt{\overline{V_n^2}}/\sqrt{f_2 - f_1}$ approaches the Johnson noise value when short bias pulses are used. This Johnson noise can be less than the thermal noise density, with the result that the $D^*$ computed using Eq. (95) with pulsed bias can be greater than the "maximum theoretic $D^*$" (1.8E-10 cm$\sqrt{\text{Hz}}$/w at 300 K) usually quoted for room temperature sensors (Rogalski, 1994). The reason for this apparent paradox is that the method used to calculate the maximum theoretic $D^*$ implicitly assumes that the noise bandwidth is lower than the thermal noise bandwidth $1/2\pi\tau$; that is, assumes a near dc-bias condition.

Although $D^*$ computed in this way may exceed the maximum theoretic $D^*$ under pulsed bias, use of the expression NEP = $\sqrt{A}\sqrt{f_2 - f_1}/D^*$ shows that the NEP of a microbolometer never becomes lower than the ideal value $g\sqrt{KT^2/c}$.

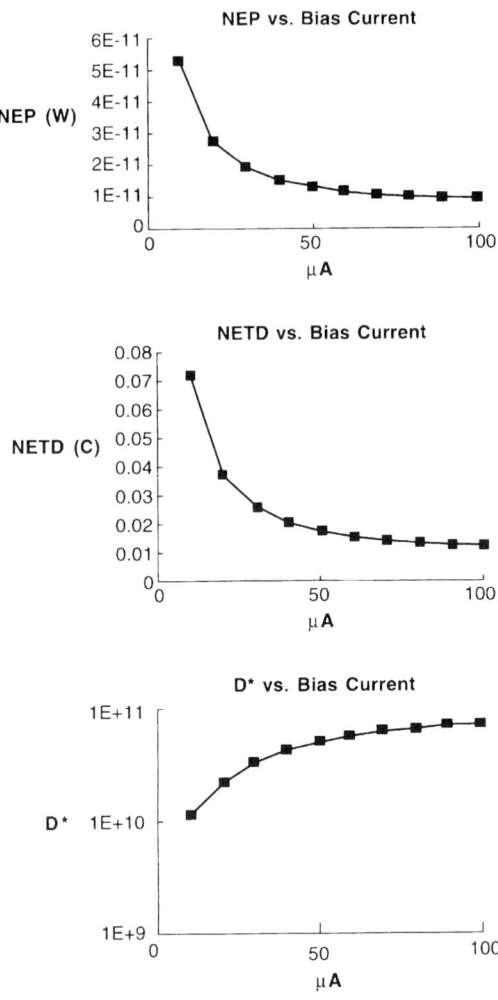

FIG. 23. Numerical calculations assuming the parameters used in Table III. NEP, noise equivalent power, NETD, noise equivalent temperature difference; $D^*$, detectivity.

Detectivity $D^*$ also can be obtained from the NETD by using the relation

$$D^* = \frac{4F_{no}^2}{A\varepsilon\pi.\text{NETD}(dL/dT_t)} \qquad (96)$$

As a numerical example, assuming the microbolometer parameters listed in Table III, the previous equations produce the following curves in Figs. 23(a) to 23(c).

## 4. COMPARISON WITH THE IDEAL BOLOMETER

We can define the *quality factor* $Q$ as the ratio of the calculated NEP (or NETD) to the *ideal NEP* (or NETD), which we define to be the ideal NEP value $g\sqrt{KT^2/c}$ (Eqs. (91) and (92)), which is attained if there is neither Johnson nor $1/f$ noise, but only thermal noise. The parameter $Q$ indicates how closely the microbolometer performance approaches the thermodynamic limit set by the thermal noise of the microbolometer.

From Eq. (92) for the NEP with current signals, we find that $Q$ can be written as

$$Q = \sqrt{1 + \frac{4KT\frac{1}{R+R_L}(f_2 - f_1)\overline{f(\beta)} + V_b^2 k \ln\left(\frac{f_2}{f_1}\right)\left\{\frac{R}{(R+R_L)^2}\right\}^2}{\left\{V_b \frac{R\alpha}{(R+R_L)^2}\right\}^2 \frac{kT^2}{c}}} \quad (97)$$

For constant-voltage bias — which is a convenient read-out technique that has been used for many bolometer arrays to date — $R_L = 0$ and $V_b/R = I$, and assuming the full electrothermal effect,

$$Q = \sqrt{1 + \frac{4KTR(f_2 - f_1)\{1 + \beta\}^2 + I^2 R^2 k \ln\frac{f_2}{f_1}}{(IR\alpha)^2 \frac{KT^2}{c}}} \quad (98)$$

Figure 24 plots the computed values of $Q$, assuming the parameters in Table IV, representing a microbolometer with a high TCR value. We see from Fig.

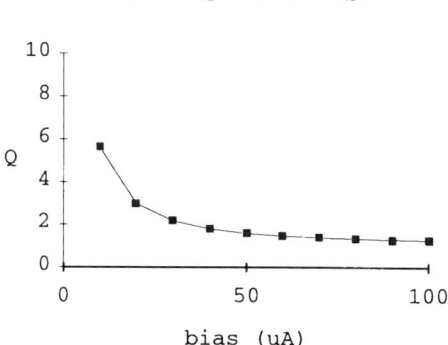

FIG. 24. Computed values of $Q$, assuming the parameters used in Table III.

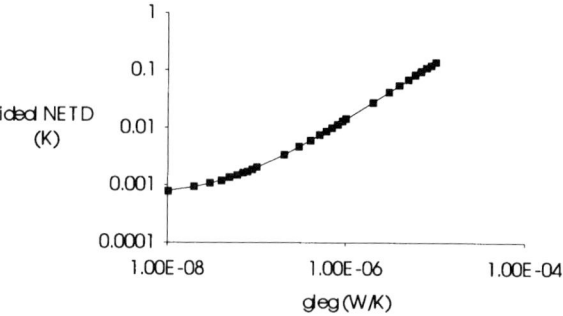

FIG. 25. Plot of the ideal noise equivalent temperature difference (NETD) versus $g_{leg}$ (thermal conductance of microbolometer legs) for microbolometer parameters of Table IV.

24 that an array with the parameters in Table IV operates efficiently, coming within a factor of two of the ideal (thermal limit) NETD, even at a low bias current of 35 μA ($\Delta T \approx 0.3°C$). In Fig. 24. $Q = 1$ corresponds to the ideal (thermodynamic limit) NETD of

$$\text{NETD}_{ideal} = \frac{4F_{no}^2}{AF_f \varepsilon \pi (dL/dT_t)} (g_{leg} + g_{rad}) \sqrt{\frac{KT^2}{c}} \qquad (99)$$

At first it is surprising that the array operates efficiently with a noise bandwidth much higher than the minimum value of $1/2\pi\tau$. The efficient

TABLE IV
ASSUMED MICROBOLOMETER PARAMETERS

| Parameter | Value | Parameter | Value |
|---|---|---|---|
| $R$ | 20 KΩ | $A_c$ | 50 μm square |
| $R_L$ | 0 (constant voltage bias) | $F_f$ | 0.75 |
| $V_b$ | 3 V | $\varepsilon$ | 0.8 |
| $t_{pulse}$ | 35 μsec | $f_2$ | 15 KHz[b] |
| Bias pulse period | 33 msec[a] | $f_1$ | 0.001 Hz |
| $T_s$ | 300 K | $k$ | 1E-13 |
| $\alpha$ | $-0.10$ K$^{-1}$ at 300 K | $F_{no}$ | 1.0 |
| $c$ | 3E-9 J/K | $dL/dT_t$ | 8E-5 W/cm².sr.K[c] |
| $g$ | 1.5E-7 W/K | | |

[a] 30 Hz frame rate.
[b] $1/(2*35 \mu sec)$, equation 70.
[c] 300 K blackbody target, 8 to 12 μm, Table I.

operation is due to the fact that the bias current can be made much higher when using short bias pulses.

The minimum possible value of $g$ is the radiation-limited value $g_{rad} = 8\sigma A\varepsilon T^3$. Figure 25 shows a plot of the ideal NETD versus $g_{leg}$ value for the microbolometer parameters in Table IV.

5. JOHNSON NOISE APPROXIMATION

Here we make the approximation that all noise terms are negligible except Johnson noise. We also assume that the read-out system bandwidth is set by an integrator with noise bandwidth $\Delta f = 1/2\Delta t$, that the microbolometer is operated at constant voltage ($R_L = 0$), and that current signals are measured. Then Eqs. (92) and (94) give

$$\text{NETD} = \frac{4F_{no}^2}{A\varepsilon\pi(dL/dT_t)}\sqrt{2KT}\frac{g_{eff}}{\alpha}\frac{1}{\sqrt{\Delta T}} \qquad (100)$$

In this approximation we see that the NETD is inversely proportional to the square root of the microbolometer temperature increase $\Delta T$ in each bias pulse. Typical values of $\Delta T$ used to date (Wood and Stelzer, 1992) are a few degrees K.

## V. Electronic Read-Out Circuits for Two-Dimensional Microbolometer Arrays

The read-out circuit for two-dimensional microbolometer arrays may be fabricated in the underlying Si substrate, providing a monolithic IR-sensitive focal plane. The basic function of the monolithic read-out circuit is to apply a short bias pulse to each bolometer in the array in turn, while simultaneously measuring the signal (either the bolometer voltage $V$ or the bolometer current $I$) from that microbolometer. As discussed, this method of read-out is quite efficient (Eq. (97)).

The general circuit principle is shown in Fig. 26. Row and column conductors are fabricated in the underlying Si, with a microbolometer and an on–off electronic "pixel switch" lying at each intersection of the rows and columns. Peripheral circuits are placed in the Si to control the voltage on each row and column. The control voltages on each row and column are used to open and close the electronic pixel switches at each microbolometer, allowing a bias pulse to be applied to any selected microbolometer(s). The

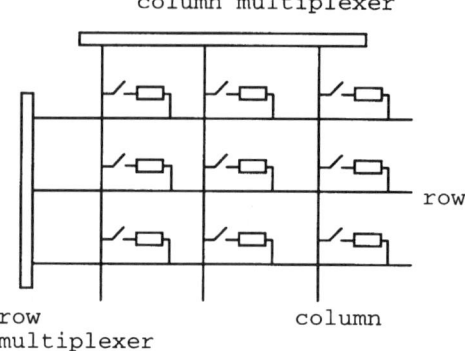

FIG. 26. General form of a microbolometer read-out circuit.

resulting bias currents or voltages comprise the signals, which are routed to an output signal line.

A "small" array of dimension 64 × 64 with a frame rate of 30 Hz implies a "pixel time" of $1/64 \times 64 \times 30 = 8.1$ μsec. If a single microbolometer is read out at any instant (called a *fully serial read-out*), each microbolometer can be biased for $\Delta t = 8.1$ μsec. A suitable circuit is shown in Fig. 27. In this circuit, each bolometer is biased "on" for 8.1 μsec by application of a bias voltage $V_b$ between the appropriate row and column metallization. This applies a bias of $V_b - 0.7$ to a microbolometer (the forward bias voltage decrease across the series diode is assumed to be 0.7 V). The diode at each microbolometer acts as a "self-closing" pixel switch, which only conducts

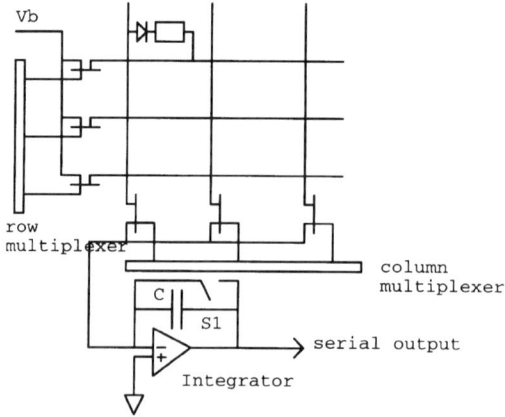

FIG. 27. Serial read-out circuit. $V_b$, bias voltage to load resistor; C, capacitance, S1, a switch.

when the proper voltage is applied. The column metallization of the selected microbolometer is held at ground potential by the action of the preamplifier, which integrates microbolometer current on capacitor $C$ during the interval of each bias pulse. The integration operation automatically limits the noise bandwith to the optimum value $1/2\Delta t \approx 62$ KHz. At the end of each 8.1-$\mu$sec interval, the charge on $C$ is discharged by momentarily closing switch S1.

In practice, some fraction of the pixel time of 8.1 $\mu$sec would not be available for signal integration. This is because the switching circuits cause transient disturbances, which experience has shown to be noisy, and must be allowed to decay to a negligible level before commencing signal integration. For larger arrays, the pixel time for fully serial read-out becomes progressively shorter, so that the transient decay time becomes a more significant fraction of the pixel time. Also, as pixel time is reduced, examination of the performance equations shows that bias amplitude must be increased to maintain a given level of performance (a good rule-of-thumb is that $\Delta T$ must be held constant for a given performance, Eq. (100)). The bias requirement for efficient fully serial read-out of larger arrays may approach monolithic circuit voltage or current limits.

To avoid these difficulties with larger arrays, several microbolometers may be read out simultaneously (called a *parallel read-out*), to allow a sufficiently long pixel time. Early 240 × 336 arrays used 14 microbolometers biased simultaneously, with 14 parallel analog signals brought out of the focal plane to 14 external integrators. This multiple-analog-output scheme provided efficient read-out but suffered from the practical difficulty that small temperature changes in the camera caused changes in the different integrator gains and offsets, resulting in pattern noise developing on the video display. To overcome this pattern noise, matched monolithic integrators can be used, and monolithic multiplexers can output a single serial channel of integrated signals from the focal plane.

With modern integrated circuits, parallel read-out of 240 × 336 arrays can be extended to whole rows of pixels (Fig. 28), and frame rates increased to 60 Hz, providing a pixel time of up to $1/240 \times 60 = 70 \mu$sec. In Fig. 28, all microbolometers in each row of the array, in turn, are biased. Since integrators hold their input terminals at ground potential, the bias arrangement in Fig. 28 provides the microbolometers with a constant-voltage bias equal to $V_{substrate}$. Microbolometer current signals flow down columns and are integrated in an integrator at the base of each column. Integrator outputs are transferred to a storage register before being multiplexed out serially, while current signals from the next row of the array are being integrated. In this arrangement, the previous numerical calculations have shown that the noise of a typical microbolometer (Table III) would be about 70 pA rms. Monolithic circuitry should ideally contribute negligible noise compared with microbolometer noise. This is not difficult to obtain with

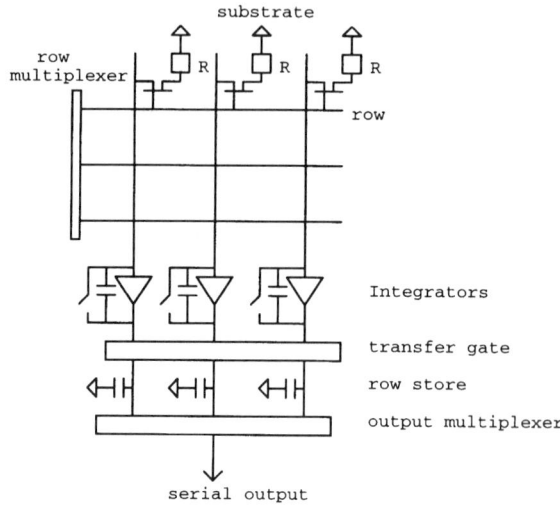

FIG. 28. Read-out circuit for larger arrays. R, resistance.

multiplexers and pixel switches, which are simply on–off switches: FET switches have negligible noise because they operate in either cut-off or saturation, where gate noise does not influence their characteristics. Complementary metal-oxide semiconductor (CMOS) multiplexers are therefore noiseless. CMOS amplifiers of adequately low noise, however, are difficult to achieve in CMOS, because they operate in linear regions where gate 1/f noise influences their performance. Adequately low noise preamplifiers are relatively easy to design with bipolar devices, so that mixed bipolar-CMOS (BICMOS) is an attractive Si technology for low-noise monolithic read-out (Unewisse et al., 1995a).

An important practical consideration for microbolometer focal plane operation is the relative magnitude of the noise and the microbolometer-to-microbolometer relative signal level variations. This ratio sets the required dynamic range of the camera circuitry. The variation in microbolometer resistance across an array is termed the *nonuniformity* of the array. The microbolometer resistances differ primarily because of resistance variations between individual microbolometers due to fabrication process nonuniformities.

If the variation in room temperature resistance between different microbolometers is $dR$, then at the start of a bias pulse $V_b$ the instantaneous voltage across a microbolometer in the array is

$$V = \frac{R}{R + R_L} V_b \qquad (101)$$

and the voltage offset between different microbolometers is

$$dV = \frac{V_b R_L}{(R + R_L)^2} dR \qquad (102)$$

The current offset between different microbolometers is similarly found to be

$$dI = -\frac{V_b}{(R + R_L)^2} dR \qquad (103)$$

In addition to the built-in microbolometer room temperature resistance variation due to fabrication differences, a temperature increase $\Delta T$ occurs during each individual bias pulse, producing an additional microbolometer resistance change of about $R\alpha\Delta T$ during each bias pulse.

The magnitudes of the resultant microbolometer offsets can be estimated using the preceding small-signal equations, or can be computed with more precision using the numerical approach described previously. For the parameters in Table III, a numerical computation shows that a microbolometer resistance nonuniformity of 1% rms produces an overall voltage offset nonuniformity of 20 mV rms. This may be compared with a noise level (Fig. 22) of about 2.6 $\mu$V rms. The dynamic range of the analog circuits therefore needs to be 20E-3/2.6E-6 = 7700. Digitization of the analog signals is similarly required to be performed with a high-number dynamic range. The rms noise of an ideal analog–digital (A–D) converter due to its finite least-significant-bit (lsb) quantization intervals (Bennett, 1956) is $1/2\sqrt{3} = 0.3$ lsb rms. Setting this equal to the rms noise (2.6 $\mu$V in this numerical example) and taking the peak-to-peak offset variation to be five times the rms variation, we find that we need a 14-bit A–D converter to accept the expected range of signal levels from an array with 1% rms resistance nonuniformity.

Since the array offset nonuniformity is expected to be fixed, it is possible to compensate for the nonuniformity by a variety of circuit techniques that are well known to A–D designers. Methods of compensating uniformities caused by bolometer heating also have been suggested by Jansson et al. (1995).

## VI. Offset Compensation Schemes

As has been noted, because of microbolometer resistance variations within an array, the output signals have fixed offsets that are very large compared with the noise level. Removal of these offsets is essential in order to produce an image that is viewable by a human observer. Offset correction in microbolometer cameras has been demonstrated by two basic methods:

1. Choppers. If the microbolometer is read out at a 60-Hz field rate and alternate fields are obscured by a rotating chopper blade, then alternate fields may be subtracted to remove pixel offsets, yielding an offset-corrected image at a 30-Hz frame rate.
2. Shutter. If the array temperature is held fixed, a few initial reference frames may be acquired while a shutter is held closed, stored, and used to remove offsets from all following frames by subtraction.

In both techniques, several reference frames should be averaged, to ensure that noise on the reference frame is negligible compared with the time-dependent noise.

Each scheme has advantages and disadvantages, and either a chopper or a shutter may be more desirable in different applications. A serious disadvantage of a chopper is that the field rate must be twice the frame rate: This, in turn, requires half the $\tau$ value, and since $c$ is usually difficult to reduce, this implies $g$ must be doubled, resulting in a reduction in responsivity by a factor of about two (Eq. (39)). An advantage of a chopper is the reduction of drift and 1/f noise. The chopper must produce a "shadow," which travels across the array at the same speed as the array is read out.

A major disadvantage of a shutter is that 1/f noise is fully present, and shutter closures are intermittently required to prevent 1/f noise from building up "fixed pattern" noise. The operational *staring time* of the system can be defined as $T_{stare} \approx 1/4f_1$, where $f_1$ is the low-frequency bandwidth limit that produces a rms 1/f noise equal to the rms time-dependent Johnson noise. For measurement of current signals, this can be written using Eq. (86):

$$4KT\frac{1}{R+R_L}(f_2 - f_1) = V_b^2 k \ln\left(\frac{f_2}{f_1}\right)\left\{\frac{R}{(R+R_L)^2}\right\}^2 \overline{f(\beta)} \qquad (104)$$

Figure 29 shows a numerical computation of $T_{stare}$ (as defined by Eq. (104) versus bias current for the assumed parameters in Table III, with three assumed values $k = $ 2E-13, 5E-13, and 1E-12.

Most microbolometer cameras reported to date employ a shutter to avoid the NETD increase associated with a faster field rate. For chopperless operation, the array temperature may be conveniently stabilized by a thermoelectric (TE) device, or by holding the array at an elevated temperature with a heater.

The minimum power required to stabilize an array at a given temperature is the power loss–gain due to thermal radiation, thermal conductance through the bond wires, and the bias joule power. For a 240 × 340 array this is typically on the order of 100 mW for a temperature difference of 25°C.

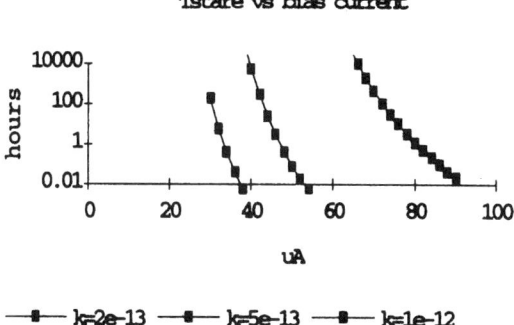

FIG. 29. "Starting" time versus bias for $k = $ 2E-13, 5E-13, and 1E-13.

In practice, the heat load may be dominated by heat lost through the TE stabilizer itself: a small commercially available TE stabilizer draws about 150 mW for a 25°C temperature difference.

The precision to which microbolometer temperature must be maintained is set by the appearance of pattern noise, due to different changes in microbolometer properties with temperature. Measurements of pattern noise versus temperature in 240 × 340 arrays for gain-correction purposes show that the required temperature stability is about 0.05°C (Wood and Stelzer, 1992).

## VII. Gain Correction

Figure 30 shows the measured responsivity histogram of a 240 × 336 array (Wood, 1992). In normal imager operation, this responsivity nonuniformity is reduced by multiplying each pixel by a gain-correction number, found by imaging a uniform blackbody at two temperatures. After gain correction, pattern noise has been found (Wood and Stelzer, 1992) to remain less than time-dependent noise for target temperature changes of about 10°C from the calibration temperatures. Since the bolometer temperature change is typically 1/200 of the target temperature change (Eq. (8)), the bolometer temperature stability required for chopperless operation is about 1/20°C. For radiometric applications, the requirements are stricter: If target temperatures are required to be measured to 1°C, then the temperature regulation requirement is about 1/200°C (equation 8). Thermistor temperature sensors can measure array temperature to better than 1 mK, and feedback circuits can be used to regulate TE stabilizers to maintain temperature regulation.

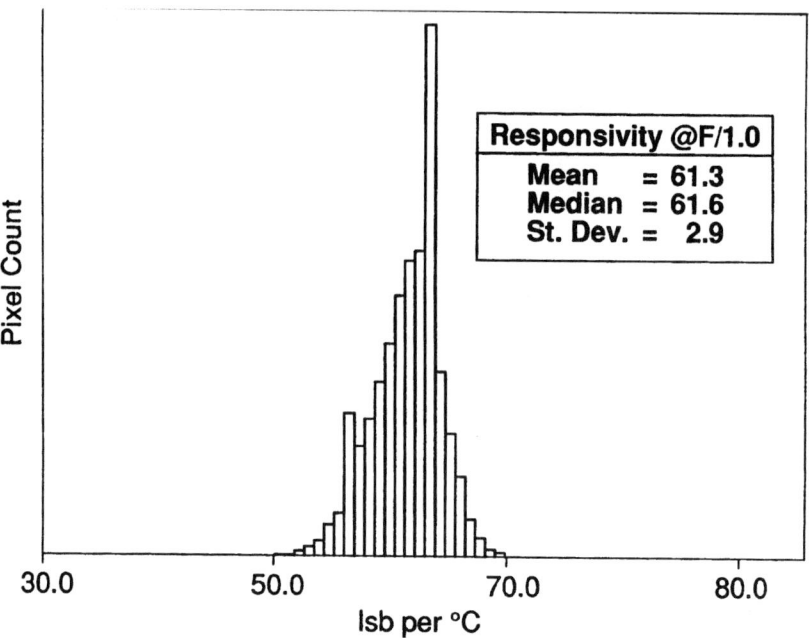

FIG. 30. Histogram of responsivity measured on a 240 × 340 microbolometer array. (Reprinted from Wood et al. (1993) with permission from the authors.)

## VIII. Modulation Transfer Function (MTF)

The modulation transfer function (MTF) of a microbolometer array is determined by the optical system, cell dimensions, and fill factor (Lloyd, 1979). There is no measurable cross talk of signals between adjacent microbolometers because each microbolometer is individually supported by a large thermal mass (the Si substrate). Figure 31 (see color plate section) shows a magnified section of a video display from a microbolometer array imaging a point target that is very hot: no cross talk between the pixel viewing the point target and adjacent pixels is visible above the noise level.

## IX. Microbolometer Physical Design, Fabrication, and Packaging

The basic technologic advance that made microbolometer arrays practical is called *micromachining*, which is the technology of constructing microscopic thermally isolated structures (or *microbridges*) in Si-based materials by selectively etching certain materials, crystalline planes, or both. Micro-

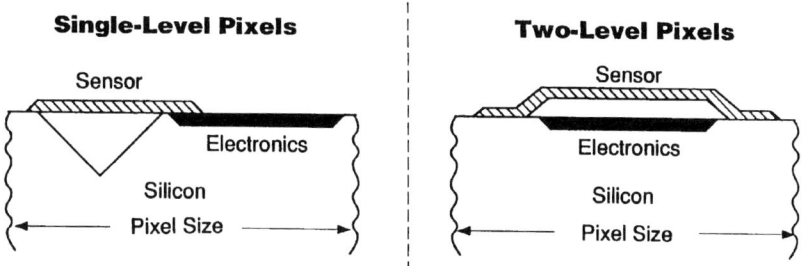

FIG. 32. Illustration of one-level and two-level microbolometer designs.

machining has grown to be a large technical area that is well-described in many other publications (e.g., Petersen, 1982; Middelhoek and Andet, 1989) and will not be repeated here, except to describe simply the particular micromachining techniques and materials (Table V) used to fabricate microbolometers.

The micromachined microbolometers reported to date may be classified into two broad design categories (Fig. 32). "One-level" microbolometers consist of a microbridge that is essentially level with the original Si surface and is thermally isolated by an etched "pit" in the underlying Si. "Two-level" microbolometers consist of a microbridge elevated above the original Si surface so that the Si is intact underneath.

TABLE V

Reported Parameters of Materials used in Microbolometer Fabrication (Higashi 1997).

| Material | Thermal conductivity W/cm.K | Thermal capacity J/cc.K |
|---|---|---|
| Al (2000A) | 1.32 | 6.65 |
| Cr (500A) | 0.29 | 3.31 |
| NiFe (80:20, 500A) | 0.25 | 3.91 |
| Si3N4: | | |
| sputtered 8000A | 0.020 | 2.16 |
| lpcvd 30,000A | 0.032 | 2.1 |
| pecvd 5000A | 0.045 | 3.33 |
| SiO2 | 0.014 | 2.27 |
| VOx | 0.05 | 3.0 |
| ZrO2 | 0.01 | 2.8 |

Pecvd, plasma enhanced chemical vapor deposition; lpcvd, low pressure chemical vapor deposition.

FIG. 33. Microscope photograph of a prototype single-level microbolometer (75 × 75 μm) suitable for infrared imaging.

1. ONE-LEVEL MICROBOLOMETERS

Early microbolometer arrays suitable for IR imaging (Wood, 1983, 1986) were one-level types, consisting of 75-μm square microbridges of $Si_3N_4$, thermally isolated from the underlying Si by anisotropic etching (Fig. 33). The microbridge supported a thin-film serpentine Ni–Fe resistor with a TCR of $+0.0023\,K^{-1}$. Infrared absorbtion was enhanced by a coating of carbon black. Quantitative measurements on these single-level microbolometers were made by Arch and Heisler (1982), who used a dc-bias current to demonstrate a $D^*$ of $1E9\,cm\sqrt{Hz}/w$, using microbridges about 250 μm square. The very small mass of microbolometers (about 1E-9 gm) allows great shocks to be withstood. Single-level microbolometers of the type shown in Fig. 33 have been subjected to accelerations of 14,000 g without damage (Wood, 1986).

Infrared imaging with one-level microbolometers was obtained with 1 × 16 linear arrays (Wood, 1986), and later with 64 × 128 arrays of

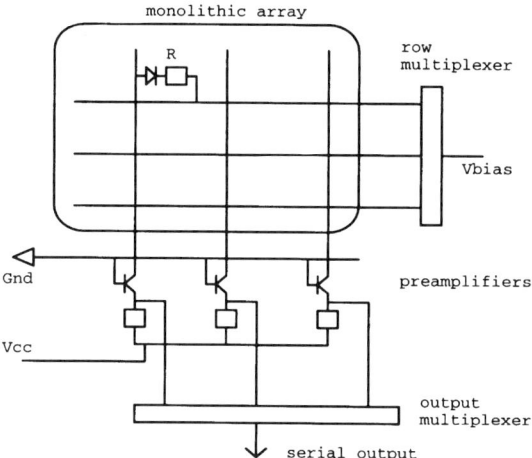

FIG. 35. Read-out used for 64 × 16 microbolometer arrays. $V_b$, bias voltage to load resistor; Gnd, zero volts; Vcc, supply voltage.

microbolometers. These microbolometers consisted of $Si_5N_4$ microbridges that were about 50 μm square and 0.8 μm thick, with serpentine Ni-Fe metallic sensing resistors (TCR +0.0023 $K^{-1}$). Thermal conductance was provided by L-shaped supporting legs of $Si_3N_4$ that were 100 μm long, 5 μm wide, and 0.5 μm thick, with 500-Å Ni-Fe metal conductors (Fig. 34; see color plate section). This design provided $c$ = 2.5E-9 J/K, $g$ = 5E-7 W/K, and $\tau$ = 5 msec. The thermal conductance of the legs was approximately equally contributed to by the $Si_3N_4$ and the leg metallizations. A 50% IR absorbtance was provided by a 50-Å metal film of about 189 ohms per square (Nishioka et al., 1978) deposited across the microbridge area. The arrays were fabricated with a monolithic read-out consisting of a single p-n diode adjacent to each pixel, connected as in Fig. 35.

Figure 35 provides a pulsed bias, biasing one complete row of 16 microbolometers at a time with a constant voltage bias. Current signals flowed down the 16 columns to 16 common-base bipolar transistor integrators. The arrays were operated in an evacuated package with a Ge window with no thermal stabilization. A traveling-shadow toothed-wheel chopper was used to provide a shadow that moved across the array at the same speed at which the rows were read out, using an array scan rate of 60-Hz. Alternate frames were digitally subtracted to give a camera frame rate of 30 Hz. The array size was later increased to 128 × 64 and pixel dimensions reduced to 50 μm. An NETD ($F_{no}$ = 1, 30 Hz frame rate) of 0.25°C, including camera noise, was measured (Wood et al., 1986) using fully serial pulsed bias read-out. The characteristics of these arrays are summarized in Table VI, and the pixel design is shown in Fig. 34.

TABLE VI

SUMMARY OF 64 × 16 AND 64 × 128 SINGLE-LEVEL MICROBOLOMETER ARRAY PARAMETERS

| Parameter | Value |
|---|---|
| Array size | 64 × 128 |
| Pixel size | 50 × 60 $\mu$m |
| Design | Single layer |
| Bridge | 50 $\mu$m × 50 $\mu$m × 0.8 $\mu$m $Si_3N_4$ |
| Legs | 50 $\mu$m × 4 $\mu$m × 0.8 $\mu$m $Si_3N_4$ |
| Fill factor | 0.18 |
| Package | Vacuum |
| Thermal stabilization | None |
| Nominal operating temperature | 25°C |
| Thermal capacity | 2.5E-9 J/K |
| Thermal conductance in vacuum | 5E-7 W/K |
| Time constant | 5 msec |
| Absorber | 180 $\Omega$/square nickel |
| TCR material | Serpentine 500 A nickel–iron, TCR +0.0023 $K^{-1}$ |
| Read-out | Pulsed constant-voltage, fully serial read-out, diode switch alongside each pixel (Fig. 35) |
| Pixel resistance | 2 K$\Omega$ |
| Field rate | 60 Hz |
| Frame rate | 30 Hz |
| Offset compensation | 60 Hz chopper |
| Bias | 1 mA pulses |
| Measured camera NETD | 0.25°C, $F_{no} = 1$, 8–14 $\mu$m, 300 K target |

$Si_3N_4$, silicon nitride; TCR, temperature coefficient of resistance; NETD, noise equivalent temperature difference.

Liddiard (1983) reported the one-layer microbolometer structure shown in Fig. 36. The bolometer used a $Si_3N_4$ pellicle microbridge, with through-the-wafer etching. Liddiard achieved a $D^*$ of 1.6E8 cm$\sqrt{Hz}$/w with a fast response time (<1 msec) and 50 $\mu$m pixels. Downey et al. (1984) reported a one-level microbolometer, consisting of a large area (5 × 5 mm) 5-$\mu$m-thick Si plate suspended on a Si frame, fabricated by an etching process.

Many other reports of single-level microbolometer designs are found in the literature cited in this chapter.

2. TWO-LEVEL MICROBOLOMETERS

The principal disadvantage of the one-level microbolometer is that, since the underlying Si is removed, read-out electronics must be placed adjacent to each microbridge. This necessarily produces a poor fill factor

K. C. LIDDIARD

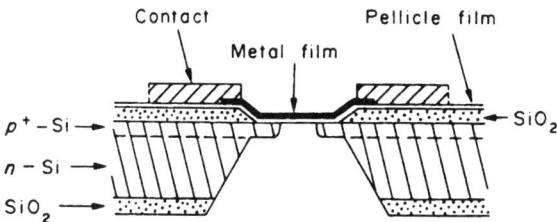

FIG. 36. Single-level microbolometer constructed using the through-the-wafer etching technique. (Reprinted from Liddiard (1984). Thin film resistance bolometer IR detectors. *Infrared Phys.* **24**, 57; with permission from the author.)

for pixel sizes less than about 75 μm. In Fig. 34, the fill factor is about 0.2.

A two-level design is shown in Fig. 37 (Fig. 37b; see color plate section). It is an improvement over the one-level microbolometer design because it allows greater fill factor (read-out circuitry may be placed in the Si underneath the microbridge) and greater IR absorptance (the underlying cavity produces a resonant optical cavity) (Wood, *et al.*, 1990).

Two-level microbolometers may be produced in several ways using different variants of micromachining. Figure 38 illustrates the typical fabrication steps of a two-level microbolometer (Wood 1993c). Fabrication begins with implantation of the required read-out electronics and conducting metallizations in the Si wafer. The wafer is then planarized with a material, such as spun-on polyimide, which can be photolithographically patterned to form sacrificial mesas (Unewisse *et al.*, 1995b). Silicon nitride layers are sputtered over the sacrificial mesas, together with TCR material

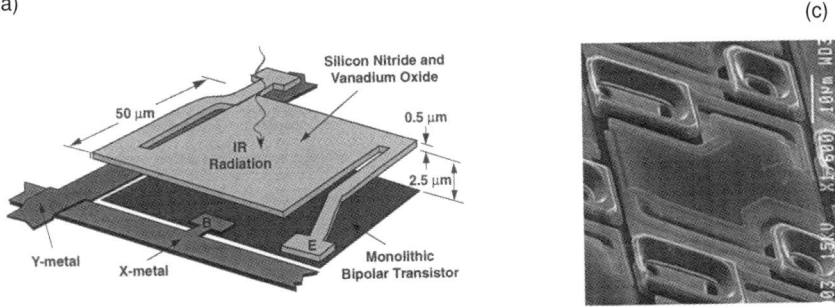

FIG. 37. Two-level microbolometer construction (a) and microscope photographs (b, c).

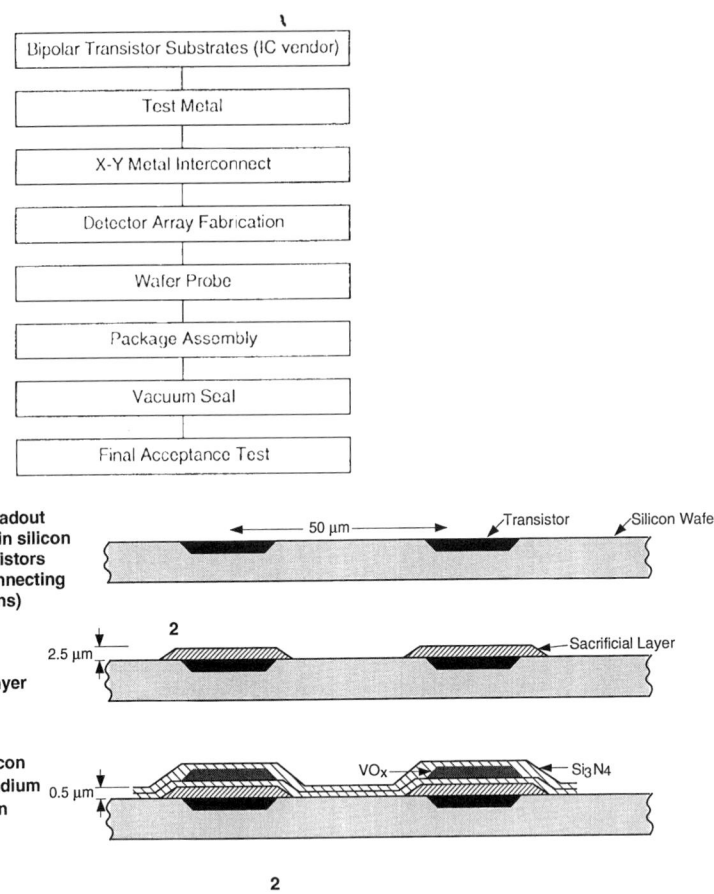

FIG. 38. Fabrication sequence for two-level microbolometers.

and connecting metallizations. As a final step, the sacrificial mesas are removed by a material-selective etch, for example, oxygen plasma etching, to leave a self-supporting two-level structure.

Figure 37 shows a $35 \times 35 \times 0.8\ \mu$m $Si_3N_4$ microbridge suspended over the Si substrate, with a gap of approximately 2.5 $\mu$m between the bridge and the Si. The parameters in Table VI show that this gives a thermal capacity of 3E-9 J/K. The supporting legs are $Si_3N_4$, with 500-Å Ni–Cr conductive films, giving $g = 2E\text{-}7$ W/K (Fig. 39) and $\tau = c/g = 20$ msec. The sensing material is a 500-Å layer of vanadium oxide, with TCR $= -0.023\ K^{-1}$ at

## TABLE VII
### Summary of 240 × 336 Two-Level Microbolometer Array Parameters

| Parameter | Value |
| --- | --- |
| Array size | 240 × 336 |
| Pixel size | 50 μm |
| Design | Two layer |
| Bridge | 35 μm × 35 μm × 0.8 μm $Si_3N_4$ |
| Legs | 50 μm × 2 μm × 0.8 μm $Si_3N_4$ |
| Fill factor | 0.70 |
| Package | Vacuum |
| Thermal stabilization | Thermo-electric stabilizer |
| Nominal operating temperature | 25°C |
| Thermal capacity | 3E-9 J/K |
| Thermal conductance | 2E-7 W/K (in vacuum) |
| Time constant | 15 msec (in vacuum) |
| Absorber | 80% mean, 8 to 14 μm (Fig. 40) |
| TCR material | 500 Å VOTCR $-0.023\,K^{-1}$ |
| Pixel resistance | 20 KΩ |
| Read-out | Pulsed constant-voltage, bipolar transistor under each pixel, 14 pixels in parallel, 14 bipolar pre-amplifiers |
| Field rate | 30 Hz |
| Frame rate | 30 Hz |
| Offset compensation | Intermittent shutter |
| Bias | 5 μsec 250 μA pulses |
| NETD | 0.039°C, $F_{no} = 1$, 8–14 μm, 300 K target (Fig. 45) |

$Si_3N_4$, silicon nitride; TCR, temperature coefficient of resistance; VO, vanadium oxide; NETD, noise equivalent temperature difference.

25°C, protected from etchants by being sandwiched between layers of $Si_3N_4$.

The use of a vacuum gap of approximately 2.5 μm, together with a thin-film metal reflector layer on the underlying substrate, can produce a quarter wave resonant cavity between the bolometer and the underlying reflector for wavelengths near 10 μm. The IR absorption of this multilayer structure can be computed using computer programs adapted from those used for dielectric multilayer interference filters. A calculated absorption curve versus wavelength is shown in Fig. 40 (Cole, 1995). Absorption calculations such as those in Fig. 40 have been verified by experimental measurements on large-area thin-film experimental structures, using dielectric films of refractive index near one to simulate the vacuum gap (Cole, 1995), and by multiwavelength responsivity measurements in microbolometers (Wood, 1994).

An alternative method of obtaining high IR absorption is to use semiconductor TCR-material structures illustrated in Fig. 41, where the semiconductor TCR material itself is made thick enough to act as a quarter wave

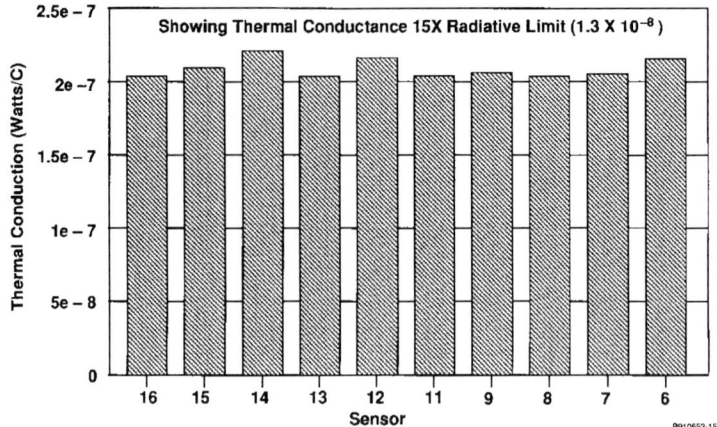

FIG. 39. Measured $g$ values for two-level microbolometers.

absorber, producing 90% absorption (Unewisse et al., 1994, 1955a). Thin metal films of about 189 ohms per square also provide close to 50% absorption (Nishioka et al., 1978), which may be increased in selected wavebands by quarter wave reflectors (Liddiard, 1994). Gold black absorbers also have been demonstrated to give close to 100% absorption on

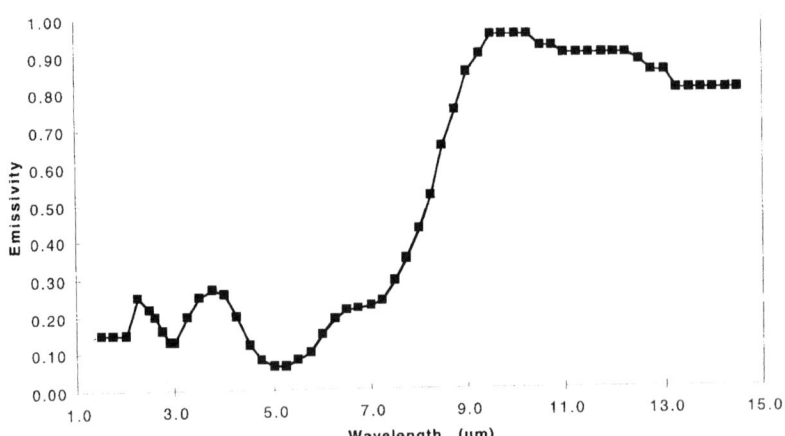

FIG. 40. Calculated absorption coefficient of a two-level microbolometer. (Reprinted from Cole (1995). Honeywell Research Memorandum; Honeywell Inc., Plymouth, Minnesota; with permission.)

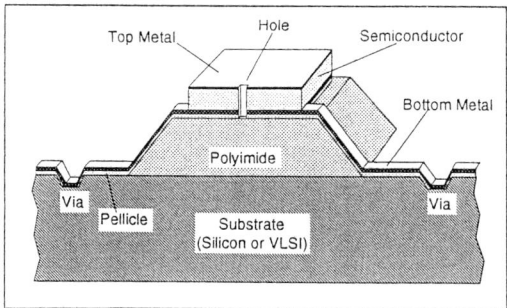

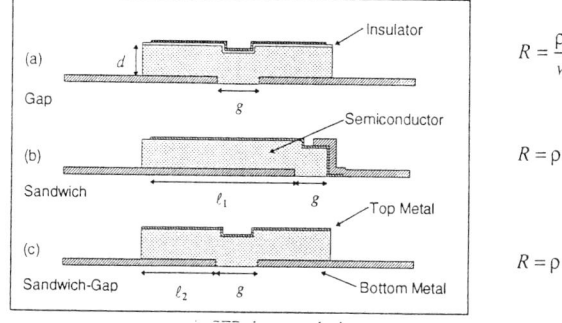

FIG. 41. Microbolometer absorber structure. (Reprinted from Unewisse et al. (1995b). Proc. SPIE **2554**; with permission from the SPIE.)

micromachined bolometers (Kimura, 1981; Carlson, 1988; Lang et al., 1990).

Figures 42 (Fig. 42a; see color plate section) and 43 show microphotographs of completed two-dimensional arrays, and a 4-inch diameter wafer of two-level microbolometers, complete with monolithic read-out electronics. These arrays employ a bipolar read-out circuit that applies constant-voltage pulses to pixels and allows current signals to flow simultaneously through 14 adjacent pixels, to 14 off-chip bipolar preamplifiers. Bipolar monolithic read-out and pre-amplifiers were selected for early arrays, rather than CMOS. This was due to the lower 1/f noise of bipolar circuits but also to the inadequate capabilities of dealing with the CMOS topography that existed in 1982, when multiyear program plans were defined. A prototype camera was constructed to demonstrate IR imaging with these arrays. This camera used the chopperless mode of operation, with a thermally stabilized array scanned at 30 Hz. Two-level microbolometer parameters are summarized in Table VII, and Figure 44

FIG. 42. (a) A 240 × 336 two-level microbolometer array (microscope photograph).

FIG. 42. (b) Packaged 240 × 336 two-level microbolometer array.

shows a typical IR image. Figure 45 shows measurements of noise, responsivity, and NETD obtained with these microbolometers (Wood *et al.*, 1993). Figure 46 shows a two-level pixel design using a Ti resistor material (Tanaka *et al.*, 1995), with a representative image from a 128 × 128 array.

3. PACKAGING

The thermal conductivity of standard temperature and pressure (STP) air is 2.5E-4 $Wcm^{-1}K^{-1}$, and therefore a two-level bolometer of 50-$\mu m^2$ area suspended 2.5 $\mu$m above an underlying substrate has a thermal conductance in STP air of about 2.5E-5 W/K. This is much greater than the thermal conductance of typical supporting legs (Fig. 39), and therefore, if operated in air, the microbolometer responsivity is greatly reduced. As air pressure is reduced, the number of gas molecules decreases, and the mean free path increases at the same rate. Thus the conductivity remains almost independent of air pressure, until the mean free path becomes limited by the physical gap distance of about 2.5 $\mu$m. This occurs at an air pressure of about

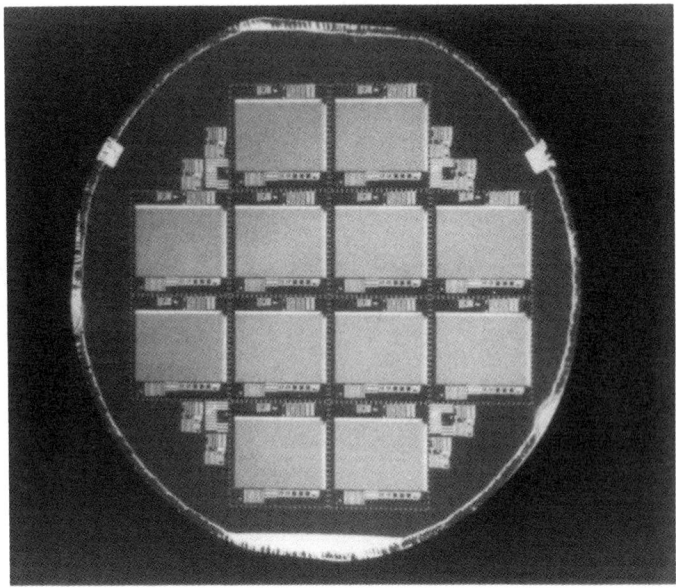

FIG. 43. Four-inch-diameter silicon wafer with 12 completed 240 × 336 two-level microbolometer arrays.

FIG. 44. Infrared image from 240 × 336 two-level microbolometer array.

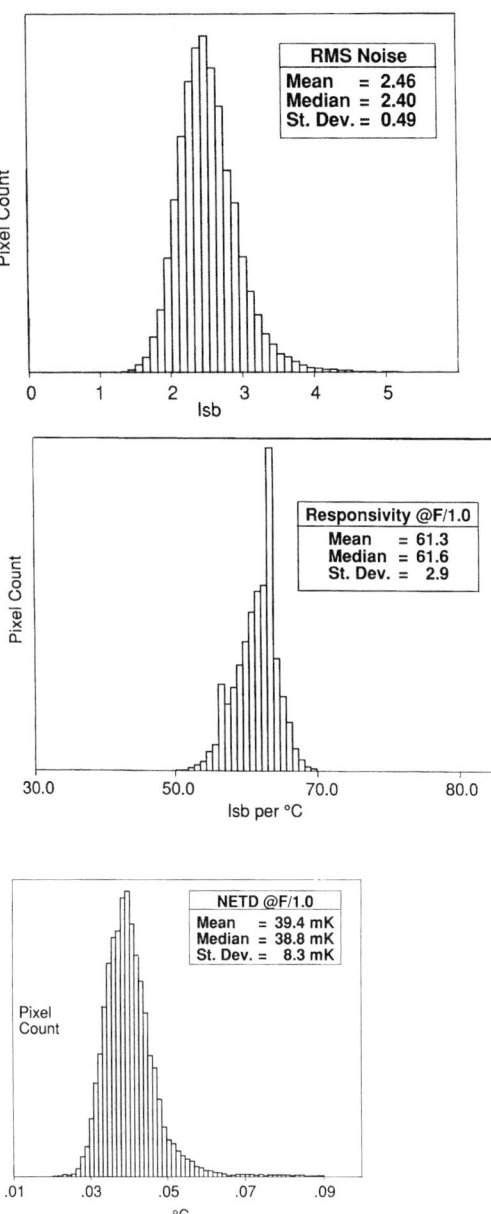

FIG. 45. Measured responsivity, noise, and noise equivalent temperature difference of 240 × 336 two-level microbolometer arrays of Table VII (ISb, least significant bit of A/D converter).

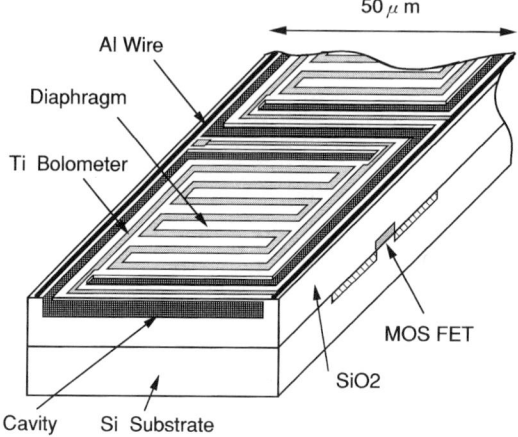
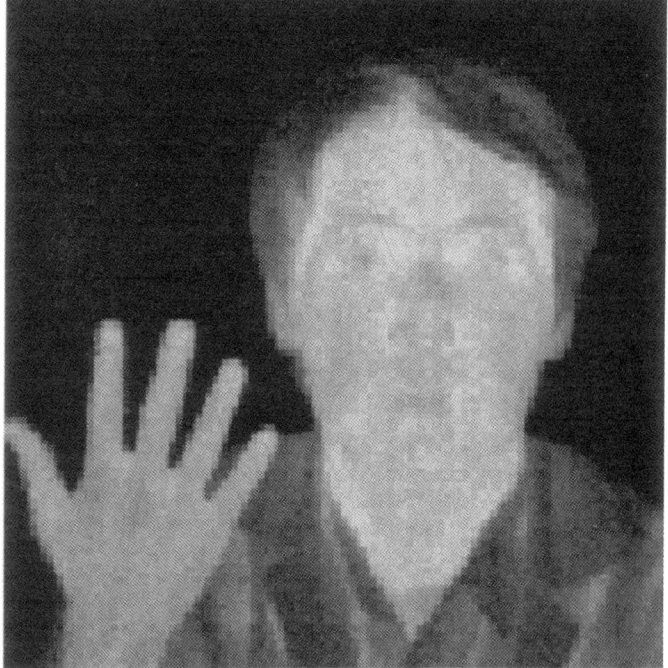

FIG. 46. Two-level microbolometer with titanium resistor temperature coefficient of resistance (TCR) material, with representative infrared image. (Reprinted from Tanaka *et al.* (1995). Silicon IC process compatible bolometer infrared focal plane array. *Proc. Int. Conf. Solid State Sens. Actuators. 8th*, Stockholm, Sweden, pp. 632–635; with permission from the authors.)

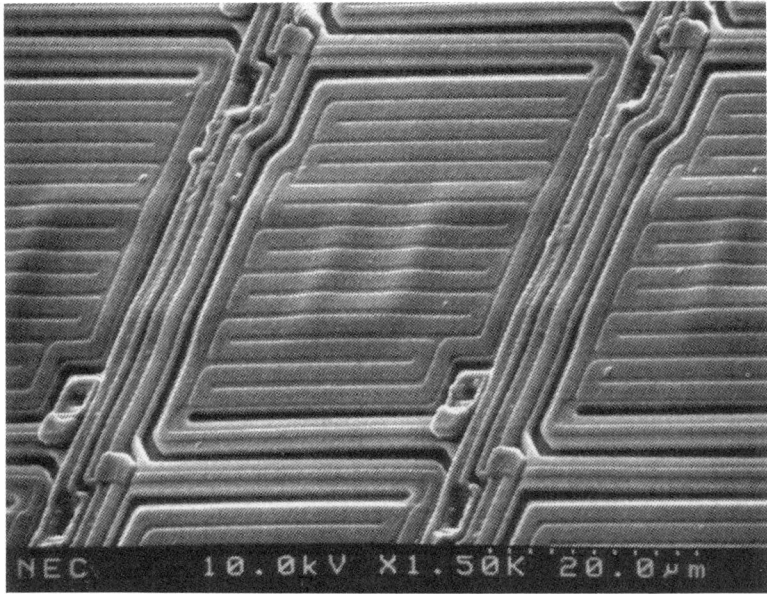

FIG. 46. (Continued)

| | |
|---|---|
| Sensor Scheme | Titanium Bolometer |
| Pixel Size | 50 x 50 $\mu m^2$ |
| Optics f/no. | 0.8 |
| Thermal Conductance | 0.35 $\mu W \cdot K^{-1}$ |
| Pixel Fill Factor | 0.59 |
| Thermal Time Constant | 5 msec |
| Bolometer Resistance | 2.6 K$\Omega$ |
| Bolometer TCR | 0.2 %$K^{-1}$ |
| Pixel Number | 128 x 128 |
| Integration Time | 1 $\mu$sec |
| Frame Rate | 30 Hz |
| NETD | 0.7 °C (I = 1 mA) |

FIG. 46. (*Continued*)

200 mTorr. Although the effective mean free path is fixed below this pressure, the number of gas molecules decreases with pressure and therefore the thermal conductance of air decreases approximately linearly with pressure (van Herwaarden, 1994). The thermal conductance of a typical microbolometer eventually becomes limited by the leg thermal conductance, which typically occurs at an air pressure of about 50 mTorr. Further reduction in air pressure provides negligible reduction in microbolometer $g$ value. Because the responsivity of a microbolometer is proportional to $1/g$, full sensitivity requires a typical air pressure of 50 mTorr or less. Little improvement in sensitivity is attained by operating at lower pressures.

Although this air pressure is easy to obtain with air pumps, maintaining this pressure for many years in a small-volume package requires attention to outgassing phenomenon. Long-lived (multiyear) sealed vacuum packages have been demonstrated using packages constructed of brazed and soldered materials that were carefully cleaned and baked before sealing to produce low outgassing. Some package designs use internal getters to absorb outgassing from internal components, providing a longer package vacuum life. Figure 47 shows the construction of a sealed vacuum package designed for 240 x 336 arrays operating in chopperless mode. Thermal stabilization was provided by a TE stabilizer mounted inside the package.

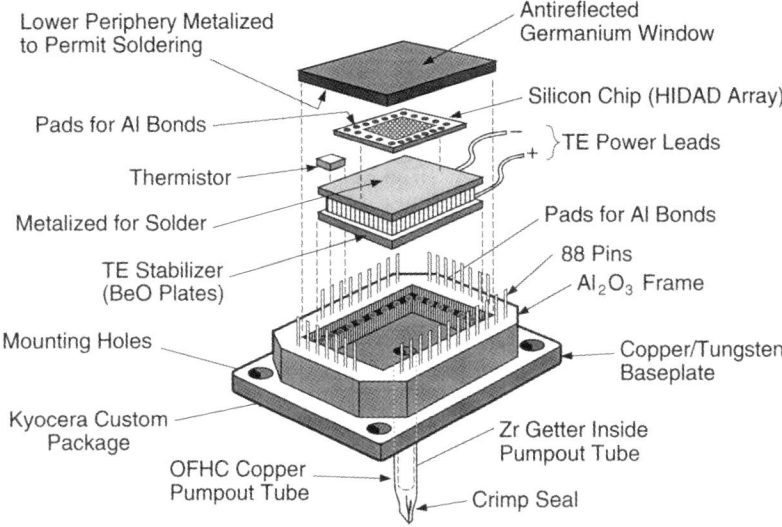

FIG. 47. Construction of a sealed vacuum package for 240 × 336 microbolometer arrays.

To reduce packaging costs, packages filled with heavy (that is, low thermal conductivity) gases have been used (Unewisse *et al.*, 1995a). Except for microbolometers with higher $g$ values, gas-filled packages do not allow the same performance as do vacuum packages. Figure 48 shows the measured responsivity of microbolometers in different gases.

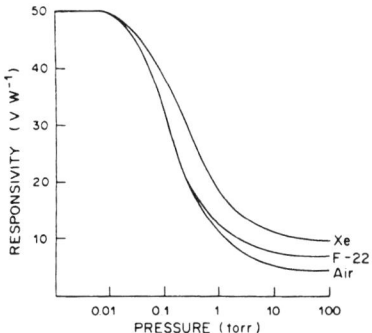

FIG. 48. Measured responsivity variation of microbolometers with gas pressure and composition. (Reprinted from Liddiard (1984). Thin film resistance bolometer IR detectors. *Infrared Phys.* **24**, 57; with permission from the author.)

## X. Practical Camera Development

Camera size has decreased dramatically from 1982 to 1996. Figure 49 shows the form of prototype microbolometer cameras used at Honeywell for most early microbolometer development. Figure 50 shows the first portable camera using 240 × 336 arrays (with parameters listed in Table VII) produced by Honeywell in 1990 (Wood et al., 1993a–d) which achieved an NETD of 39 mK ($F_{no} = 1$). Figure 51(a) shows a compact 240 × 340 microbolometer camera produced by Amber Engineering, Inc. in 1995, with a high-quality IR image (Figs. 51b,c).

### Acknowledgments

The author thanks his many colleagues at Honeywell, both present and past, who played major roles in the achievement of uncooled IR imaging with microbolometer arrays, and whose work is described herein. A large part of the development work at Honeywell was directed by the staff of the US Night Vision and Electronic Sensors Directorate and the Defense Advanced Research Projects Agency. Government funding has been provided to Honeywell under the following programs: ASP Sensor Development, contract DAAL01-85-C-0153; High Density Array Development, contract DAAB07-87-C-F024; Low Cost Uncooled Sensor Prototype, contract DAAB07-90-C-F300; and the Technology Reinvestment Program, contract MDA972-95-3-0022.

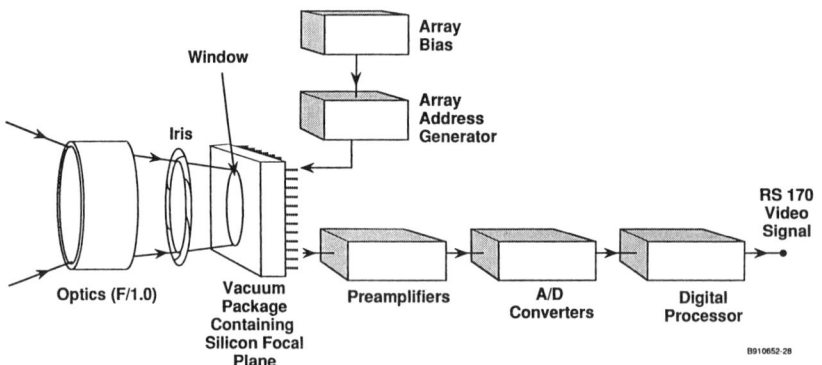

FIG. 49. A microbolometer camera block diagram. (Courtesy of Honeywell Inc., Plymouth, Minnesota, with permission.)

3 MONOLITHIC SILICON MICROBOLOMETER ARRAYS    117

FIG. 50. Portable 240 × 336 microbolometer camera. (Courtesy of Honeywell Inc. Plymouth, Minnesota; with permission.)

(a)

FIG. 51. Compact 240 × 340 microbolometer camera (a) and infrared images (b, c). (Courtesy of Raytheon Amber (1995). Sentinel Operator's Reference Manual, October 1995. Amber Engineering, Goleta, Georgia; with permission.)

Fig. 51b.

Fig. 51c.

## REFERENCES

Andrew, V. N., Chudnovskii, F. A., Petrov, A. V., and Terukov, E. I. (1978). *Phys. Stats Solidi A* **48**, K153–K156.
Arch, D., and Heisler, M. (1982). Honeywell Research Memorandum.
Baliga, S. B., Rost, N. R., and Doctor, A. P. (1994). *Proc. SPIE—Int. Soc. Opt. Eng.* **2225**, 72–78.
Barr, S. (1961). *Infrared Phys.* **1**, 1–4.
Barr, S. (1963). *Infrared Phys.* **3**, 195–206.
Bennett, W. R. (1956). *Proc. IRE* **44**, 609–638.
Brunetti, L. (1992). *Sens. Actuators, A: Phys.* **32**(1–3), 423–427.
Brunetti, L., and Monticone, E. (1993) *Meas. Sci. Technol.* **4**(11), 1244–1248.
Carlson, R. A. (1988). Honeywell Research Memorandum.
Cole, B. E. (1995). Honeywell Research Memorandum.
Dagge, K., Frank, W., Seeger, A., and Stoll, H. (1996). *J. Appl. Phys.* **68**(9), 1198–1200.
Dereniak, E. L., and Crowe, D. G. (1984). "Optical Radiation Detectors." Wiley (Interscience), New York.
Downey, P. M., Jeffries, A. D., Meyer, S. S., Weiss, R., Bachner, F. J., Donnelly, J. P., Lindley, W. T., Mountain, R. W., and Silversmith, D. J. (1984). *Appl. Opt.* **23**, 910.
Estill, S., and Brozel, M. R. (1994). *Proc. Spring Meet. Mater. Res. Soc., Infrared Detec.—Mater. Process, Devices Mat. Res. Soc. Symp. Proc.*, *1993*, vol. 299.
Fisher, B. (1982). *J. Chem. Phys. Solids* **43**, 205–211.
Flannery, R. E., and Miller, J. E. (1992). *SPIE* **1689**, 379.
Higashi, R. (1996). Honeywell Research Memorandum.
Higashi, R. (1997). Honeywell Research Memorandum.
Horn, S. B., and Buser, R. G. (1993). *Proc. SPIE—Int. Soc. Opt. Eng.* **2020**, 304–321.
Jansson, C., Ringh, U., and Liddiard, K. C. (1995) *Proc. SPIE* **2552**.
Jensen, A. S. (1993). *Proc. SPIE—Int. Soc. Opt. Eng.* **2020**, 340–349.
Jerominek, H., Picard, F., and Vincent, D. (1993). *Opt. Eng.* **32**(9), 2092–2099.
Johnson, B. R., Ohnstein, T., Han, C. J., Higashi, R. E., Kruse, P. W., Wood, R. A., Marsh, H., and Dunham, S. B. (1993). *IEEE Trans. Appl. Superconduc.* **3**(1, pt. 4), 2856–2859.
Johnson, R. G. (1978). Honeywell Research Memorandum.
Jones, R. C. (1953). *J. Opt. Soc. Am.* **43**: 1.
Katsan, I. I., Lupina, B. I., Pochtar, V. I., and Frolov, G. A. (1994). *Instrum. Exp. Techn. (Engl. Transl.)* **37**, 526–527.
Katsan, I. I., Lupina, B. I., Pochtar, V. I., and Frolov, G. A. (1994). *Prib. Tekh. Eksp.* **4**, 199–202.
Kawakuba, T., and Kakagawa, T. (1964). *J. Phys. Soc. Jpn.* **19**, 517.
Kimura, M. (1981). *Proc. Sens. Symp. 1st*, pp. 227–232.
Klein, C. A. (1976). *Infrared Phys.* **16**, 663–664.
Kruse, P. W. (1982). Honeywell Research Memorandum.
Kruse, P. W. (1995). *Infrared Phys. Technol.* **36**, 869–882.
Lang, W., Kuehl, K., and Obermeier, E. (1990). *Sens. Actuators, A: Phys.* **22**(1–3, pt. 3), 473–477.
Lang, W., Steiner, P., Schaber, U., and Richter, A. (1994). *Sens. Actuators, A: Phys.* **43**(1–3), 185–187.
Langley, S. P. (1881). *Proc. Am. Acad. Arts Sci.* **16**, 342.
Lewis, W. B. (1947). *Proc. Phys. Soc. London* **59**, 34–40.
Liddiard, K. C. (1984). *Infrared Phys.* **24**, 57.
Liddiard, K. C. (1986). *Infrared Phys.* **26**, 43–49.

Liddiard, K. C., Unewisse, M. H., and Reinhold, O. (1994). *Proc. SPIE—Int. Soc. Opt. Eng.* **2225**, 62–71.
Ling, C. C., and Rebeiz, G. M. (1991). *IEEE Trans. Microwave Theory Techn.* **39**(8), 1257–1261.
Ling, C. C., Landry, J. C., Davee, H., Chin, G., and Rebeiz, G. M. (1994). *IEEE Trans. Microwave Theory Techn.* **42**(4, pt. 2), 758–760.
Lloyd, J. M. (1979). "Thermal Imaging Systems." Plenum, New York.
MacDonald, M. E., and Grossman, E. N. (1995). *IEEE Trans. Microwave Theory Techn.* **43** (4, pt. 1), 893–896.
Marasco, P. L., and Dereniak, E. L. (1993). *Proc. SPIE—Int. Soc. Opt. Eng.* **2020**, 363–378.
Mather, J. C. (1982). *Appl. Opt.* **21**, 1125.
Middelhoek, S., and Audet, S. A. (1989). "Silicon Sensors." Academic Press, San Diego, CA.
Mori, T., Udoh, T., Komatsu, K., and Kimura, M. (1994). *Proc. IEEE MicroElectroMech. Syst., 1994*, pp. 257–262.
Motchenbacher, C. D., and Fitchen, (1973). Low Noise Electronic Design." Wiley, New York.
Neikirk, D. P., Lam, W. W., and Rutledge, D. B. (1984). *Int. J. Infrared Millimeter Waves* **5**, 245–476.
Newman, E. A., and Hartline, P. H. (1982). *Sci. Am.* **246**, 116–127.
Nishioka, N. S., Richards, P. L., and Woody, D. P. (1978). *Appl. Opt.* **17**, 1562–1567.
Petersen, K. E. (1982). *Proc. IEEE* **70**(5), 420–457.
Richards, P. L. (1994). *J. Appl. Phys.* **76**, 1–23.
Rogalski, A. (1994). *Infrared Phys. Technol.* **35**(1), 1–21.
Schnelle, W., and Dillner, U. (1989). *Phys. Status Solidi A* **115**(2), 505–513.
Scott, R. S., and Fredericks, G. E. (1976). *Infrared Phys.* **16**, 619–626.
Shie, J. S., and Weng, P. K. (1992). *Sensors and Actuators* **A33**, 183–189.
Slater, P. N. (1980). "Remote Sensing". Addison-Wesley, Reading, MA.
Smith, R. A., Jones, F. E., and Chasmar, R. P. (1958). " The Detection and Measurement of Infrared Radiation". Clarendon Press, Oxford.
Tanaka, A., Matsumoto, S., Tsukamoto, N., Itoh, S., Endoh, T., Nakazato, A., Kumazawa, Y., Hijikawa, M., Gotoh, H., Tanaka, T., and Teranashi, N. (1995). *Proc. Int. Conf. Solid State Sens. Actuators, 8th*, Stockholm, pp. 632–635.
Umadevi, P., Nagendra, C. L., Thutupalli, G. K., and Mahadevan, K. (1991). *Proc. SPIE— Int. Soc. Opt. Eng.* **1484**, 125–135.
Unewisse, M. H., Passmore, S. J., Liddiard, K. C., and Watson, R. J. (1994). *Proc. SPIE—Int. Soc. Opt. Eng.* **2269**, 43–52.
Unewisse, M. H., Liddiard, K. C., Craig, B. I., Passmore, S. J., Watson, R. J., Clarke, R. E., and Reinhold, O. (1995a). *Proc. SPIE* **2552**, 000–000.
Unewisse, M. H., Craig, B. I., Watson, R. J., Reinhold, O., and Liddiard, K. C. (1995b) *Proc. SPIE* **2554**.
Van Der Ziel, A. (1986). "Noise in Solid State Devices and Circuits". Wiley, New York.
Van Herwaarden, A. W. (1994). Ph.D. Thesis, Delft University of Technology, Delft, The Netherlands.
Voss, R. F., and Clarke, J. (1976). *Phys. Rev. B* **13**, 556–573.
Wentworth, S. M. and Neikirk, D. P. (1992). *IEEE Trans. Microwave Theory Tech.* **40**(2), 196–201.
Wood, R. A. (1983). Honeywell Research Memorandum.
Wood, R. A. (1993a). *Proc. SPIE*, San Diego, *1993*, p. 322.
Wood, R. A. (1993b). *Proc. SPIE* **2020** (Infrared Technol. XIX), 329.
Wood, R. A. (1993c). *Proc. IEDM*, Washington, DC, *1993*.
Wood, R. A. (1994). Honeywell Research Memorandum.
Wood, R. A., and Foss, N. A. (1993). *Laser Focus World*, June, pp. 101–106.

Wood, R. A., and Stelzer, E. (1992). "High Density Array Development," Final Report DAAB07-87-C-F024. NVEOD, Fort Belvoir, VA.
Wood, R. A., Johnson, R. J., Higashi, R. E., and Foss, N. A. (1986). *Proc. IRIS DSG, 1986.*
Wood, R. A., Johnson, R., Higashi, R., and Foss, N. A. (1987a). *Proc. Nat. IRIS, 1987.*
Wood, R. A., Johnson, R., Higashi, R., and Foss, N. A. (1987b). *Proc. IRIS DSG, 1987.*
Wood, R. A., Carney, J., Higashi, R. E., Ohnstein, T., and Holmen, J. (1988). *Proc. IRIS DSG 1988.*
Wood, R. A., Carlson, R. A., Higashi, R. E., and Foss, N. A. (1989). *Proc. IRIS DSG, 1989.*
Wood, R. A., Cole, B. E., Foss, N. A., Han, C. J., Higashi, R. E., and Lubke, R. (1990). *Proc. IRIS DSG, 1990.*
Wood, R. A., Cole, B. E., Han, C. J., and Higashi, R. E. (1991a). *Proc. IRIS DSG*, Boulder, *1991.*
Wood, R. A., Cole, B. E., Han, C. J., Higashi, R. E., Nielsen, D., Weinstein, A., and Miller, J. E. (1991b). *Proc. GOMAC*, Orlando, FL, *1991.*
Wood, R. A., Han, C. J., and Kruse, P. W. (1992a). *IEEE Solid State Actuator Workshop*, Hilton Head Island, SC, *1992*, pp. 132–135.
Wood, R. A., Han, C. J., Cole, B. E., Higashi, R. E., Holmen, J., Ridley, J., Johnson, B., and Nielsen, D. (1992b). *Proc. IRIS Passive Sens., 1992.*
Wood, R. A., Han, C. J., Cole, B. E., and Higashi, R. E. (1992c). *Proc. IRIS Detec. Spec. Group, 1992.*
Wood, R. A., Cole, B. E., Han, C. J., and Higashi, R. E. (1993). *Proc. IRIS DSG*, Bedford, MA, *1993.*
Zwerdling, S., Smith, R. A., and Theriault, J. P. (1968). *Infrared Phys.* **8**, 271.

FIG. 31. (Chapter 3) Magnified area of a display screen from an infrared camera viewing a point hot object, illustrating no cross talk observable above the random noise level. (Reprinted from Wood et al. (1987b); with permission from the authors.)

FIG. 34. (Chapter 3) Microscope photograph of single-level microbolometers with "L-shaped" legs for improved thermal isolation.

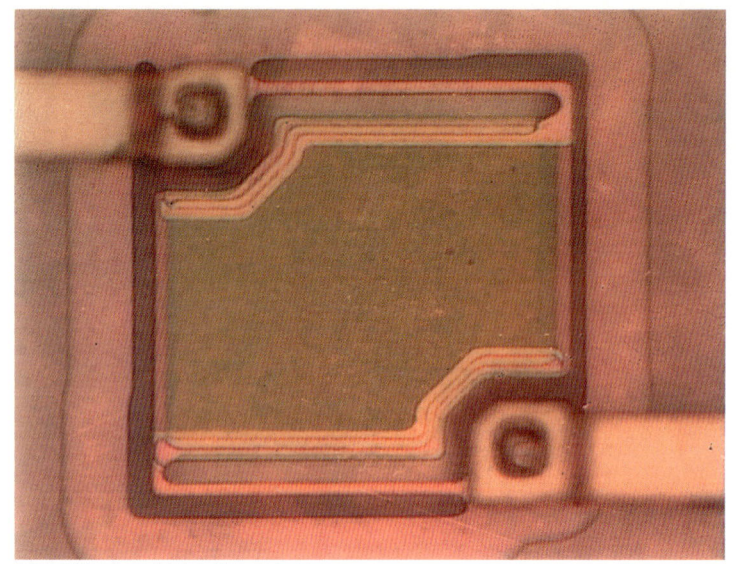

FIG. 37(b). (Chapter 3)

FIG. 42(a). (Chapter 3)

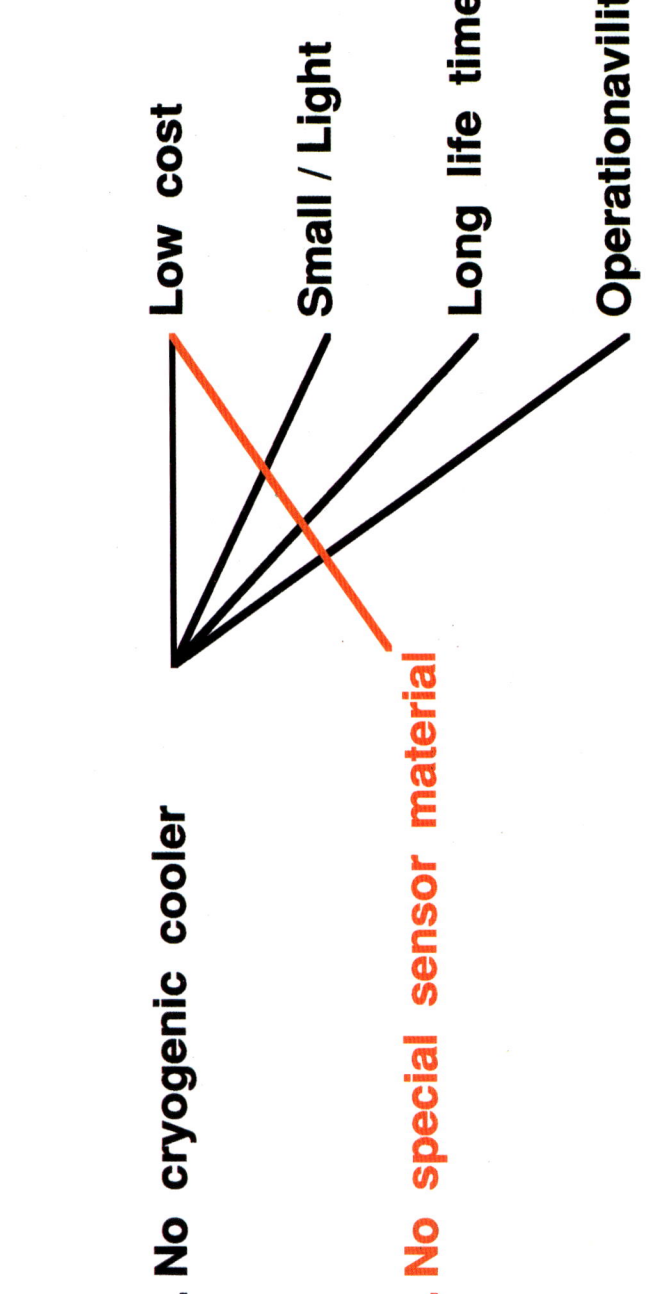

FIG. 1. (Chapter 6) Merit for uncooled infrared (IR) detector.

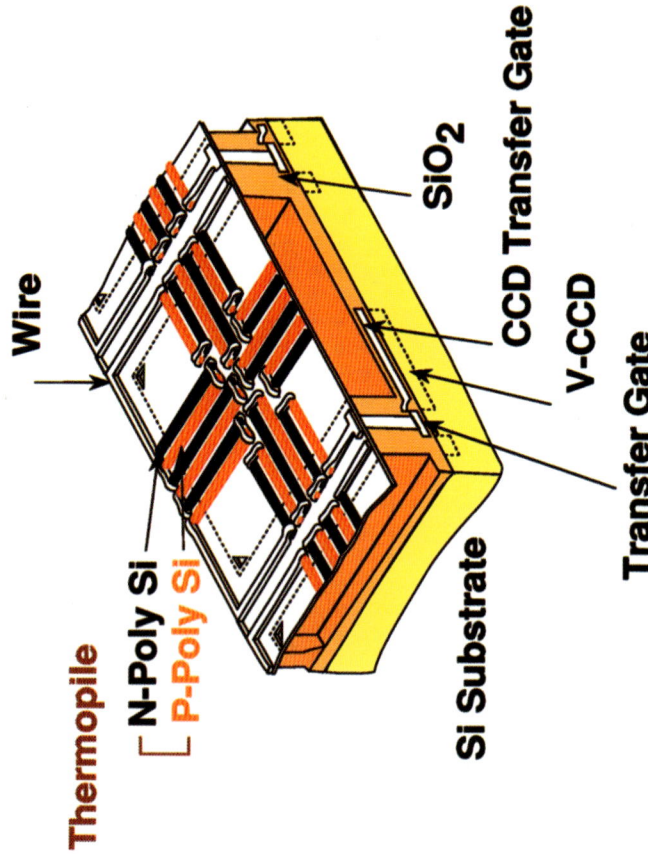

FIG. 5. (Chapter 6) Pixel structure of the thermopile infrared focal plane array. $SiO_2$, silicon dioxide; V-CCD, charge-coupled device.

CHAPTER 4

# Hybrid Pyroelectric–Ferroelectric Bolometer Arrays

*Charles M. Hanson*

UNCOOLED INFRARED SYSTEMS DEPARTMENT
TEXAS INSTRUMENTS, INC.
DALLAS, TEXAS

| | |
|---|---|
| I. INTRODUCTION . . . . . . . . . . . . . . . . . . . . . . . . . . . . . . | 123 |
| II. PRINCIPLES OF PYROELECTRIC DETECTORS . . . . . . . . . . . . . . . . . | 124 |
|     1. *Pyroelectricity and Ferroelectric Materials* . . . . . . . . . . . . . . . | 124 |
|     2. *Modes of Operation* . . . . . . . . . . . . . . . . . . . . . . . . | 139 |
|     3. *Signal and Noise* . . . . . . . . . . . . . . . . . . . . . . . . . | 144 |
| III. PRACTICAL CONSIDERATIONS AND DESIGNS . . . . . . . . . . . . . . . . . | 154 |
|     1. *Ferroelectric Material Selection* . . . . . . . . . . . . . . . . . . . | 154 |
|     2. *Thermal Isolation* . . . . . . . . . . . . . . . . . . . . . . . . . | 156 |
|     3. *Modulation Transfer Function (MTF)* . . . . . . . . . . . . . . . . . | 158 |
|     4. *Read-out Electronics* . . . . . . . . . . . . . . . . . . . . . . . | 159 |
|     5. *System Electronics* . . . . . . . . . . . . . . . . . . . . . . . . | 161 |
|     6. *Choppers* . . . . . . . . . . . . . . . . . . . . . . . . . . . . | 162 |
| IV. SYSTEMS IMPLEMENTATIONS . . . . . . . . . . . . . . . . . . . . . . | 169 |
|     References . . . . . . . . . . . . . . . . . . . . . . . . . . . . . | 173 |

## I. Introduction

This chapter explores the physics and applications of a class of uncooled infrared (IR) detectors distinguished by two features: They employ in some manner the pyroelectric properties of ferroelectric materials, and they comprise separate detection and electronic read-out means that are bonded together toward the end of the fabrication process.

This technology began in the 1970s at a time when bulk ferroelectric materials were easily synthesized, but thin films were not, a time when a bump-bond process for manufacturing hybrid IR photodetectors was rapidly evolving. The high thermal response of ferroelectric materials adequately compensated for what was lacking in thermal isolation structures.

Hybrid pyroelectric–ferroelectric technology has matured greatly during its 20 years. Thermal isolation structure improvements and reduced detector thicknesses have evolved together to yield performance levels unsurpassed even by detectors having thermal isolation orders of magnitude greater. This highlights the superiority of the pyroelectric–ferroelectric detection mechanism.

This technology, pioneered independently in the mid-1970s by Hopper (1978, 1979) and McCormack (1979) at Texas Instruments (Dallas, Texas) and by Iwasa and Butler (Butler *et al.*, 1988) at Loral (at the time, the Honeywell Radiation Center), in the 1990s, has become the foundation for rapid expansion of thermal imaging markets in both the commercial and military sectors. Applications that could not conceivably afford the cost of thermal imaging in the 1970s now have choices. The Loral hybrid approach has since been discontinued; however, very early Putley (1980) and Watton (1986) of the Royal Signals and Radar Establishment, and Whatmore (1986) and Porter (1981) of GEC Marconi (at the time, Plessey), began an independent effort in the United Kingdom. Over the intervening years several companies have made attempts at hybrid pyroelectric detectors; however, only Texas Instruments and GEC Marconi have maintained a commitment that has led to salable products.

Unfortunately, most of the work done in the United States in the 1970s and 1980s was classified information, and therefore little open literature exists describing the details, except for a few patents. For this reason specific references to the work are limited.

This chapter explains the principles of operation of pyroelectric detectors and gives a basis for detector design decisions. A discussion of the fundamentals of ferroelectric materials from a phenomenologic point of view begins the chapter, and it continues with an exposition regarding signal and noise. Many of the practical issues of system implementation—the options and the rational basis of successful programs—are addressed. The chapter concludes with a brief description of some of the more important systems that recently have become available.

## II. Principles of Pyroelectric Detectors

### 1. Pyroelectricity and Ferroelectric Materials

Every material becomes polarized with the application of an electric field. The degree to which this is true for a given material is reflected in its permittivity or dielectric constant. Some materials are polarized even in the absence of an applied electric field; that is, they are spontaneously polarized. Ferroelectric materials are those whose spontaneous polarization can be switched or reversed by application of an electric field.

In general, a material's polarization, whether spontaneous or induced by an electric field, is temperature-dependent. This temperature dependence is the pyroelectric effect. Of the 32 possible crystallographic systems, 21 are piezoelectric; of those 21 systems, 10 are pyroelectric. Ferroelectric materials are a subset of the pyroelectric materials. Thus all ferroelectric materials are pyroelectric, but not all pyroelectric materials are ferroelectric. In practice, ferroelectric materials make much better IR detectors than do other pyroelectric materials, and we shall limit our discussion primarily to ferroelectric detectors. First, a brief review of ferroelectric physics is in order.

A material in an external electric field is generally described by the equation

$$D = \varepsilon_0 E + P \tag{1}$$

where $D$ is the electric displacement vector, $\varepsilon_0$ is the permittivity of free space, $E$ is the electric field vector, and $P$ is the polarization vector.

For many materials, $P$ is proportional to $E$:

$$P = \chi \varepsilon_0 E \tag{2}$$

and therefore

$$\begin{aligned} D &= (\chi + 1)\varepsilon_0 E \\ &= \varepsilon \varepsilon_0 E \end{aligned} \tag{3}$$

where $\varepsilon = \chi + 1$ is the relative permittivity or relative dielectric constant, usually called simply the dielectric constant, and $\chi$ is the electric susceptibility. These materials are called linear dielectrics because of the linear relationship between $D$ and $E$. Equation (3) is not valid for ferroelectrics, which are highly nonlinear. For nonlinear materials a dynamic permittivity is defined by

$$\varepsilon \varepsilon_0 = \frac{\partial D}{\partial E} \tag{4}$$

The following analysis generally follows that of Lines and Glass (1977), except that we have generalized to explicitly include electric field effects. The fundamentals of ferroelectric behavior can be described by a one-dimensional electric Gibbs free energy expressed as a power series in electric displacement:

$$G = \tfrac{1}{2}\beta(T - T_0)D^2 + \tfrac{1}{4}\gamma D^4 + \tfrac{1}{6}\delta D^6 - ED \tag{5}$$

where $G$ is the electric Gibbs free energy; $T$ is the temperature of the ferroelectric; $T_0$ is the Curie-Weiss temperature, a characteristic of the material; $E$ is the applied electric field; $D$ is the magnitude of the electric displacement; and $\beta$, $\gamma$, and $\delta$ are material parameters. In the absence of an electric field, the free energy is an even function of electric displacement because it is independent of direction of polarization. The material parameters $\beta$, $\gamma$, and $\delta$ may be functions of temperature, and the series expansion may be carried to higher order but Eq. (5), as written, contains the essence of ferroelectricity. Also, $D$ and $E$ in Eq. (5) should, strictly speaking, be vectors rather than scalars, and the parameters $\beta$, $\gamma$, and $\delta$ should be tensors; however, these complications are unnecessary for the present analysis, and we shall henceforth treat $D$, $E$, and $P$ as scalars. For ferroelectrics $P$ is generally much greater than $\varepsilon_0 E$, and to a good approximation,

$$D \approx P \tag{6}$$

Because of this near equality, many authors substitute $P$ for $D$. We shall consistently use $D$, which has, from Gauss' law, the rigorous property

$$Q = DA \tag{7}$$

where $Q$ is the charge on one of the plates of a parallel-plate capacitor and $A$ is the area of one of the capacitor plates. The variables $D$, $E$, and $T$ are thermodynamic variables of the ferroelectric material, much like pressure, volume, and temperature in a gas, and each is expressible as a function of the other two as an equation of state.

The permissible states for a given field are given by the solutions to the equation

$$\left(\frac{\partial G}{\partial D}\right)_{E,T} = 0 \tag{8}$$

or

$$E = \beta(T - T_0)D + \gamma D^3 + \delta D^5 \tag{9}$$

This is the equation of state for a ferroelectric material. For each value of $E$, there may be multiple values of $D$. The value of $G$ for each value of $D$

specifies which states are stable, which are unstable, and which are metastable. This describes hysteresis in ferroelectrics.

The dielectric constant is given by

$$\frac{1}{\varepsilon\varepsilon_0} = \left(\frac{\partial E}{\partial D}\right)_T = \beta(T - T_0) + 3\gamma D^2 + 5\delta D^4 \tag{10}$$

Note that

$$\left(\frac{\partial E}{\partial D}\right)_T = \frac{\partial^2 G}{\partial D^2} \tag{11}$$

and therefore a negative dielectric constant indicates a maximum $G$, which is an unstable state.

The pyroelectric coefficient, defined by

$$p = \left(\frac{\partial D}{\partial T}\right)_E \tag{12}$$

is the fundamental parameter for current response (short circuit) of a pyroelectric detector. The equivalent parameter for voltage response (open circuit) is

$$\left(\frac{\partial E}{\partial T}\right)_D = -\left(\frac{\partial D}{\partial T}\right)_E \bigg/ \left(\frac{\partial D}{\partial E}\right)_T = -\frac{p}{\varepsilon\varepsilon_0} \tag{13}$$

From Eqs. (9) and (13),

$$\left(\frac{\partial E}{\partial T}\right)_D = -\frac{p}{\varepsilon\varepsilon_0} = \beta D \tag{14}$$

Equation (14) is important because it indicates that increasing the magnitude of the polarization of a given material improves the voltage responsivity. Pyroelectric detectors have a finite impedance of their own, and therefore they never operate in a purely open-circuit or short-circuit configuration. However, as we shall see later, both response parameters are important figures of merit for real detectors.

Ferroelectric materials can be broadly classified by the order of their phase transition from a high-temperature paraelectric phase to a lower-temperature ferroelectric phase. A first-order thermodynamic phase transition has a discontinuity in the first temperature derivative of its free energy, and a second-order transition has a discontinuity in the second derivative. The order of the transition has important implications for the interpretation of experimental results as they apply to pyroelectric detectors. Materials of both types are of interest for use as pyroelectric detectors, and therefore we will consider both.

### a. Second-order Ferroelectrics

The second-order ferroelectric phase transition is simpler, and we consider it first. Equation (5) well-represents a second-order transition if $\beta$ and $\gamma$ are positive. The parameter $\delta$ adds nothing to the model and can be set to zero. The important equations become

$$G = \tfrac{1}{2}\beta(T - T_0)D^2 + \tfrac{1}{4}\gamma D^4 - ED \qquad (15)$$

$$E = \beta(T - T_0)D + \gamma D^3 \qquad (16)$$

$$\frac{1}{\varepsilon\varepsilon_0} = \beta(T - T_0) + 3\gamma D^2 \qquad (17)$$

When $E = 0$, Eq. (16) yields the solutions for spontaneous polarization

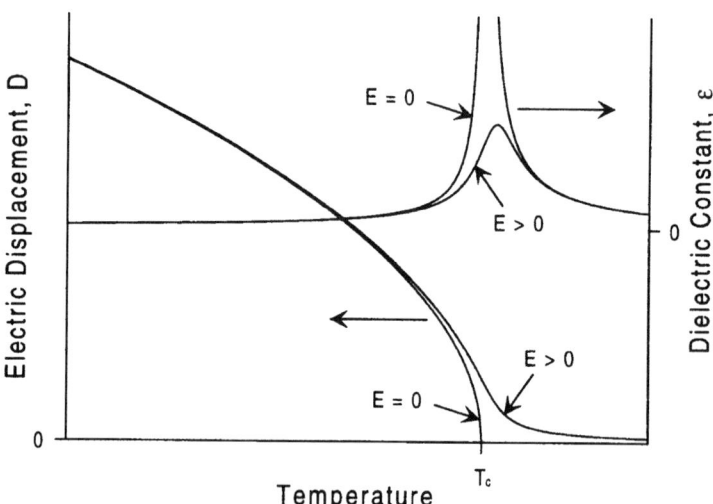

FIG. 1. Polarization and dielectric constant, with and without an applied field, as a function of temperature for a second-order ferroelectric. $T_c$, transition temperature.

(noting that $D = P$ when $E = 0$)

$$P_s = \begin{Bmatrix} 0 \\ \pm \sqrt{\dfrac{\beta}{\gamma}(T_0 - T)} \end{Bmatrix} \qquad (18)$$

The first solution, $P_s = 0$, is, from Eq. (17), unstable when $T < T_0$ and stable when $T > T_0$. Conversely, the other two solutions are stable below $T_0$ and unstable above. Thus the spontaneous polarization for a second-order ferroelectric is as shown in Fig. 1.

When $E \neq 0$, the solutions are still closed form but are somewhat more complex. This situation is also shown in Fig. 1. Note that the pyroelectric coefficient (the slope of the $D$ versus $T$ curve) is suppressed by the field; however, according to Eq. (14), the voltage response is improved (because of increased $D$). This results from the fact that the dielectric constant is suppressed to a greater extent than is the pyroelectric coefficient.

The temperature $T_0$ clearly demarks the transition from a polar phase at low temperature to a nonpolar phase at high temperature. This is shown in Fig. 2, where the minima of the $G$ versus $D$ curves represent stable states. Figure 2 shows that as the temperature is increased from below $T_0$, the $D \neq 0$ phase disappears at $T_0$, and the $D = 0$ phase appears at $T_0$. Therefore, there is a sharp transition; that is, there is no temperature at which the two phases coexist, and there is no temperature at which neither phase exists.

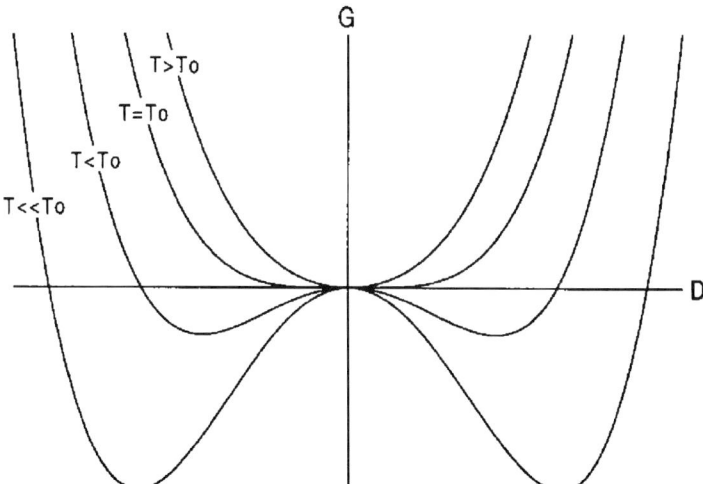

FIG. 2. Free-energy diagrams for both phases of a second-order ferroelectric. $T_0$, temperature.

The transition temperature, or Curie temperature, for a second-order transition is

$$T_c = T_0 \tag{19}$$

The pyroelectric coefficient is the magnitude of the slope of $P$ versus $T$. Therefore, there is no pyroelectric effect above $T_c$ in the absence of an applied field. It is clear from Fig. 1 that the pyroelectric coefficient becomes very large as $T$ approaches $T_c$ from below.

From Eqs. (17) and (18), the dielectric constant above and below the transition is given by

$$\frac{1}{\varepsilon \varepsilon_0} = \begin{cases} \beta(T - T_c) & \text{for } T > T_c \\ 2\beta(T_c - T) & \text{for } T < T_c \end{cases} \tag{20}$$

and so the dielectric constant diverges as $T$ approaches $T_c$ from either direction, but more sharply from below. Since $D$ generally increases with applied field, Eq. (17) shows that the peaking of the dielectric constant near the phase transition becomes more diffuse with increasing field.

Equation (16) implies a hysteresis loop, as shown in Fig. 3. The loop is best understood by considering the $T < T_c$ energy diagrams in Fig. 4. Start tracing the hysteresis loop at large negative $E$ and $D$. When $E$ is increased to zero, we have the situation shown in Fig. 4(a). There are two possible equal and opposite values of $D$, and both have the same energy. Because it

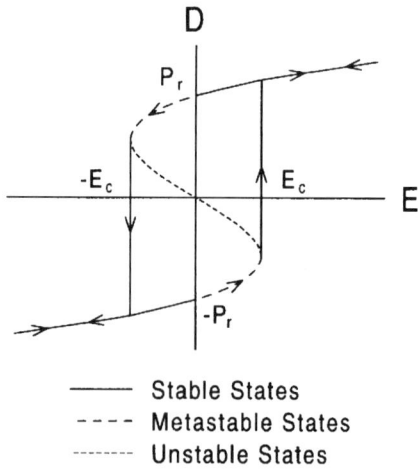

FIG. 3. Hysteresis loop for a second-order ferroelectric. $P_r$, remanent polarization; $E_c$, coercive field.

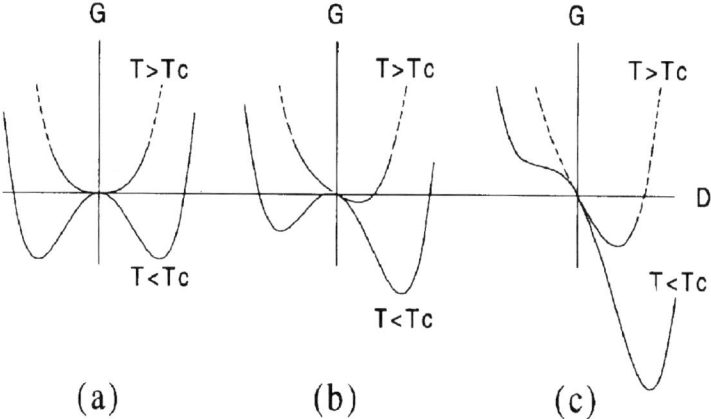

FIG. 4. Gibbs' free energy for a second-order phase transition for (a) $E = 0$, (b) $E > E_c$, and (c) $E \gg E_c$. $E_c$, coercive field; $T_c$, transition temperature.

is already in the negative $D$ state, there is no driving force to cause a polarization reversal. The magnitude of $D$ at this point is the remanent polarization $P_r$. As we proceed to positive $E$, the situation shown in Fig. 4(b) pertains. The energy of the negative $D$ state is a local minimum, but not the absolute minimum. It is therefore metastable and will eventually decay into the lower-energy positive $D$ state, given a decay mechanism. In this region we have partial polarization reversal, and the net negative polarization begins to decrease in magnitude. As we further increase $E$, there comes a point at which the secondary energy minimum disappears, as in Fig. 4(c). Then there is only one stable state, and the polarization reverses completely. The value of the electric field at this point is the coercive field $E_c$. The coercive field is temperature-dependent, tending to zero as $T$ approaches $T_c$. Note that the dielectric constant is the slope of the $D$ versus $E$ curve, and so it too experiences hysteresis.

We can see that the phase transition is indeed second order by examining the temperature derivatives of Eq. (15) above and below the transition temperature. The entropy is related to the first derivative by

$$S = -\frac{\partial G}{\partial T}$$

$$= \begin{cases} \frac{1}{2} \frac{\beta^2}{\gamma} (T - T_c) & \text{for } T < T_c \\ 0 & \text{for } T > T_c \end{cases} \quad (21)$$

which is continuous at $T_c$. The specific heat is related to the entropy and to the second temperature derivative of the free energy by

$$c_p = T\frac{dS}{dT}$$

$$= -T\frac{\partial^2 G}{\partial T^2}$$

$$= \begin{cases} \frac{1}{2}\frac{\beta^2}{\gamma} & \text{for } T < T_c \\ 0 & \text{for } T > T_c \end{cases} \tag{22}$$

which is discontinuous at $T_0$. Thus the specific heat increases discontinuously as the material experiences a transition from its paraelectric phase to its ferroelectric phase. This is characteristic of a second-order transition.

b. *First-Order Ferroelectrics*

Ferroelectric materials having a first-order transition behave in a more complicated fashion than do second-order transition materials. In order for Eq. (5) to represent a first-order transition, $\gamma$ must be negative, and $\beta$ and $\delta$ must be positive. Letting $\gamma' = -\gamma$, we have

$$G = \tfrac{1}{2}\beta(T - T_0)D^2 - \tfrac{1}{4}\gamma'D^4 + \tfrac{1}{6}\delta D^6 - ED \tag{23}$$

The corresponding equation of state is

$$E = \beta(T - T_0)D - \gamma'D^3 + \delta D^5 \tag{24}$$

and the dielectric constant is given by

$$\frac{1}{\varepsilon\varepsilon_0} = \beta(T - T_0) - 3\gamma'D^2 + 5\delta D^4 \tag{25}$$

The spontaneous polarization is determined by Eq. (24) with $E = 0$. The result is

$$P_s = \begin{Bmatrix} 0 \\ \pm\sqrt{\dfrac{\gamma' + \sqrt{\gamma'^2 - 4\delta\beta(T - T_0)}}{2\delta}} \\ \pm\sqrt{\dfrac{\gamma' - \sqrt{\gamma'^2 - 4\delta\beta(T - T_0)}}{2\delta}} \end{Bmatrix} \quad (26)$$

and the complexity is obvious. In order to determine the stability of the various solutions, we must again turn to the energy diagrams (Fig. 5) where the minima of the G versus D curves represent stables states. The first obvious difference from second-order transitions is that there is a temperature range in which polar and nonpolar phases coexist. This makes the definition of the transition a little trickier. The transition temperature $T_c$ is the temperture below which the polar phase is more stable and above which the nonpolar phase is more stable. At $T_c$, therefore, the two phases have exactly the same energy. The transition temperature $T_c$ is given by

$$T_c = T_0 + \frac{3\gamma'^2}{16\beta\delta} \quad (27)$$

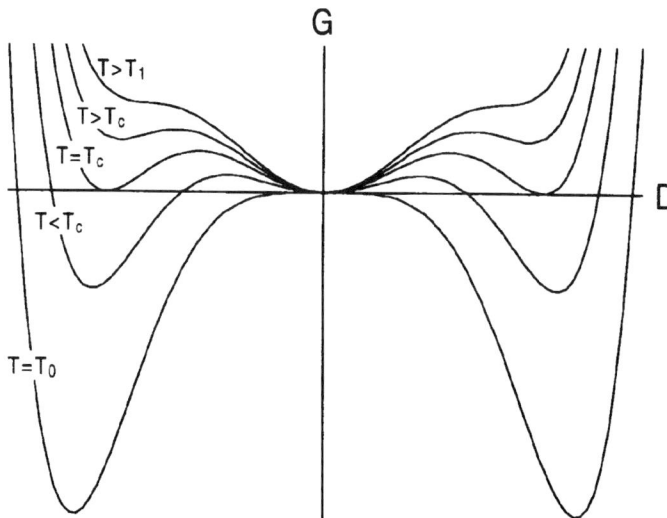

FIG. 5. Free-energy diagrams for a first-order ferroelectric. $T_c$, transition temperature.

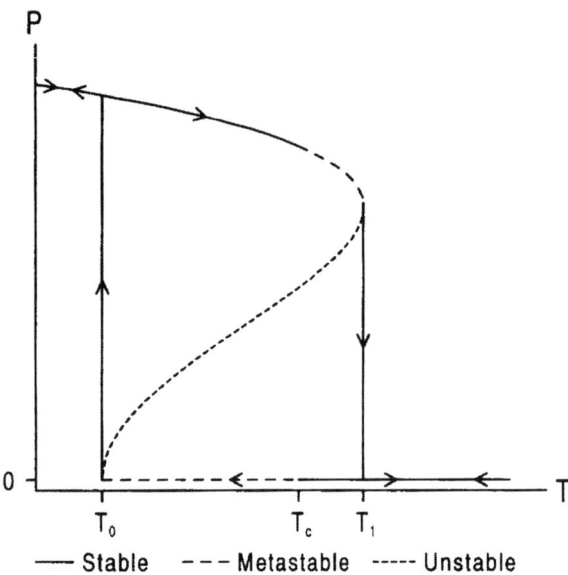

FIG. 6. Ideal thermal hysteresis loop for a first-order ferroelectric. $T_c$, transition temperature.

Below $T_0$ the nonpolar phase becomes unstable. There is a temperature $T_1$ above which the polar phase becomes unstable, given by

$$T_1 = T_0 + \frac{\gamma'^2}{4\beta\delta} \qquad (28)$$

These considerations lead to the thermal hysteresis loop show in Fig. 6. As the temperature increases from the high-polarization region below $T_0$, the polarization gradually diminishes. At $T_0$ the nonpolar phase is no longer unstable, but the material is already in the energetically most favorable polar phase, and no depolarization occurs. The polar phase becomes metastable at $T_c$. At this point the nonpolar phase becomes energetically more favorable, and some depolarization is likely to occur. When the temperature reaches $T_1$ the polar phase is unstable, and the material completely and discontinuously depolarizes. Retracing, as the temperature decreases to $T_1$, the polar phase is no longer unstable. The material is already in the energetically more favorable nonpolar phase, and therefore no spontaneous polarization occurs. As the temperature decreases further to $T_c$, the nonpolar phase becomes metastable and the polar phase becomes

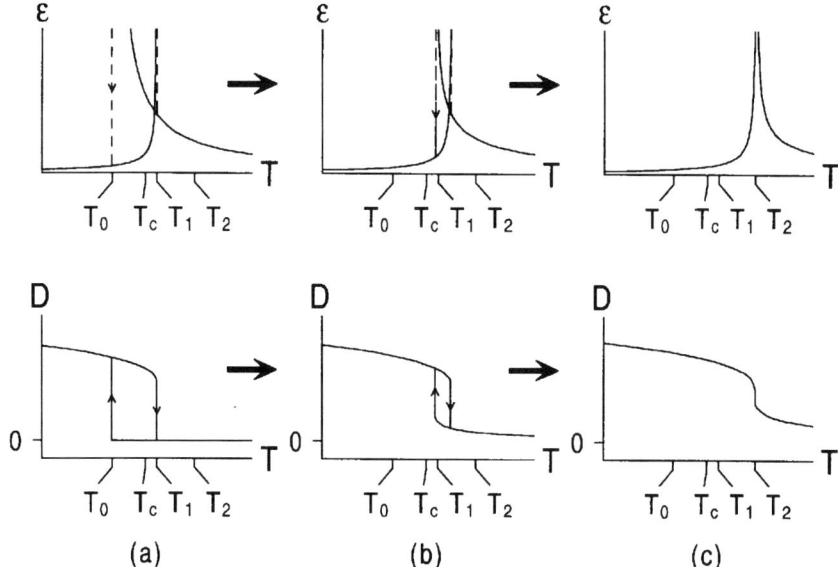

FIG. 7. Effect of increasing applied electric field on thermal hysteresis of electric displacement (*bottom*) and dielectric constant (*top*), beginning with $E = 0$ in (a). $T_c$, transition temperature; $T_2$, maximum temperature.

energetically more favorable. Some spontaneous polarization begins to occur. When the temperature reaches $T_0$, the nonpolar phase becomes unstable, and the material polarizes spontaneously and discontinuously. The spontaneous polarization generally is not directly measurable, because, in the absence of external influences, domains form in such a manner that the net polarization is zero. The effects are evident, however, in the dielectric constant and other material properties that depend on polarization.

Application of an electric field suppresses the thermal hysteresis, as shown in Fig. 7. The field improves the favorability of the polar phase, thereby raising the temperature at which that phase becomes unstable. At the same time, the nonpolar phase experiences a field-induced polarization. These effects suppress hysteresis and reduce and eventually eliminate the discontinuity in polarization. The electric field at which thermal hysteresis disappears is

$$E_2 = \frac{6}{25}\sqrt{\frac{3\gamma'^5}{10\delta^3}} \qquad (29)$$

and the temperature of the inflection point at that field is

$$T_2 = T_0 + \frac{9\gamma'^2}{20\beta\delta} \tag{30}$$

Thus $E_2$ is the maximum field at which there is a phase transition, and $T_2$ is the maximum temperature at which the ferroelectric phase can be induced by an electric field.

This importance of thermal hysteresis for pyroelectric detectors is not widely appreciated. The effective pyroelectric coefficient is the slope of the $D$ versus $T$ curve. In the vicinity of the transition, therefore, there is ambiguity in its value. The simplest and most common means of measuring the pyroelectric coefficient is the Byer-Roundy method, wherein the temperature is steadily increased and the corresponding release of electric charge is measured. Measurement with decreasing temperature yields similar results, but with a temperature shift that readily can be confused with thermal lag due to the mass of the apparatus. The operation of a pyroelectric detector, however, is very different. In operation, the temperature fluctuates only very slightly, thus traversing an "inner" loop, as shown in Fig. 8. As a result the slope of the $D$ versus $T$ curve is reduced, giving a lower pyroelectric coefficient. This is an important reason for discrepancies between predicted and actual responsivities of pyroelectric detectors. For this reason it is preferable to measure the pyroelectric coefficient using the Chynoweth (1956) method, which approximates a detector configuration.

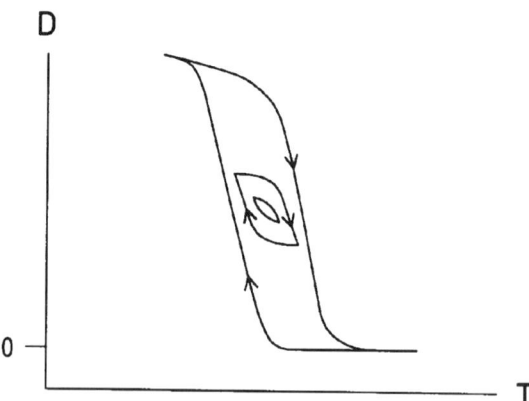

FIG. 8. Conceptual sketch showing the effect of reducing the thermal excursion within the limits of the thermal hysteresis loop.

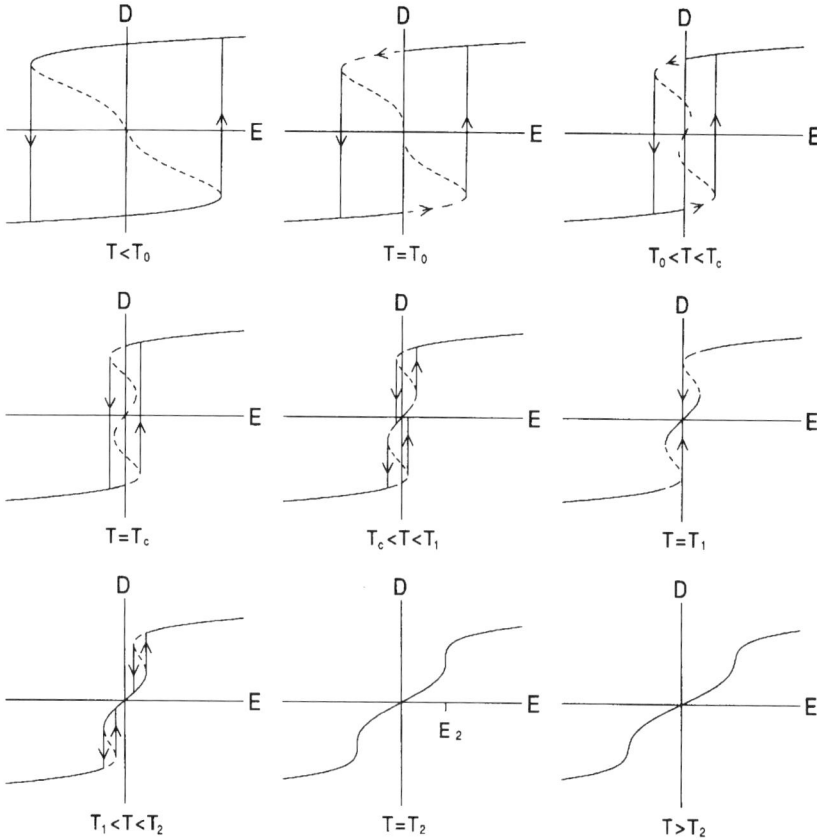

FIG. 9. Hysteresis loops for first-order ferroelectric materials at selected temperatures.

Field dependence of the pyroelectric behavior of first-order ferroelectrics is quite different from that of second-order materials. In first-order materials the pyroelectric coefficient near the transition actually increases with field for $E < E_2$. This results from a sharpening of the transition as the thermal hysteresis diminishes with increasing field. The increase in polarization with field indicates that the voltage response increases as well, regardless of the fact that the dielectric constant also increases near the transition when $E < E_2$.

Electrical hysteresis in first-order ferroelectrics has complications not present in second-order materials. Figure 9 shows hysteresis loops for a variety of temperatures. At low temperature the hysteresis is much like that of a second-order ferroelectric. As the temperature increases past $T_0$, a metastable nonpolar phase appears at low fields. At $T_c$ a stable nonpolar phase appears at low fields. At some temperature between $T_c$ and $T_1$, the stable nonpolar phase is sustained at sufficiently high field to provide an intermediate state between the large positive polarization state and the large negative polarization state, and the hysteresis loop begins to be squeezed in the middle. At $T_1$ the loop disappears momentarily, but it reappears as a double loop above $T_1$. At $T_2$ the double loop disappears, and the dielectric constant diverges at field $E_2$. Above $T_2$ there is no more switching behavior, although the dielectric properties are highly nonlinear.

The order of the transition is evident from the discontinuity in spontaneous polarization at the transition temperature $T_c$. From Eqs. (26) and (27), the value of $D$ at $T_c$ is

$$P_c = \begin{cases} \sqrt{\dfrac{3\gamma'}{4\delta}} & \text{for the polar phase} \\ 0 & \text{for the nonpolar phase} \end{cases} \quad (31)$$

The entropy at $T_c$ is therefore

$$S = -\left(\frac{\partial G}{\partial T}\right)_E = -\frac{1}{2}\beta P_c^2 = \begin{cases} -\dfrac{3\beta\gamma'}{8\delta} & \text{for the polar phase} \\ 0 & \text{for the nonpolar phase} \end{cases} \quad (32)$$

Thus there is a discontinuity in the first temperature derivative of the free energy, and the transition is first-order. There is latent heat associated with the transition, given by

$$\Delta Q = T_c \Delta S = \frac{1}{2} T_c \beta P_c^2 = \frac{3\beta\gamma'}{8\delta} T_c \quad (33)$$

This means that the ferroelectric material can absorb heat without changing temperature. At first glance this would seem to destroy its potential use as an IR detector. However, under these conditions the polarization changes with absorbed heat rather than with temperature change. Furthermore, increasing an applied electric field reduces and eventually eliminates the latent heat. This effect is related to the thermal hysteresis discussed earlier and can result in misleading pyroelectric measurements using the Byer-Roundy technique.

## 2. MODES OF OPERATION

The most basic problem of processing thermal images results from the fact that signals of interest are small fractions of the total detectable radiation flux. Most detectors compound the problem by producing a signal level (dark current) even in the absence of incident flux. These phenomena produce noise that limits the sensitivity of the detectors; however, more important for staring imagers, these phenomena produce image nonuniformities that dominate system performance. For single-element detectors, the effect is of relatively little consequence. For large arrays of detectors, small pixel-to-pixel variations in responsivity, resistivity, or other pertinent parameters produce pixel-to-pixel variations in output levels that dwarf signals of interest. Solving this problem requires introduction of system electronic complexities that increase cost, weight, and power dissipation, and even then long-term or even short-term drift presents a limit to the achievable sensitivity.

For scanned near-linear detector arrays, this problem has traditionally been solved by ac-coupling each detector pixel at its output. For two-dimensional staring arrays this is not practical because of the limited available space for circuitry and because lower operating frequencies require larger ac-coupling capacitors. Pyroelectric detectors, however, are inherently ac-coupled, and as such, they provide an elegant solution to the problem.

Pyroelectric detectors are capacitors, usually simple parallel plate capacitors. There are three different but related ways that ferroelectric materials can be used as IR detectors. In the *poled* mode the ferroelectric material is prepared by application of an electric field, and the material remains polarized after removal of the field. The temperature dependence of the polarization is inherent pyroelectricity. In the *induced pyroelectric* mode an electric field remains applied during operation, inducing a polarization in addition to the spontaneous polarization. The temperature dependence of the induced polarization is induced pyroelectricity. In the *pulsed* (or ac) mode the applied electric field is modulated, releasing the (temperature-dependent) charge on the capacitor.

The first two modes are inherently ac-coupled, which considerably simplifies system implementation, but which requires a chopper. The pulsed mode does not require a chopper, but it is dc-coupled and presents the same problems as most nonpyroelectric detectors, as discussed briefly earlier. The three modes of operation are discussed more fully in the following sections.

### a. Inherent Pyroelectric Mode

When a ferroelectric material is initially prepared, it comprises domains—small regions, each of which is coherently polarized within itself. The energy of interaction among these domains is such that they are generally aligned to produce no net polarization in a ferroelectric capacitor. Application of an electric field greater than the coercive field tends to align the domains so that they contribute cooperatively to a net polarization. Figure 10 shows this effect as a partial tracing of a hysteresis loop. The remanent polarization, that is, the polarization that remains after removal of the electric field, is temperature-dependent. It decreases with increasing temperature so that essentially no spontaneous polarization exists above the transition temperature. Note from Fig. 10 that the final polarization state depends on the maximum applied electric field. At sufficiently large field, a further increase will not expand the loop. This is the saturation condition.

The inherent pyroelectric mode requires no bias field, and therefore the electrical conductivity of the material is not very critical, provided the capacitor is not so leaky that its electrical RC time constant is comparable to the signal integration time. Operation in this mode requires that the operating temperature be sufficiently below the transition temperature that spontaneous depoling does not occur, although some depoling is tolerable

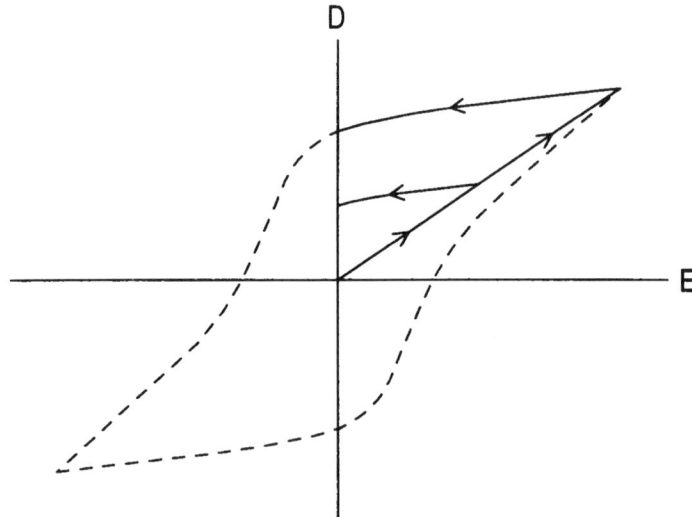

FIG. 10. Polling process, in the context of the hysteresis loop.

if the original state is periodically restored by poling. This limits the range of suitable material compositions to those whose transition is well above normal ambient operating temperatures. As we saw in Figs. 1 and 6, well below the transition temperature the electric displacement $D$ is large, but the pyroelectric coefficient $p$ is small. A low value of $p$ implies low current-mode responsivity; however, from Eq. (14), a high value of $D$ implies high voltage-mode responsivity.

### b. Induced Pyroelectric Mode

In the induced pyroelectric mode of operation the inherent pyroelectric effect is enhanced by application of an external biasing electric field (Hopper, 1978). An applied field always increases $D$ and therefore always increases voltage-mode responsivity. For a second-order ferroelectric material an applied field degrades current-mode responsivity. For a first-order material it may improve or degrade current-mode responsivity, depending on the operating temperature and on the strength of the applied field.

The presence of a field stabilizes operation so that detectors can operate near the ferroelectric transition temperature without risk of depoling. This broadens the range of suitable material compositions to include those having transitions near ambient operating temperatures. It also permits optimization of properties by manipulation of operating field and temperature, especially in first-order ferroelectrics. Associated with the applied field is a leakage current. Although leakage currents are small, generally much less than $10^{-12}$ A for 50-$\mu$m square pixels, they can be large enough to be significant compared with typical low-level signals. Furthermore, leakage is not typically uniform over a detector array. Fortunately, artifacts resulting from leakage are of constant polarity, whereas chopped signals are of alternating polarity. Therefore, subsequent signal processing automatically eliminates the offsets without any additional effort. However, it remains important to limit leakage in order to control the dynamic range requirements of the electronic preprocessor. Even at its worst, leakage-induced offset nonuniformities are far more subtle than are those present in photodiodes, photoconductors, or bolometers.

Perhaps the greatest benefit of operating in the induced pyroelectric mode occurs when operating near the ferroelectric phase transition. Or, to put it another way, perhaps the greatest benefit of using an applied field is that it enables operation near the ferroelectric phase transition. Such operation eases material requirements, partly because crystalline orientation becomes relatively unimportant. Below the phase transition the material is polar and the crystal structure loses a degree of symmetry. In this region polar alignment is critical. Above the transition the material is nonpolar and more symmetric. Here alignment is unimportant, but there is little or no inherent pyroelectric effect. Slightly below the transition the symmetry is broken, but only by a small degree. In this region an applied field readily supplies sufficient energy to dominate the polarization. For materials having cubic crystal structure in the nonpolar phase, the contribution of each domain will be no less than $1/\sqrt{3}$ of its maximum possible contribution. For a matrix of randomly oriented grains, the net polarization is almost 80% of the maximum. Thus the performance penalty for using a ceramic instead of a single crystal is relatively minor.

Responsivity is temperature-dependent, with a peak near the phase transition. Optimum operation requires stabilization near the responsivity peak, which makes temperature control necessary. This requirement is not very stringent in practice. Figure 11 shows $(\partial E/\partial T)_D$ for a first-order transition ferroelectric as a function of temperature, for two different applied electric fields. Increasing the bias field not only increases responsivity, it also

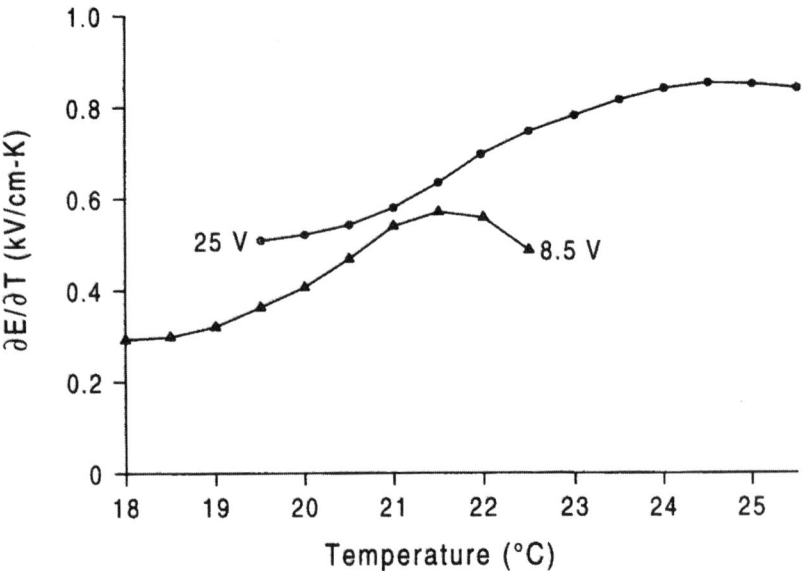

FIG. 11. Pyroelectric material responsivity figure of merit $[(\partial E/\partial T)_D = p/\varepsilon\varepsilon_0]$ for a ceramic $Ba_{0.67}Sr_{0.33}TiO_3$ (barium strontium titanate) sample 25-$\mu$m thick, at two different bias voltages. (Courtesy of Texas Instruments, Dallas, Texas; with permission.)

shifts the temperature dependence and broadens the peak. Performance changes little within a temperature range of about 0.1°C, which permits use of relatively simple control electronics.

The induced pyroelectric mode is sometimes referred to as the dielectric bolometer mode. A bolometer is a device whose impedance changes with temperature; a dielectric bolometer is a bolometer for which the impedance is purely capacitive. From this point of view, a simple material figure of merit for responsivity is

$$\alpha = \frac{1}{\varepsilon} \frac{d\varepsilon}{dT} \qquad (34)$$

In this model the voltage change for a unit change in detector temperature is $\alpha V$, where $V$ is the applied potential. For typical ferroelectric materials, $\alpha$ can range from about 5%/°C to more than 100%/°C. This picture is easy to understand, but is not a good model for ferroelectric materials.

### c. Pulsed Pyroelectric Mode

The pulsed pyroelectric mode of operation makes use of the temperature dependence of the total polarization of a ferroelectric material (Hanson, 1988a,b; Pines, 1991), as seen in Fig. 9. Application of an electric field at a given temperature generates a polarization that depends on the field and the temperature. If $D$ is the electric displacement, then the charge on the ferroelectric capacitor is $Q = DA_d$, where $A_d$ is the area of one of the capacitor plates. Reversal of the field changes the polarization to an equal but opposite value, thus releasing an electric charge $2DA_d$. Since $D$ is temperature-dependent, so also is the released charge, and its variation with temperature is proportional to the pyroelectric coefficient. The net result is a responsivity that is twice that of a detector operating in the induced pyroelectric mode, with corresponding increase in the Johnson noise of only $\sqrt{2}$.

Pulsed-mode operation offers the advantage of operation without a chopper. However, as with all chopperless IR sensors, it has the concomitant dynamic range, uniformity, and drift problems that make impractical truly chopperless staring IR sensors. This mode sacrifices the system-level advantages of the ac-coupling inherent in the other pyroelectric modes. Because of this, practical implementation of pulsed-mode pyroelectric detector arrays has not yet been done and is not likely to be done in the foreseeable future.

## 3. SIGNAL AND NOISE

### a. Responsivity

The responsivity of any thermal detector consists of two parts: The detector must absorb incident IR radiation and convert it to a temperature change, then it must sense the temperature change. The first component is common to all thermal detectors and will be discussed only briefly. In a lump-parameter model, the detector is described by two parameters. The thermal capacitance or thermal mass, given by

$$C_{th} = c_p A_d z_d \tag{35}$$

affects only the temporal response, whereas the thermal conductance $G_{th}$ affects both the temporal response and the steady-state response. The equation describing the temperature change that results from an incident radiant flux is

$$C_{th} \frac{dT}{dt} + G_{th}(T - T_0) = \eta_{abs} \Phi(t) \tag{36}$$

where $T_0$ is the equilibrium temperature in the absence of radiant flux, $\eta_{abs}$ is the IR absorption efficiency, and $\Phi(t)$ is the (time-dependent) incident radiant power. It is convenient to let

$$\Phi(t) = \Phi_0 + \Delta\Phi(t) \tag{37}$$

where $\Phi_0$ is the steady-state background radiant power, so that Eq. (36) becomes

$$C_{th} \frac{dT}{dt} + G_{th}(T - T_0) = \eta_{abs}(\Phi_0 + \Delta\Phi(t)) \tag{38}$$

If the input radiation is constant; that is, if $\Delta\Phi = 0$, then

$$T = T_1 = T_0 + \frac{\eta_{abs}\Phi_0}{G_{th}} \tag{39}$$

If we let

$$T = T_1 + \Delta T(t) \tag{40}$$

then Eq. (38) becomes

$$C_{th} \frac{d\Delta T}{dt} + G_{th}\Delta T = \eta_{abs}\Delta\Phi(t) \qquad (41)$$

Although $\Delta\Phi(t)$ is usually a trapezoidal wave, we assume for convenience that a sinusoidal wave is a reasonable approximation. Thus

$$\Delta\Phi(t) = \tfrac{1}{2}\delta\Phi e^{i\omega t} \qquad (42)$$

where $\delta\Phi$ is the difference in incident radiant power between the open and closed phases of the chopper. The solution of Eq. (41) is, apart from a phase shift and transient effects,

$$\Delta T(t) = \tfrac{1}{2}\delta T e^{i\omega t} \qquad (43)$$

where

$$\begin{aligned}\delta T &= \frac{\eta_{abs}}{G_{th}} \frac{1}{\sqrt{1+\omega^2\tau_{th}^2}} \delta\Phi \\ &= \frac{\eta_{abs}}{C_{th}} \frac{\tau_{th}}{\sqrt{1+\omega^2\tau_{th}^2}} \delta\Phi \end{aligned} \qquad (44)$$

and

$$\tau_{th} = \frac{C_{th}}{G_{th}} \qquad (45)$$

is the thermal time constant.

The second component of responsivity is the conversion of a temperature change into a measurable electrical signal. A pyroelectric detector is a nonlinear capacitor with properties described by Eq. (9), or even more generally,

$$E = E(D, T) \qquad (46)$$

A typical circuit for implementation of a pyroelectric detector is shown in Fig. 12. The small-signal capacitance of the detector is $C_d$, and $R_d$ encompasses both leakage and dielectric loss (to be discussed later). The load impedance may be deliberate or parasitic. The relevant equations are

$$V_0 - V = z_d E(D, T) = I_d R_d \qquad (47)$$

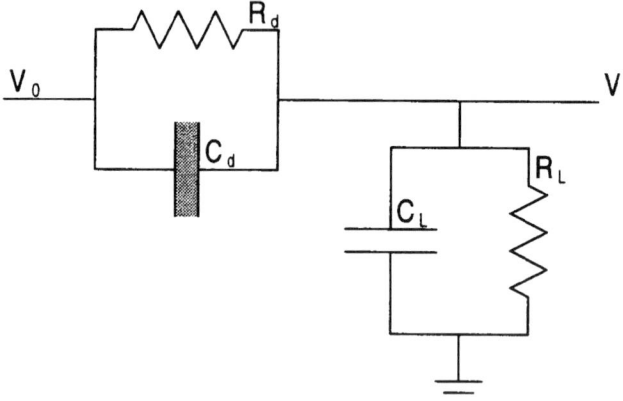

FIG. 12. Equivalent electrical circuit for a pyroelectric detector. $C_d$, small-signal capacitance of the detector; $R_d$, detector resistance; $C_L$, load resistor capacitance; $R_L$, load resistance; $V_0$, applied constant bias voltage.

and

$$V = I_L R_L = \frac{Q_L}{C_L} \tag{48}$$

where $V_0$ is the applied constant bias voltage, $z_d$ is the detector thickness, $I_d$ is the detector leakage current, $I_L$ is the load resistor current, $Q_L$ is the charge on the load capacitor, and $V$ is the output signal. Appropriately combining Eqs. (47) and (48) yields

$$V_0 = z_d E(D, T) + \frac{Q_L}{C_L} \tag{49}$$

Taking the time derivative of this gives

$$\begin{aligned} 0 &= z_d \left[ \left(\frac{\partial E}{\partial D}\right)_T \frac{dD}{dt} + \left(\frac{\partial E}{\partial T}\right)_D \frac{dT}{dt} \right] + \frac{1}{C_L} \frac{dQ_L}{dt} \\ &= \frac{A_d}{C_d} \left[ \frac{dD}{dt} - p \frac{dT}{dt} \right] + \frac{1}{C_L} \frac{dQ_L}{dt} \end{aligned} \tag{50}$$

where we have made use of Eqs. (10) and (13), where

$$C_d = \frac{\varepsilon \varepsilon_0 A_d}{z_d} \tag{51}$$

is the detector dynamic (small-signal) capacitance, and where $A_d$ is the area of a detector capacitor plate.

Conservation of charge demands that

$$A_d \frac{dD}{dt} + I_d = \frac{dQ_L}{dt} + I_L \tag{52}$$

Combining Eqs. (47), (48), (50), and (52),

$$C \frac{dV}{dt} + \frac{1}{R} V = \frac{V_0}{R_d} + pA_d \frac{dT}{dt} \tag{53}$$

where

$$C = C_d + C_L \tag{54}$$

and

$$R = \frac{R_d R_L}{R_d + R_L} \tag{55}$$

With a temperature variation described by Eq. (43), the solution to Eq. (53) is, apart from a phase shift,

$$\Delta V(t) = \frac{R_L}{R_d + R_L} V_0 + \frac{1}{2} \delta V e^{i\omega t}$$

where

$$\delta V = \frac{pA_d}{C_d + C_L} \frac{\omega \tau_e}{\sqrt{1 + \omega^2 \tau_e^2}} \delta T \tag{56}$$

and

$$\tau_e = RC \tag{57}$$

is the electrical time constant for the inherent high-pass filter. The time-independent term is the offset voltage created by leakage current through the capacitor. Thus the responsivity is

$$\mathcal{R} = \frac{\delta V}{\delta \Phi} = \frac{\eta_{\text{abs}}}{C_{\text{th}}} \frac{pA_d}{C_d + C_L} \frac{\omega \tau_e}{\sqrt{1 + \omega^2 \tau_e^2}} \frac{\tau_{\text{th}}}{\sqrt{1 + \omega^2 \tau_{\text{th}}^2}} \tag{58}$$

Note that for fixed frequency and fixed electrical and thermal capacitances, a long thermal time constant does not degrade responsivity, but actually improves it.

From Eq. (58) we see that if $C_d \ll C_L$, a convenient material figure of merit (for current responsivity) is

$$F_I = \frac{p}{c_p} \qquad (59)$$

and if $C_d \gg C_L$, a convenient figure of merit (for voltage responsivity) is

$$F_V = \frac{p}{c_p \varepsilon \varepsilon_0} \qquad (60)$$

In general, a more useful figure of merit, which includes the effect of input capacitance of the circuit with which the detector will be used, is

$$F'_V = \frac{1}{C_d + C_L} \frac{p}{c_p} \qquad (61)$$

This reduces to $F_I$ or $F_V$ when $C_L$ is comparatively small or large, respectively.

### b. *Noise*

Three independent noise sources set the sensitivity limits of most thermal detectors. These are temperature fluctuation noise, Johnson noise, and preamplifier noise. The fundamental limit is set by temperature fluctuation noise, which results from the random exchange of energy between the detector and its environment. Much of detector design is devoted to reducing Johnson noise and preamplifier noise relative to temperature fluctuation noise. In the final analysis it is unimportant which noise mechanism dominates, provided the total noise is the minimum for a given responsivity.

In principle, preamplifier noise is independent of detector parameters. In practice, the detector impedance can load the preamplifier input node and alter its noise characteristics. This effect should not dominate and, to a first approximation, the preamplifier noise referred to the preamplifier input node is usually independent of detector impedance. Given this assumption, the sensitivity figure of merit for a device limited by preamplifier noise is the same as the responsivity figure of merit, Eq. (61).

Johnson noise is generated by the real power-dissipating part of an impedance (Callen and Welton, 1951). Since an ideal capacitor has no real component to its impedance, one might expect that it would not exhibit Johnson noise. However, there are at least three ways in which capacitors can have a real component of impedance. An ideal capacitor generates a 90-degree phase shift between current and voltage so that there is no power dissipation. Any phenomenon that causes a different phase shift causes power dissipation and effectively generates a real impedance. The difference in phase angle between an ideal capacitor and an actual device is commonly denoted by $\delta$, and $\tan\delta$ is the ratio of the real (effective, series) impedance to the pure imaginary component. Capacitors exhibiting high $\tan\delta$ are referred to as *lossy*, and power dissipated as a result is know as *dielectric loss*.

A leaky capacitor is effectively an ideal capacitor in parallel with a resistor. The impedance of such an $R-C$ combination is

$$Z_p = \frac{R_p}{1 + i\omega R_p C_d} \tag{62}$$

and so

$$\tan\delta = \frac{1}{\omega R_p C_d} \tag{63}$$

Note that $\tan\delta$ is frequency-dependent if $R_p$ is constant. Conversely, if we have a measured $\tan\delta$, then the effective parallel resistance is

$$R_p = \frac{1}{\omega C_d \tan\delta} \tag{64}$$

Except for especially poor quality ferroelectric materials, electrical leakage is not usually a significant contributor to dielectric loss.

A series resistance can also contribute to dielectric loss. The impedance of this arrangement is

$$Z_s = R_s + \frac{1}{i\omega C_d} \tag{65}$$

and so

$$\tan\delta = \omega R_s C_d \tag{66}$$

Again, $\tan\delta$ is frequency-dependent if $R_s$ is constant; however, the frequency dependence differs from the parallel case. Series resistances would have to exceed several M$\Omega$ to contribute significantly to dielectric loss, and so ordinary lead resistances are not generally very important. However, poorly formed electrical contacts can have a dramatic effect on the measured $\tan\delta$.

A third loss mechanism results from time delays in dielectric polarization following an applied oscillating electric field. This can result from many causes, but most affect only very high frequencies and are inconsequential for pyroelectric detectors. In ferroelectric materials, polarization changes in response to an electric field by two means. The field alters the polarization of ferroelectric domains already aligned with the field, and the field causes alignment of domains not previously aligned. Most field-induced alignment occurs by expansion of the already aligned domains at the expense of the nonaligned domains. Such domain growth by motion of the domain walls is dissipative, and this is a major source of dielectric loss. These losses are strongly dependent on the quality and stoichiometry of the ferroelectric materials. Even in this case $\tan\delta$ is frequency-dependent, but generally to a much lesser extent than ordinary resistances in combination with the capacitor.

The spectral density of Johnson noise, also known as dielectric-loss noise or $\tan\delta$ noise is, referred to the preamplifier input,

$$v_J^2 = 4kT \frac{R}{1 + \omega^2 R^2 C^2} \tag{67}$$

where

$$R = \frac{R_L}{1 + \omega R_L C_d \tan\delta} \tag{68}$$

Eq. (54) gives $C$, and we have assumed that the leakage resistance $R_d$ has been included in $\tan\delta$. At sufficiently high frequency (or large $R_L$), we have

$$\begin{aligned} v_J^2 &= 4kT \frac{\tan\delta}{\omega} \frac{C_d}{C^2 + C_d^2 \tan^2\delta} \\ &\approx 4kT \frac{\tan\delta}{\omega} \frac{C_d}{(C_d + C_L)^2} \end{aligned} \tag{69}$$

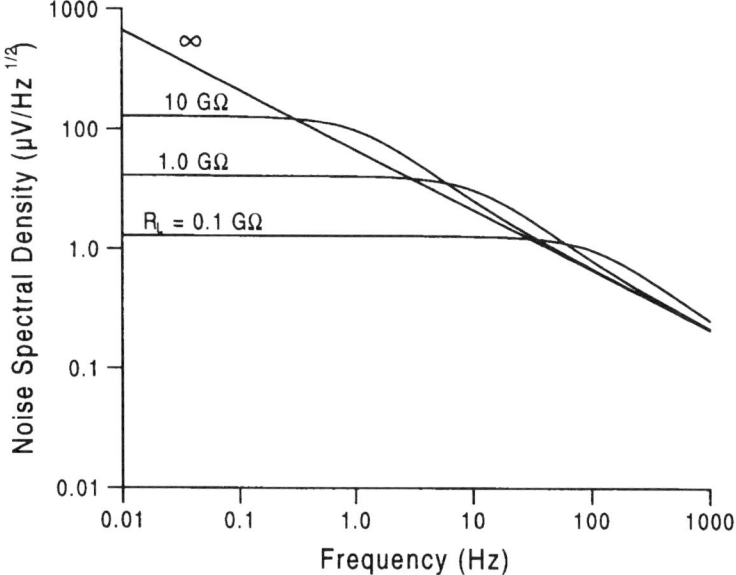

FIG. 13. Dielectric loss noise spectral density for typical parameters and various load resistances, $R_L$.

Figure 13 shows the noise spectral density for dielectric loss noise. Three points are worthy of note from the equations and the figure:

1. Dielectric loss noise is $1/f$ noise at higher frequencies. At low frequencies $R_L$ dominates the circuit impedance and the spectrum becomes flat.
2. Dielectric loss noise is attenuated by charge-sharing between the detector and the load capacitor.
3. Dielectric loss noise can be suppressed by the appropriate choice of $R_L$ matched to subsequent filters.

It is also worthy of note that if $\tan\delta$ is frequency-independent, then it has no effect on the frequency response of the circuit. This is evident from the fact that both resistances derived from Eqs. (63) and (66) are inversely proportional to $\omega$.

The conventional sensitivity figure of merit $D^*$ is of little practical use because of frequency dependencies and filter factors. However, its analytical expression is useful for examining the relative importance of the various parameters. The expression for $D^*$ at the chop frequency is

$$D^* = \frac{\mathcal{R}}{v_J}\sqrt{A_d} = \frac{\eta_{abs}}{c_p}\frac{p}{\sqrt{z_d\varepsilon\varepsilon_0\tan\delta}}\sqrt{\frac{\omega_c}{4kT}}\frac{\omega_c\tau_e}{\sqrt{1+\omega_c^2\tau_e^2}}\frac{\tau_{th}}{\sqrt{1+\omega_c^2\tau_{th}^2}} \quad (70)$$

From this, a good figure of merit for devices limited by Johnson noise is

$$F_J = \frac{p}{c_p\sqrt{\varepsilon\varepsilon_0\tan\delta}} \quad (71)$$

Temperature fluctuation noise sets the ultimate limit of performance for thermal detectors. It results from the random exchange of heat between the detector and its environment. Heat exchange may occur via radiation (photons), conduction (phonons), or any other means of heat transfer. It is usually assumed that the formalism appropriate for analysis of noise due to radiation exchange can be generalized to apply as well to other heat exchange mechanisms. There is some rationale for this assumption, considering that phonons and photons are both bosons, and therefore the same statistical arguments apply to both.

The power fluctuation spectral density is given by

$$\delta\Phi^2 = 4kT^2 G_{th} \quad (72)$$

which is valid to very high frequencies at ordinary temperatures. The equivalent temperature fluctuation spectrum is, according to Eq. (44),

$$\delta T^2 = \frac{4kT^2}{C_{th}}\frac{\tau_{th}}{1+\omega^2\tau_{th}^2} \quad (73)$$

where the factor $\eta_{abs}$ has been dropped because Eq. (72) represents the absorbed power. When integrated with infinite bandwidth (i.e., the total temperature fluctuation bandlimited only by the thermal time constant) the result is

$$\delta T = \sqrt{\frac{kT^2}{C_{th}}} \quad (74)$$

The voltage fluctuation spectrum equivalent to Eq. (73) is

$$v_T^2 = \frac{4kT^2}{C_{th}}\left(\frac{pA_d}{C_d+C_L}\right)^2\frac{\tau_{th}}{1+\omega^2\tau_{th}^2}\frac{\omega^2\tau_e^2}{1+\omega^2\tau_e^2} \quad (75)$$

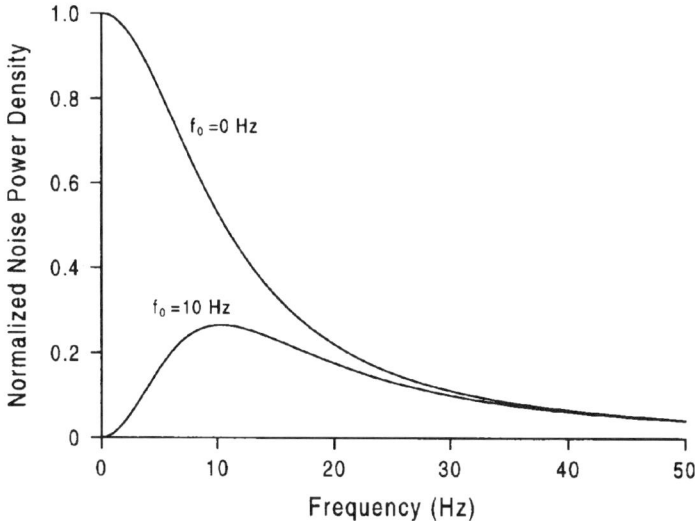

FIG. 14. Temperature fluctuation noise power spectrum with and without self-filtering. The thermal time constant is 15 msec.

and the voltage equivalent to Eq. (74) is

$$\delta V_T = \frac{pA_d}{C_d + C_L} \sqrt{\frac{kT^2}{C_{th}} \frac{\tau_e^2}{\tau_{th}^2 + \tau_e^2}} \qquad (76)$$

If $\tau_e$ is sufficiently large, then the high-pass filter has little effect on this noise. However, if the high-pass filter characteristic frequency $f_0 = 1/2\pi\tau_e$ is comparable to the thermal bandwidth, then the noise is dramatically suppressed with minimal effect on responsivity. Figure 14 shows the temperature fluctuation noise power spectrum with and without self-filtering. The higher the value of $f_0$, the more effectively it suppresses temperature fluctuation noise, and the more it degrades responsivity. The signal-to-noise ratio increases monotonically with $f_0$ whenever $\omega_c \tau_{th} > 1$, as shown in Fig. 15, and decreases monotonically with $f_0$ whenever $\omega_c \tau_{th} < 1$ (not usually the case). Of course, once the temperature fluctuation noise is suppressed to the point that it no longer dominates, the analysis is no longer relevant. Also, subsequent filters in the sensor can modify the illustrated behavior somewhat.

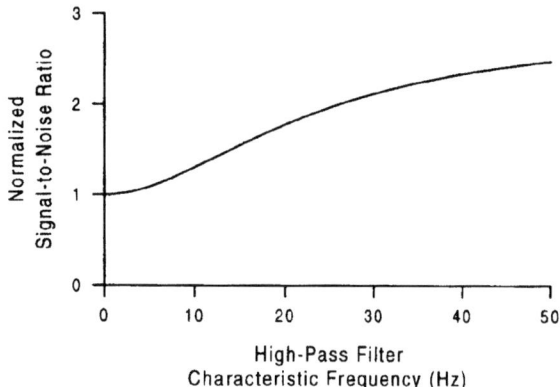

FIG. 15. Effect of the intrinsic high-pass filter characteristic frequency on the signal-to-noise ratio limited by temperature fluctuation noise. The chop frequency as shown is 30 Hz, and the thermal time constant is 15 msec.

For a device dominated by temperature fluctuation noise,

$$D_T^* = \frac{\mathcal{R}}{v_T}\sqrt{A_d} = \eta_{\text{abs}}\sqrt{\frac{A_d}{4kT^2 G_{\text{th}}}} \qquad (77)$$

which, as expected, is independent of material parameters. Of course this is not fully indicative of the potential of pyroelectric detectors because it does not include the intrinsic low-frequency filtering that is usually so difficult to achieve in staring IR detector arrays.

## III. Practical Considerations and Designs

1. Ferroelectric Material Selection

The common wisdom concerning selection of materials for pyroelectric detectors is to maximize the values of the figures of merit, Eqs. (59), (60), and (71) (Mansingh and Arora, 1991). These generally are considered the appropriate figures of merit for current responsivity, voltage responsivity, and Johnson noise sensitivity, respectively. Figures of merit are useful only if they reliably measure the "goodness" of a material in some way. A

TABLE I
Typical Properties[a] of Ferroelectric Materials

| Material | $p$ (nC/cm$^2$ K) | $\varepsilon$ | tan$\delta$ | $c_p$ (J/cm$^3$ K) | $F_V$ (m$^2$/Coul) | $F_J$ (Pa$^{-1/2}$) | NETD[b] (K) |
|---|---|---|---|---|---|---|---|
| TGS | 35 | 30 | 0.01 | 2.30 | 0.573 | $9.34 \times 10^{-5}$ | 0.123 |
| LiTaO$_3$ | 20 | 60 | 0.003 | 3.30 | 0.114 | $4.80 \times 10^{-5}$ | 0.258 |
| BST [Ba$_{0.67}$Sr$_{0.33}$TiO$_3$] | 350 | 5000 | 0.01 | 2.55 | 0.031 | $6.52 \times 10^{-5}$ | 0.066 |
| PST[c] [Pb(Sc$_x$Ta$_{1-x}$)$_2$O$_6$] | 500 | 8000 | 0.01 | 2.70 | 0.026 | $6.96 \times 10^{-5}$ | 0.061 |

NETD, noise equivalent temperature difference; TSG, triglycine sulfate; LiTaO$_3$, lithium tantalate; BST, barium strontium titanate; PST, lead scandium tantalate.
[a]Unless otherwise referenced, the values listed are approximate, broadly published values.
[b]Calculated using typical device and system parameters with f/1 optics and 50-$\mu$m square pixels.
[c]Shorrocks et al. (1990).

responsivity figure of merit is valuable in selecting a material with responsivity sufficiently high that preamplifier noise is small compared with temperature fluctuation noise. A Johnson noise sensitivity figure of merit is valuable in selecting a material whose Johnson noise is small compared with the temperature fluctuation noise. Thus both figures of merit must be large to ensure temperature fluctuation noise-limited performance.

The traditional responsivity figures of merit, $F_I$ and $F_V$, readily fail this test of usefulness. While it is generally true that a large pyroelectric coefficient and a small dielectric constant are desirable, it is also true that these two parameters are not independently adjustable. Thus we find that materials having a high pyroelectric coefficient also have a high dielectric constant, and materials having a low dielectric constant also have a low pyroelectric coefficient. If, for example, we hold $p/\varepsilon$ constant and allow $\varepsilon$ (or $p$) to vary, we find, according to Eq. (60), that responsivity is degraded at a low values of $\varepsilon$ (or $p$). This is because of the effect of the preamplifier input capacitance. This means that different detector–preamplifier sizes and configurations will be optimized with different materials (Whatmore, 1986). Thus Eq. (61) is a better responsivity figure of merit, assuming one knows the pixel geometry and the circuit with which the detector material will be used. This is why we have seen better success with materials such as BST (barium strontium titanate) and PST (lead scandium tantalate) than with materials such as TGS (triglycine sulfate) and $LiTaO_3$ (lithium tantalate). Table I shows the parameter values and traditional figures of merit for typical materials of the types listed. The traditional figures of merit indicate, for example, that TGS should be a much better material than is BST; however, sensor system results indicate the contrary.

Another consideration in using published properties to select a ferroelectric material is the method used to evaluate the pyroelectric coefficient. As mentioned earlier, the commonly used Byer-Roundy technique can lead to a pyroelectric coefficient measurement that far exceeds that which can be realized in an IR detector. This is a particular danger with materials operated near a first-order phase transition.

2. THERMAL ISOLATION

It has long been recognized that performance of a thermal detector improves with increased thermal isolation provided there is a concomitant decrease in thermal mass. Various means have been proposed and implemented for thermally isolating from the environment the elements of hybrid pyroelectric detector arrays. Early efforts sometimes simply made use of the low thermal conductance of most ferroelectric materials to create a thermal gradient within the material itself (Fig. 16(a)). The ferroelectric material layer was relatively thick, and it was attached to a read-out integrated circuit (ROIC) by means of soft metal bumps that provided little thermal isolation (Shorrocks and Edwards, 1992). The bottom side of the capacitor was essentially fixed at the temperature of the ROIC, while the top side was somewhat isolated by the thickness of the capacitor. Thermal isolation was poor, and only about half the ferroelectric material effectively contributed to generation of signal. Pixels were generally large, however, and so the modest thermal isolation gave reasonably sensitive results.

Later efforts improved the situation somewhat. A layer of organic material of low thermal conductivity separated the ROIC from the electrical connections to the ferroelectric capacitors (Fig. 16(b)). Long, thin metal runs on top of the organic layer led to vias etched in the layer and provided electrical connections while minimizing thermal loss. This method supported pixels as small as 75-$\mu$m square and 75-$\mu$m thick, with thermal conductances as low as 40 $\mu$W/K.

As pixel pitch shrank to improve sensor resolution, greater demands were placed on the thermal isolation structures. A simple organic layer was no longer adequate because thermal spreading within the layer increased the effective conductance cross section. This was solved by photolithographi-

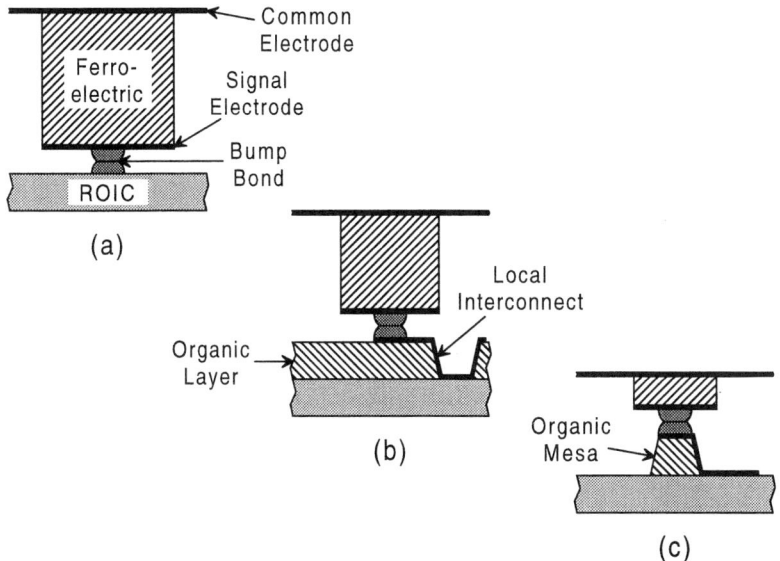

FIG. 16. Evolution of hybrid thermal isolation structures. ROIC, read-out integrated circuit.

cally removing portions of the organic layer to leave organic mesas to which the detector pixels were attached (Fig. 16(c)). This structure is currently used in modern pyroelectric detector arrays of the highest performance (Augustine and Evans, 1991; Meissner et al., 1991; Watton, 1986). Deposition of thin, conformal, over-the-edge metal runs provides electrical connection between the top of the mesas and the ROIC. Such structures are suitable for detector pixels smaller than 50-$\mu$m square and as thin as 10 $\mu$m. They provide thermal conductances from 10 $\mu$W/K to as low as 2 $\mu$W/K. Performance of these mesas is limited by the ability to consistently define tall organic mesas of small cross sections, and by the ability to maintain electrical integrity in thin over-the-edge metals.

3. MODULATION TRANSFER FUNCTION (MTF)

Initial efforts produced detector arrays from a single slab of ferroelectric material, and pixels were defined only by their electrical contacts. With this design, thermal energy absorbed at one pixel would not only flow through the thermal isolation to the substrate, it would also flow through adjacent pixels. This severely degraded the modulation transfer function (MTF). A first consideration might lead one to believe that thinning the ferroelectric material would reduce lateral thermal conduction, thereby improving MTF. However, if thickness is reduced, then thermal isolation must be accordingly increased to maintain the thermal time constant. Because the thermal MTF degradation depends on the ratio of pixel-to-pixel thermal conductance to the pixel-to-substrate thermal conductance, this approach yields no improvement in MTF.

The first-order solution to the MTF loss is reticulation of the pixels. The reticulation techniques that have gained the widest acceptance are ion milling, laser cutting, and laser-assisted chemical etching. The viability of each of these method depends on the ferroelectric material being used. Reticulation leaves pixels connected to one another only by the common electrode and IR absorber, and therefore MTF improves dramatically.

Further improvements in MTF result from careful attention to reducing lateral thermal conduction in the common electrode and IR absorber. Early designs used a conductive absorber, such as gold black, platinum black, or a co-deposited zinc sulfide–titanium mixture. These provided good absorption; however, lateral thermal conduction was excessive. Electrode–absorber layers, used by both GEC Marconi (Shorrocks and Edwards, 1992) and Texas Instruments (Hanson, 1993), make use of a tuned optical cavity (Hadley and Dennison, 1947a,b), the bottom layer of which serves as a common electrode. Figure 17 shows such a structure. The bottom metal

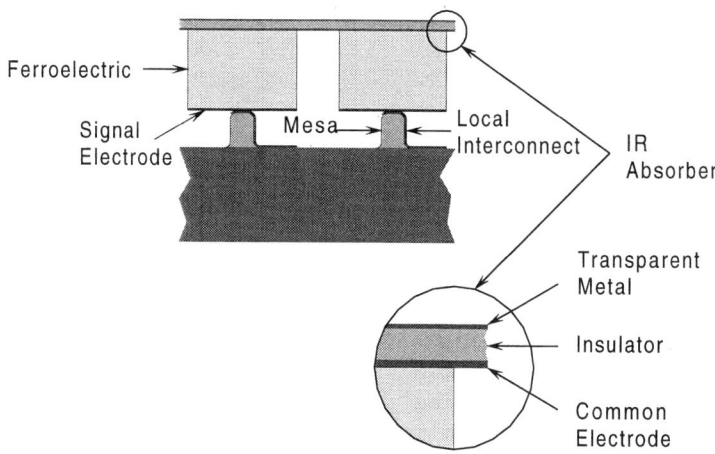

FIG. 17. Resonant infrared (IR) absorber with common electrode on reticulated detector array.

layer is sufficiently thick to be totally reflecting. The top metal layer is thin and semitransmissive, matched to the impedance of free space (377 Ω/square). The thickness of the insulating layer is adjusted to optimize absorption in the desired spectral band. Using this design, absorption efficiencies of more than 99% averaged over the 7.5 to 13 μm spectral band have been demonstrated. MTF measured at the array Nyquist frequency has been demonstrated as high as about 40%.

4. READ-OUT ELECTRONICS

Early hybrid ferroelectric IR detector arrays included an ROIC whose primary purpose was to provide accessibility to the individual detector pixels, as shown in Fig. 18. Each unit cell contained only the detector pixel and two switches. A shift register addressed the rows sequentially, closing the switches for that row and connecting its pixels to the column address lines. Another switch was used to reset the charge on the detector after each read. Amplification, bandlimiting, and multiplexing occurred external to the ROIC. The IC chip was simple, but the noise bandwidth high, as high as 40 kHz for a 245 × 328 pixel array. Because of the wide bandwidth, these devices were typically limited by dielectric loss noise.

The primary benefits of a staring array (as opposed to a scanned array) are long signal integration times and low-noise bandwidth. The early arrays

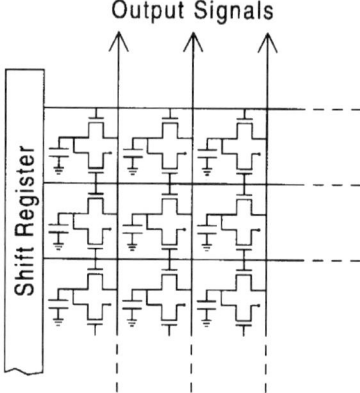

FIG. 18. Early read-out integrated circuit (ROIC) design in which the unit cell comprises only the detector pixel and a switch.

took full advantage of the long available signal integration time, but failed to take advantage of the potential for low-noise bandwidth. Recent ROIC designs have remedied this by adding signal processing to the unit cell (Hanson *et al.*, 1992; Owen *et al.*, 1994). The primary addition necessary to reduce the noise bandwidth is a low-pass filter inserted between the detector and the switch. A buffer between the filter and the switch eliminates the deleterious effects of the relatively high capacitance of the column address

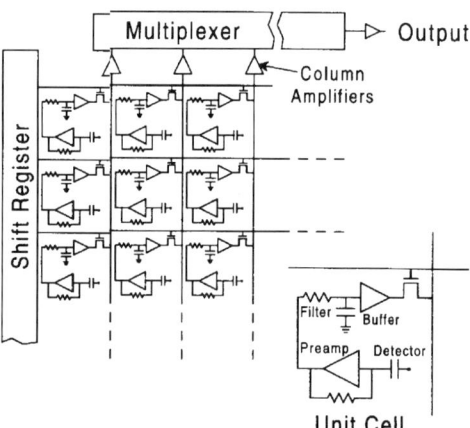

FIG. 19. State-of-the art read-out integrated circuit (ROIC), including gain and filtering in the unit cell.

lines. Because of capacitor-size limitations in the single-pole filter, the noise in such a circuit is dominated by $kTC$ noise on the capacitor. To alleviate this problem, a gain stage is inserted between the detector and the filter. The net result is suppression of the dielectric loss noise to the point that performance is limited by preamplifier noise, and performance (NETD) is improved by about a factor of two.

Current designs trade ROIC simplicity for performance. The trade is a good one because circuit geometries remain considerably larger than those allowed by state-of-the-art commercial semiconductor process design rules. Yields are high because of the design rules and because imaging arrays can tolerate small numbers of defective elements.

5. SYSTEM ELECTRONICS

The ROIC delivers the detector data row by row, and sequentially within each row, in sequence similar to a standard television camera. Alternate fields have opposite polarity because of the chopper (see §6). The data stream coming out of the detector is imperfect in two ways: The effective detector responsivities are not uniform, and neither are the effective offset voltages. Response nonuniformities result from a number of factors, including variations in pixel sizes, electrode areas, ferroelectric material properties, preamplifier gains, and column amplifier gains. Offset variations result from variations in detector pixel leakage currents, and from variations in the offsets of the preamplifiers and column amplifiers.

Figure 20 shows a block diagram for the signal processing flow in a typical system. Data arriving from the ROIC is immediately digitized, usually to 8 or 10 bits. The previous field (e.g., a closed-chopper field), which was of opposite polarity to the current field (an open-chopper field), has been stored in the dynamic field memory. The previous field is subtracted from the current field and, at the same time, the current field replaces the previous field in memory. This field subtraction effectively applies a filter function given by

$$H_\Delta(\omega) = 2\sin\left(\frac{\pi}{2}\frac{\omega}{\omega_c}\right) \quad (78)$$

where $\omega_c$ is the angular chop frequency $2\pi f_c$. This filter effectively doubles the signal and increases white noise by a factor of $\sqrt{2}$, for a net improvement in the signal-to-noise ratio of $\sqrt{2}$. For $1/f$ noise, the effect is more dramatic because of the null in the filter function at 0 Hz. Noise that is correlated from one field to the next, that is, low-frequency noise, is canceled

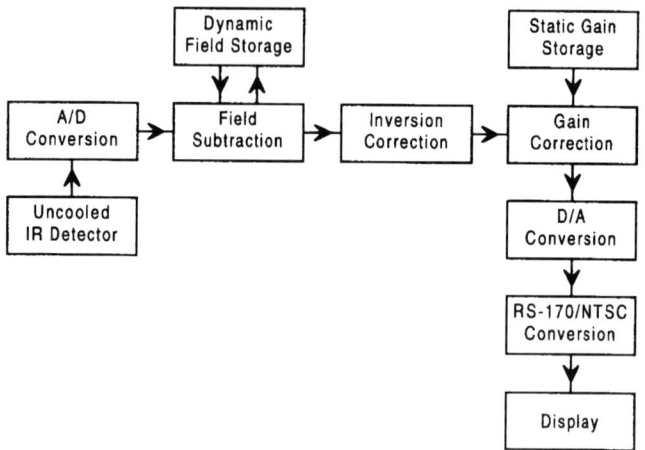

FIG. 20. Signal processing flow diagram for a typical pyroelectric imaging system. IR, infrared.

by this filter. Fixed pattern offsets are also canceled by this action. Thus there is no requirement for offset correction. The next field is of opposite polarity to its predecessor (i.e., a closed-chopper field in this example), and therefore the result of the next field-to-field subtraction will be of opposite polarity. The next step in the signal processing corrects for this by inverting alternate fields. In practice, this polarity inversion is implemented as part of the differencing operation.

Gain correction factors stored in static memory during manufacture of the detector correct unequal responsivities in the array. Any elements that are either unresponsive or outside the dynamic range of the correction algorithm are considered defective pixels, and their values are replaced by those of neighboring pixels. Defective pixels may actually be pixels that are excessively responsive or insufficiently responsive.

Subsequent to gain correction, the data stream is converted to an analog signal, and the timing signals appropriate for RS-170/NTSC or PAL standards are inserted. This signal is then available to drive any standard display.

## 6. CHOPPERS

As discussed earlier in this chapter, it is possible to operate pyroelectric–ferroelectric detectors with or without a chopper; however, we do not see the realization of chopperless implementations. Before we discuss chopper designs, let us first address the reasons use of a chopper is preferred, not only for pyroelectric imagers but for all uncooled IR imagers (with the possible exception of thermocouples–thermopiles). The fundamental issues are the electronics dynamic range and sensor stability (Hanson, 1993). The general nature of the argument is most clearly seen by the following technology-independent analysis, which ignores any nonlinearities or electrothermal interactions. The voltage output of a thermal detector can be expressed as

$$V_i = V_i(T_{scene}, T_{sub})$$
$$= V_i^0 + \mathcal{R}_l \left[ (T_{scene} - T_{scene}^0) + \frac{1}{\beta_i}(T_{sub} - T_{sub}^0) \right] \quad (79)$$

where the $i$ subscripts denote a specific pixel, the 0 superscripts denote initial set-point values, and where $V$ is the output signal, $\mathcal{R}$ is the pixel temperature responsivity ($V/K$), $T_{scene}$ is temperature of the scene, $T_{sub}$ is the temperature of the substrate, $\beta$ is the temperature increase of the detector per unit temperature increase in the scene, given by

$$\beta = \frac{\Delta T_{det}}{\Delta T_{scene}} = \frac{\eta_{abs} A_d}{4f^2 G_{th}} \left(\frac{\partial W}{\partial T}\right)_{\Delta \lambda} \quad (80)$$

For typical hybrid detectors, $\beta$ is on the order of 0.00025; however, for monolithic devices a typical value is 0.01. The offsets are determined by measuring the array output signals with the sensor viewing a uniform target of known fixed temperature $T_{scene}^0$. The detector substrate temperature at the time of this sample is $T_{sub}^0$. In photon detectors the offset variability generally results from pixel-to-pixel variations in responsivity applied to a large background flux, and from pixel-to-pixel variations in dark current. The first of these is not generally applicable to thermal detectors, but the second, or its equivalent, is. For resistive bolometers, it results from variations in the dc resistance of the elements; for capacitive bolometers, it results from variations in the large-signal capacitance and in electrical leakage; and for pyroelectric detectors, it results from variations in the remanent polarization.

Equation (79) can be rewritten to show explicitly the effects of offset and responsivity nonuniformities:

$$V_i = V_i^0 + (\mathcal{R}_0 + \Delta \mathcal{R}_i)\left(\Delta T_{scene} + \frac{\Delta T_{sub,i}}{\beta_i}\right)$$
$$= \mathcal{R}_0 \Delta T_{scene} + V_0^0 + \Delta V_i + \Delta \mathcal{R}_i \Delta T_{scene} + \mathcal{R}_0 \frac{\Delta T_{sub,i}}{\beta_i} + \Delta \mathcal{R}_i \frac{\Delta T_{sub,i}}{\beta_i} \quad (81)$$

where the 0 subscripts denote array spatial averages, and where $\Delta V_i$ is the difference between the pixel offset and the array average offset, and $\Delta\mathcal{R}_i$ is the difference between the pixel responsivity and the array average responsivity. It is informative to examine the relevance and impact on dynamic range of each term of Eq. (81):

1. $\mathcal{R}_0 \Delta T_{scene}$. This is the output of a pixel as it should be in the ideal case. If the detector element is electrically linear, as with an ordinary resistor or capacitor, then the responsivity increases linearly with bias, and therefore

$$\mathcal{R}_0 \Delta T_{scene} = \alpha V_0^0 \beta \Delta T_{scene} \tag{82}$$

where $\alpha$ is the temperature coefficient of the electrical response parameter (resistance or capacitance). When the scene temperature is equal to the noise equivalent temperature difference (NETD), then the resulting signal sets the lower end of the dynamic range requirement. This lower limit signal is

$$\Delta V_{min} = \alpha V_0^0 \beta \Delta T_{NETD} \tag{83}$$

2. $V_0^0$. This is the mean dc offset. It is a global value that can be readily subtracted from all signals. Proper design can virtually eliminate the adverse effect of this value on dynamic range requirements. The initial signal-handling capacity of the electronics, however, must be at least

$$\frac{V_0^0}{\Delta V_{min}} = \frac{1}{\alpha \beta \Delta T_{NETD}} \tag{84}$$

Even for typical monolithic devices, this dynamic range is on the order of $10^5$, or 16 to 17 digital bits. For a chopped ac-coupled system, this offset does not occur.

3. $\Delta V_i$. This is the pixel-to-pixel offset variation. It can be removed by subtracting a different fixed value from the signal of each pixel. The dynamic range must be capable of handling the offsets of virtually all the pixels, and therefore it must be approximately six times the standard deviation. Thus

$$\Delta V_{max} = 2 \times 3 \times \sigma_V \tag{85}$$

where the factor of 2 accounts for both sides of the statistical distribution and the factor of 3 roughly converts standard deviation to amplitude. The dynamic range required to handle all the offsets is

$$\frac{\Delta V_{\max}}{\Delta V_{\min}} = \frac{2 \times 3 \times \sigma_V}{\alpha \beta V_0^0 \Delta T_{\text{NETD}}} \tag{86}$$

For a responsivity nonuniformity that has a standard deviation of 5%, a typical dynamic range requirement is $1.5 \times 10^4$, or 14 bits. Therefore, the offset correction for an unchopped sensor must be calculated and applied with at least 14 bits accuracy. Offsets in a chopped ac-coupled sensor are automatically eliminated.

4. $\Delta \mathcal{R}_i \Delta T_{\text{scene}}$. This is the pixel-to-pixel gain nonuniformity. It is present in chopped and unchopped sensors; it must be corrected in either case. Note that no matter how accurately responsivity is corrected, there can always be a scene intensity that is sufficiently bright to generate noticeable nonuniformity. This is true of all sensors—cooled and uncooled, chopped and unchopped.

5. $\mathcal{R}_0 \Delta T_{\text{sub},i}/\beta_i$. This term accounts for a time-dependent offset variation due to substrate temperature drift. Note that even for a monolithic device, a drift of only about 0.01°C results in an apparent change in the scene average temperature of 1°C. This appears on the display as a phenomenon sometimes called *breathing*; the display brightness slowly increases and decreases as the temperature controller cycles. This leads to a requirement for very stringent control of the focal plane temperature for an unchopped sensor. This term also accounts for spatial nonuniformity due to temperature gradients on the focal plane. Such gradients may result from uneven heating by on-chip signal processing, and they can be quite large. A temperature difference of only 1°C across the focal plane would appear as a 100°C difference across the scene. Even when the temperature drift is uniform there is an induced nonuniformity due to variations in $\beta$, although this effect is much less important. These effects have little or no impact on a chopped sensor, unless the temperature drift rate is sufficiently high to cause substantial field-to-field temperature differences. But for an unchopped sensor, these effects demand tight temperature control and periodic recalibration.

6. $\Delta \mathcal{R}_i \Delta T_{\text{sub},i}/\beta_i$. This term accounts for reappearance of display nonuniformity even after precision gain correction. The substrate temperature drift acts as an amplification factor for residual inaccuracies in gain correction. This results in the need to recalibrate gain correction factors unless the focal plane always operates at the same temperature.

Because of the need to periodically recalibrate so-called "chopperless" sensors, occasional interpositioning of a "shutter" or "asynchronous chopper" is necessary. So the question is not "To chop or not to chop?" Rather, it is "To chop synchronously or asynchronously?" Use of an asynchronous chopper requires periodic interruption of the scene, and it requires system electronics of much greater dynamic range. These are the reasons chopped sensors are generally smaller, lighter weight, and lower power than are unchopped sensors. These are also the reasons pyroelectric sensors, which lend themselves so well to ac coupling, are not commonly used in the pulsed pyroelectric mode despite its responsivity benefits.

Chopper design has evolved considerably from its early beginnings, and the result is a mechanically simple device. The basic IR chopper concept has simple blades with straight edges, rotated about a hub positioned to the side of the focal plane (Fig. 21(a)). The rate of rotation is set to match standard television frame rates, and the size of the chopper is set so that the time to traverse adjacent rows is exactly one video line time. This design creates phase differences across a large focal plane, and different scan rates for

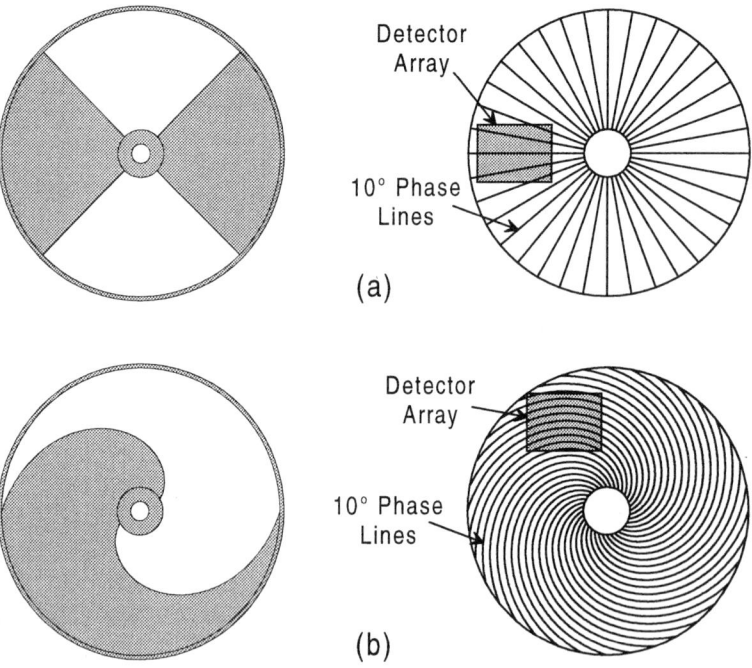

FIG. 21. Chopper with (a) straight blades and (b) Archimedes' spiral blades.

different rows. Except for the row of detectors in line with the chopper hub, the chopper blade crosses one end-of-row before the other end. This results in variability of the sampling phase along a given row. The effect becomes more pronounced at the top and bottom of the array. This results in nonuniform response and suboptimization of sensitivity.

An early solution to this problem used a continuous metal band that passed in front of and behind the detector package. Rectangular holes were cut in the band in such a fashion that the open areas were equal to the closed areas. A motor and rollers, much like the drive mechanism in a movie projector, caused the band to pass vertically over the focal plane, alternately exposing and then hiding the scene. This design suffered from mechanical complexity and difficulty in the precision manufacture of the metal bands.

Choppers currently in use are generally of the design shown in Fig. 21(b). The blade edges are Archimedes' spirals. Although residual phase shifts, occur, they are minimal and approximately the same for every row.

Operation of the detector with the chopper is as follows. As the trailing edge of the chopper blade passes over a detector pixel, the pixel is exposed to radiation from the scene. Its temperature begins to change accordingly. The temperature continues to change until the blade again obscures the scene. Immediately before the leading edge of the blade passes, the signal is sampled. As the leading edge passes over the pixel, the pixel temperature begins to return to its initial temperature. Immediately before the scene reappears, the signal is sampled again. Thus the signal is sampled twice per exposure. This sequence is illustrated in Fig. 22.

In the basic chopper design the "closed" phase is opaque and the "open" phase is transparent. The effective signal is the difference between the chopper temperature and the scene temperature. During the open phase the detector temperature increases to a (large) fraction of what it would be for

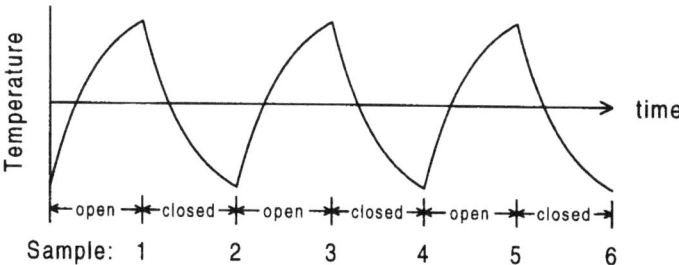

FIG. 22. Detector temperature wave generated by a chopper, showing sample points.

an unchopped system. The amplitude of the temperature swing is given by

$$\Delta T = \frac{\Phi}{G_{th}} \tanh\left(\frac{1}{4 f_c \tau_{th}}\right) \qquad (87)$$

If the detector thermal time constant is short compared with the chopper period, the temperature swing can approach that of an unchopped sensor. Lengthening the thermal integration time indefinitely achieves little added benefit.

If an opaque chopper is operating at a temperature that is substantially different from the viewed scene, the large signals generated can exceed the capability of the gain correction algorithm. The result is fixed pattern noise in the image. This effect can be eliminated by controlling the chopper temperature to minimize the array-average signal; however, this adds complexity and expense to the otherwise simple chopper. This is equivalent to the use of a fixed temperature reference in a scanned forward-looking infrared (FLIR). Another solution that has found wider acceptance in uncooled imagers is the diffusing chopper. In this design the "closed" phase of the chopper is not opaque but, rather, it is designed to diffuse the incident IR radiation so that each detector receives radiation equal to a local scene average. Figure 23 shows the effect of this type of chopper on the display. During the "open" phase, the detector receives a high-resolution image of the scene. During the "closed" phase, the detector receives a blurred image. The displayed image is the difference between the two, which effectively subtracts low-frequency objects. There are several ways of expressing the

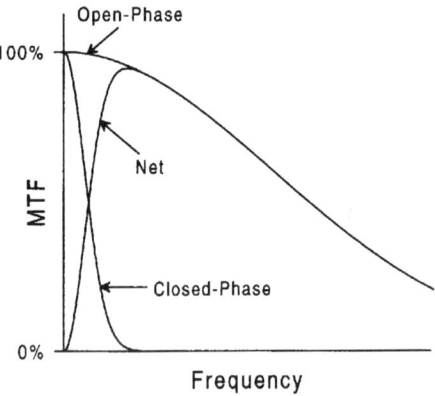

FIG. 23. Modulation transfer function (MTF) of a sensor with a diffusing chopper.

results: It applies a two-dimensional high-pass spatial filter to the scene; it differentiates the scene; or it provides edge enhancement. Bright objects in the scene appear to have a dark halo. If the diffusing phase is not sufficiently blurred, this can be a distracting artifact.

## IV. Systems Implementations

The proof of the pudding is in the eating, or in this case, in the system. The smallest, lightest weight, lowest power, lowest cost, highest performance, uncooled systems use hybrid ferroelectric detector arrays. Two companies dominate this technology, GEC Marconi in the United Kingdom and Texas Instruments in the United States. Each has a vertically integrated technology that encompasses all the major aspects of development and production of detectors as well as systems. GEC Marconi is producing $100 \times 100$ arrays with pixels on 100-$\mu$m centers. GEC Marconi has produced a variety of systems, the most intriguing of which is the fire fighter's sight. This is a very small sensor mounted on a helmet, with a separate display, also mounted on the helmet. This enables a fire fighter to see and have mobility within burning buildings while keeping both hands free. A synchronous dithering action produced by the chopper increases the image sampling beyond the stated $100 \times 100$ pixels. This system is currently being sold in the United States and elsewhere.

Texas Instruments is clearly leading in the development and production of uncooled IR systems with several commercial and military products, all built around a $328 \times 245$ array with pixels on 48.5-$\mu$m centers. NETDs less than 0.04°C have been measured on systems with f/1 optics, without correcting for system-level noise and other losses. Two products deserve particular mention. The first is the NIGHTSIGHT™ 200 Series (Fig. 24), a system designed for law enforcement and marine applications. It is a small sensor typically mounted on a pan-and-tilt head atop a police car. The display and a remote-control unit are mounted inside the car. The field of view is $12 \times 6$ degrees. These units are in production and are available to police departments and marine operators across the United States and in several other countries.

Experiences with various metropolitan police departments have been remarkable, with unexpected applications continually evolving. Drugs discarded during high-speed chases are easily recovered because of the lingering temperature difference between the recently handled drug packaging and the foliage into which it has been thrown. In a particular instance, police personnel narrowly escaped with their lives because they could see from the

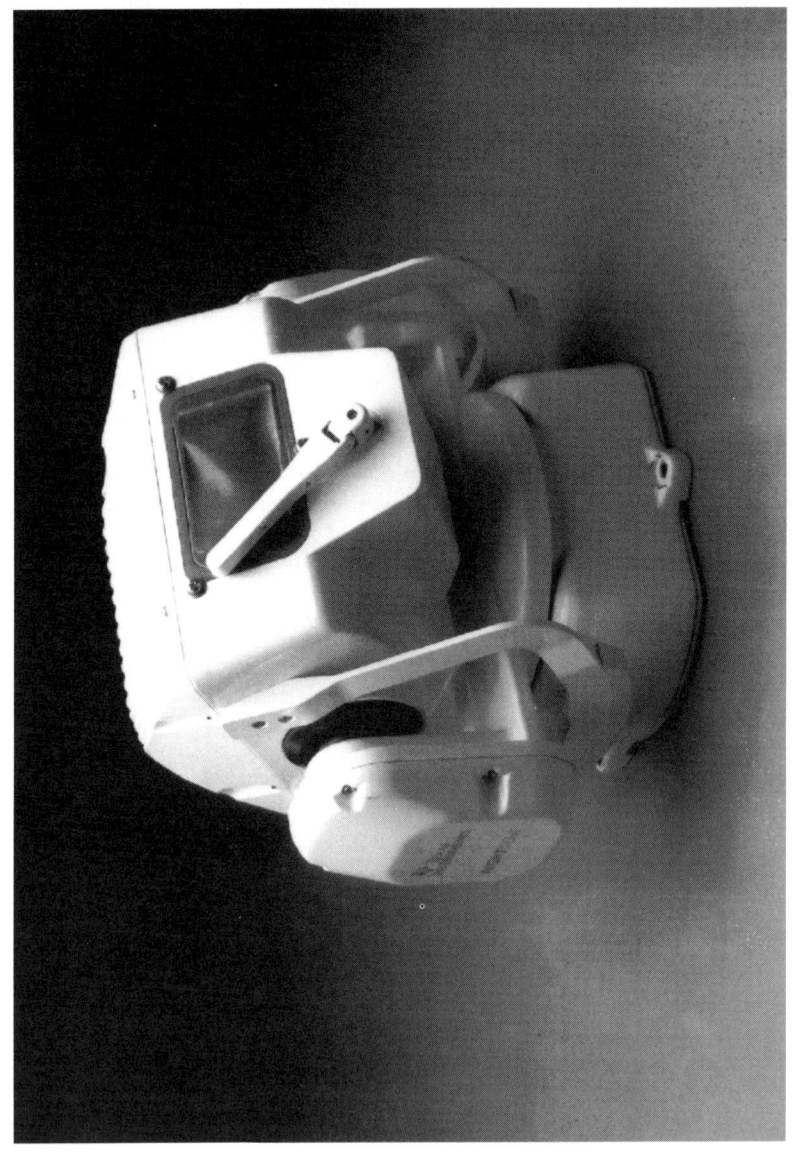

FIG. 24. Texas Instruments' NIGHTSIGHT™ 200 Series for police cars. (Courtesy of Texas Instruments, Dallas, Texas; with permission.)

FIG. 25. Texas Instruments' NIGHTSIGHT™ W1000, a multiuse, person-portable weapon sight. (Courtesy of Texas Instruments, Dallas, Texas; with permission.)

FIG. 26. Typical infrared image as viewed through a NIGHTSIGHT™ W1000 system. (Courtesy of Texas Instruments, Dallas, Texas; with permission.)

intense heat behind a door that a burning building was about to explode. Police can stealthily proceed into dark alleys without the need for headlights that would alert criminals to their presence.

The NIGHTSIGHT™ W1000 (Fig. 25) is a weapon sight for portable military weapons, including rifles, machine guns, squad automatic weapons, and light anti-aircraft missiles. Figure 26 shows typical infrared imagery created by a Texas Instruments NIGHTSIGHT™ W1000 system. This system is remarkable for its size, weight, and power. It is equipped with a 15 × 11.3 degree field-of-view lens, and a 9.3 × 7 degree lens is optional. The IFOV for the standard lens is 0.8 mrad. The system weighs less than 3.1 pounds (<1.4 kg), including a self-contained display and standard batteries sufficient for 8 to 10 h of continuous operation. Power dissipation is less than 4.1 W.

Various other uncooled ferroelectric IR sensors are gradually displacing conventional cooled IR sights for military applications. As volume builds and prices drop, commercial markets are expanding rapidly with applications only recently thought to be beyond the realm of affordability. Ferroelectric–pyroelectric sensors have been the first to exploit these markets because of their superior performance and inherent simplicity, as described earlier. The hybrid technology has the potential to achieve NETDs as low as 0.02°C with f/1 optics. However, to achieve significant performance improvements beyond that level it will be necessary to evolve to a monolithic ferroelectric detector, a technology being developed by both GEC Marconi and Texas Instruments—and by others. Monolithic ferroelectric technology is the subject of a later chapter in this book. From the point of view of cost, the distinction between the hybrid and monolithic technologies is not as significant as it first might appear. With few exceptions, the hybrid technology can be produced with standard silicon processing equipment. Although, depending on yields, a monolithic process should have fewer processes and a shorter cycle time, production cost estimates based on validated cost models project that detector cost in high volume will be limited primarily by detector packaging costs, which are not significantly different for hybrid and monolithic arrays.

### REFERENCES

Augustine, F. L., and Evans, S. B. (1991). U.S. Pat. 5,015,858.
Butler, N., McClelland, J., and Iwasa, S. (1988). *Proc. SPIE* **930** (Infrared Detect. Arrays), 151–163.
Callen, H. B., and Welton, T. A. (1951). *Phys. Rev.* **83**, 34–40.
Chynoweth, A. G. (1956). *J. Appl. Phys.* **27**, 78–74.

Hadley, L. N., and Dennison, D. M. (1947a). *J. Opt. Soc. Am.* **37**(6), 451–465.
Hadley, L. N., and Dennison, D. M. (1947b). *J. Opt. Soc. Am.* **38**(6), 483–496.
Hanson, C. M. (1988a). U.S. Pat. 4,745,278.
Hanson, C. M. (1988b). U.S. Pat. 4,792,681.
Hanson, C. M. (1993). *Proc. SPIE* **2020** (Infrared Technol. XIX), 330–339.
Hanson, C. M., Beratan, H. E., Owen, R. A., Corbin, M., and McKenney, S. (1992). *Proc. SPIE* **1735** (Infrared Detect.: State-of-the-Art), 17–26.
Hopper, G. S. (1978). U.S. Pat. 4,080,532.
Hopper, G. S. (1979). U.S. Pat. 4,162,402.
Lines, M. E., and Glass, A. M. (1977). "Principles and Applications of Ferroelectrics and Related Materials," pp. 59–86. Clarendon Press, Oxford.
Mansingh, A., and Arora, A. K. (1991). *Indian J. Pure Appl. Phys.* **29**, 657–664.
McCormack, K. (1979). U.S. Pat. 4,143,269.
Meissner, E. G., Owen, R. A., and Cronin, M. E. (1991). U.S. Pat. 5,047,644.
Owen, R. A., Belcher, J., Beratan, H. E., and Frank, S. N. (1994). *Proc. SPIE* **2225**, 79–86.
Pines, M. Y. (1991). U.S. Pat. 5,075,549.
Porter, S. G. (1981). *Ferroelectrics* **33**, 193–206.
Putley, E. H. (1980). *Infrared Phys.* **20**, 149–156.
Shorrocks, N. M., and Edwards, I. M. (1992). *Proc. IEEE Int. Symp. Appl. Ferroelect. 7th, 1990*, pp. 58–62.
Shorrocks, N. M., Whatmore, R. W., and Osbond, P. C. (1990). *Ferroelectrics* **106**, 387–392.
Watton, R. (1986). *Proc. IEEE Int. Symp. Appl. Ferroelect. 6th, 1986*, pp. 172–181.
Whatmore, R. W. (1986). *Rep. Prog. Phys.* **49**, 1335–1386.

CHAPTER 5

# Monolithic Pyroelectric Bolometer Arrays

*Dennis L. Polla and Jun R. Choi*

MICROTECHNOLOGY LABORATORY
UNIVERSITY OF MINNESOTA
MINNEAPOLIS, MINNESOTA


| | |
|---|---|
| I. INTRODUCTION | 175 |
| II. DETECTOR DESIGN METHODOLOGY | 176 |
|    1. *Materials Processing* | 178 |
|    2. *Materials Characterization* | 181 |
|    3. *Thermal Isolation Structures* | 183 |
|    4. *Micromachined Sensor Process Design* | 184 |
|    5. *Integrated Circuits* | 186 |
| III. PROCESS DESIGN | 187 |
| IV. SILICON-BASED INTEGRATED PYROELECTRIC DETECTOR ARRAYS | 189 |
|    1. *Cell Structure* | 190 |
|    2. *Circuit Operation* | 191 |
|    3. *Silicon-Based $PbTiO_3$ Array Performance* | 195 |
| V. GALLIUM ARSENIDE–BASED INTEGRATED PYROELECTRIC DETECTORS | 197 |
| VI. SUMMARY | 199 |
|    *References* | 200 |

## I. Introduction

Solid-state micromachining (Fung *et al.*, 1985; Howe, 1988; Seidel, 1987; de Rooij, 1992) has emerged over the last 10 years as both an enabling technology and value-added technology of importance to both physical microsensors and microactuators (Muller *et al.*, 1990; Gardner, 1994; Ristic, 1994; Sze, 1994). Silicon (Si) is currently the most commonly used material in these sensing and actuation applications due to its ability to support integrated circuits and due to its well-understood materials properties (Petersen, 1982; Obermeier *et al.*, 1986; Middelhoek and Audet, 1989). Solid-state micromachining offers the possibility of forming mechanical structures by fabrication methods commonly used in the manufacture of integrated circuits (Wolf and Tauber, 1986; Sze, 1988; Campbell, 1996).

Through selective process ordering of commonly used fabrication modules, such as thin-film deposition, photolithography, and etching, solid-state micromachining is routinely used to make thin mechanical diaphragms for pressure sensors (Lee and Wise, 1982; Ko et al., 1982) and microcantilever beam for accelerometers (Roylance and Angell, 1979; Chen et al., 1982; Blow et al., 1993; Chau et al., 1995). This technology is also being applied in microactuator devices such as microvalves (Jerman, 1990; Barth, 1995) and deformable micromirrors for projection displays (Hornbeck, 1989; Sampsell, 1993). In all of these applications, solid-state micromachining has been applied to form mechanical structures capable of designed deformation or controlled movement in response to an excitation such as electric field.

More generally, solid-state micromachining has enabled an entirely new technology capable of supporting sensors, signal conditioning electronics, and actuators on a common semiconductor substrate. This technology is commonly known as *microelectromechanical systems*, or *MEMS* (see *IEEE/ASME J. Microelectromechanical Systems*) in the United States and Asia and *microsystems* in Europe (Lebbink, 1994; also see *Sensors and Actuators A*).

Additionally, solid-state micromachining offers opportunities in the construction of thermal isolation microstructures where one would like to minimize heat loss from a thermally sensitive element. These structures are extremely important to the overall performance of thermal radiation microsensors such as bolometers (Lahiji and Wise, 1982) and pyroelectric microsensors (Polla et al., 1984; Ye et al., 1991a; Whatmore et al., 1995; Shorrocks et al., 1995) where minimization of conductive heat losses to the surroundings and substrate of a thermal detector is critical. This chapter focuses on the design and fabrication of pyroelectric infrared (IR) detectors, with emphasis on process integration of the following technologies: (1) solid-state micromachined thermal isolation structures, (2) pyroelectric thin films, and (3) on-chip signal processing electronics (Polla et al., 1989).

## II. Detector Design Methodology

Pyroelectric materials are attractive for uncooled IR detection due to their ability to spontaneously produce charge with a temperature change and their inherent low electrical noise. A variety of pyroelectric materials including single crystals, ceramics, polymers, and thin films have been used previously for IR detectors. The appropriate material parameters for IR detector design have been reviewed by others (Liu et al., 1975; Liu, 1976; Whatmore et al., 1990), including Hanson in Chapter 4.

In this work we have focused on the use of pyroelectric thin films due to their ability to be deposited on individual thermal isolation microstructures. The three most extensively used pyroelectric thin films in the formation of micromachined integrated sensors have been zinc oxide (ZnO) (Polla et al., 1983, 1985), lead titanate ($PbTiO_3$) (Polla et al., 1991; Ye et al., 1992), and lead zirconium titanate (PZT) (Hsueh et al., 1993; Ye and Polla, 1994; Shorrocks et al., 1995).

The ability to form precision thermal isolation structures in a semiconductor technology is a direct consequence of solid-state micromachining. This processing technique is used to minimize the unwanted thermal conduction path between a thermal sensitive element and its associated substrate (Ye et al., 1991b). Two commonly used methods for achieving thermal isolation include subtractive chemical etching and the additive

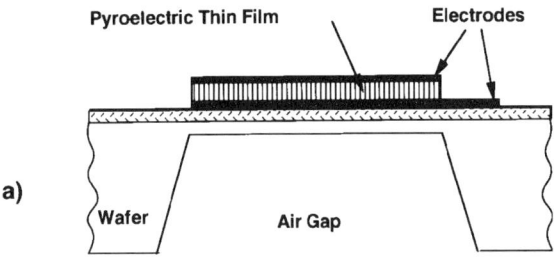

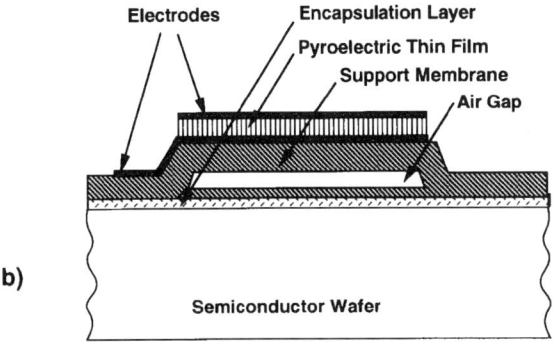

FIG. 1. Two basic solid-state micromachining methods applied in the formation of pyroelectric thin-film microsensors. (a) Bulk-micromachining in which chemical etching is used to remove a substantial portion of the semiconductor substrate and (b) surface-micromachining in which thin-film layers are added to the surface of the starting wafer with one layer selectively removed or sacrificed.

deposition of sacrificial materials and structural materials. The two processing methods are shown in cross section in Fig. 1.

The direct integration of pyroelectric elements with on-chip electronics is attractive due to the minimization of parasitic electrical interconnections. The electrical interconnection between the detector element and pre-amplifier introduces stray capacitance that reduces the detected pyroelectric voltage signal (Polla et al., 1993; Pham and Polla, 1994). Therefore, in the approaches presented in this work, the electrical interconnection distance — or more specifically the electrical interconnection capacitance — has been minimized by processing both the pyroelectric detector element and signal processing electronics on a common semiconductor substrate.

The basis for the work in this chapter makes use of pyroelectric thin films, micromachined thermal isolation structures, and on-chip electronics that are monolithically formed in a common integrated circuit (IC) fabrication process.

1. MATERIALS PROCESSING

Only a few pyroelectric materials have been successfully integrated with micromachined thermal structures. These include ZnO, PZT, and $PbTiO_3$. Of these materials, ZnO is perhaps the easiest material to prepare due the large number of deposition techniques and electrodes materials that can be used. The main disadvantage is its low pyroelectric response. Ferroelectric materials of the Perovskite family such as PZT and $PbTiO_3$ have larger pyroelectric activity but require more careful control of deposition conditions, with special attention directed toward the quality and chemical stability of electrode interfaces. Table I summarizes some of the important material properties of these pyroelectric materials.

The processing of the ferroelectric layer is of central importance to this class of uncooled IR detectors because this layer provides the desired dielectric and pyroelectric properties. In addition to achieving a thin film with the desired properties, the processing steps must be compatible with the entire fabrication sequence. Solution-gelation (sol-gel) and other solution deposition methods are well-suited for pyroelectric MEMS applications, because coatings with reproducible compositions and properties can be prepared at acceptable processing temperatures (Tamagawa et al., 1990; Hsueh et al., 1991). Uncooled IR detector structures of $PbTiO_3$ have been prepared by sol-gel deposition methods. In a typical sol-gel process, a solution typically comprised largely of metal alkoxides is synthesized and deposited by spin coating; the coating that results is heated to develop a crystalline ceramic layer. Many synthetic routes for solution deposition

TABLE I
PYROELECTRIC THIN FILM PROPERTIES

| Property | Zinc oxide | PZT (54/46) | PbTiO$_3$ |
|---|---|---|---|
| Relative dielectric constant | 10.3 | 965–1006 | 80–110 |
| Dielectric loss tangent at 1 KHz | — | 0.02–0.04 | 0.001–0.002 |
| Remanent polarization | Nonferroelectric | 35 μC/cm$^2$ | 30 μC/cm$^2$ |
| Coercive field | Nonferroelectric | 54 kV/cm | 40 kV/cm |
| Resistivity | $3 \times 10^7$ Ωcm | $>10^{11}$ Ωcm | $>10^{11}$ Ωcm |
| Dielectric breakdown strength | 0.5 MV/cm | 0.6–0.8 MV/cm | 0.4–0.7 MV/cm |
| Pyroelectric coefficient | 0.9 nC/cm$^2$ K | | |
| Piezoelectric coefficient $d_{33}$ | 12 pC/N | 190–220 pC/N | — |
| Young's modulus $E_Y$ | – | $4 \times 10^{11}$ N/m$^2$ | $4 \times 10^{11}$ N/m$^2$ |

PZT, lead zirconium tantalate; PbTiO$_3$, lead tantalate.

processing have been proposed and coatings by these routes have been characterized. In this work, the alkoxide route based on 2-methoxyethanol (2-MOE) originally developed by Budd et al. (1985) has been used. This route (Fig. 2) provides coatings with consistent microstructure and properties; however, the solution preparation is equipment- and time-intensive and the solvent is carcinogenic. Thus we continually evaluate alternative routes. One with promise is an aqueous acetate method (Lin et al., 1992). For example, preparation of PZT by this solution processing route simply involves mixing titanium isopropoxide with acetic acid and water, followed by sonication and mixing (at 100°C) with zirconium acetylacetonate and lead acetate trihydrate. Thin films prepared by this method form into the desired Perovskite phase after thermal treatment at 700°C; their microstructures are dense and finely grained (grain sizes 0.1 to 0.3 μm). Properties from this aqueous acetate route are sensitive to aging of the solution; however, this behavior is now well characterized, predictable, and stable when finally crystallized.

Because spin coating is used to prepare the ferroelectric thin films, inherent challenges arise for accommodating thermal isolation structures with variations in surface topography. To investigate these limitations, we prepared sol-gel PZT coatings (based on the 2-MOE route) on substrates with elevated surface features (step height 0.13 to 1 mm; 45 degree rise angle, Cooney et al., 1994). The distribution of PZT after spin coating depended on step height and spinning spin rate. Sol-gel coatings partially planarized

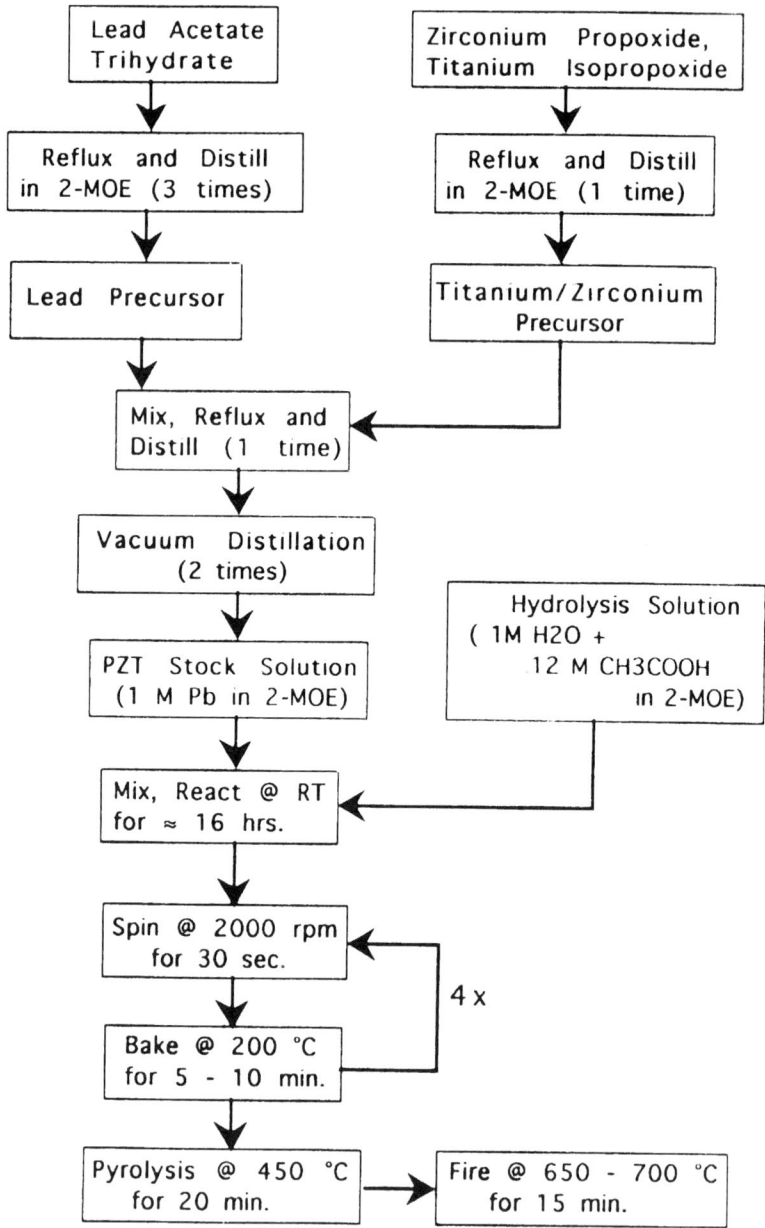

FIG. 2. Basic sol-gel processing method used for lead zirconium tantalate (PZT)-type ferroelectrics. 2-MOE, 2-methoxyethane; Pb, lead; RT, room temperature.

the features so that the PZT coatings were thinner on top of the features as compared with the surrounding substrate surface. After the high-temperature (700°C) heating, cracks and delamination developed near the top and bottom of the step when the ratio of coating thickness to step height was less than approximately 0.35. These defects originated from stresses that are enhanced by the presence of the step and by the thickness nonuniformities. To best accommodate surface features, multilayer deposition was used to build up the coating thickness prior to the higher temperature processing steps. In our investigation, features up to approximately 0.8-$\mu$m high with a rise angle of 45 degrees could be successfully coated (using multiple deposition). Planar device designs, such as the one shown in Fig. 1(b), eliminate these problems and are always preferred. For structures that must be raised (e.g., thermal isolation structures elevated above on-chip transistors), shallower rise angles lessen the thickness nonuniformity and cracking.

2. MATERIALS CHARACTERIZATION

There are three general goals of the materials synthesis described above previously:

1. Achievement of good crystallinity in the thin film as deposited on a microstructure
2. Predictable control of the pyroelectric and dielectric properties of the active thin films, and
3. Control of the composite stress associated with the different materials of the thermal isolation structure with pyroelectric thin-film capacitor.

Usually, X-ray diffractometry is used to determine the crystallinity of the pyroelectric thin film. Diagnostic depositions to assess the results of materials synthesis are carried out on metalized Si wafers. Figure 3 shows an X-ray diffraction spectrum for a 3600-Å PZT (54/46) thin film. The mixed (100) and (101) is characteristic of most of the $PbTiO_3$ and PZT thin films prepared according to the sol-gel method described previously.

The degree of crystallinity versus evidence of any nonpyroelectric pyrochlore phase has direct implications on the expected responsivity of thin films for IR detector applications. Therefore, both pyroelectric and dielectric properties are routinely characterized after solutions have been synthesized. Figure 4 shows the pyroelectric response for ZnO, $PbTiO_3$, and PZT (54/46) as a function of temperature obtained by thermal ramping experiments. The data shown in uncorrected for secondary and tertiary pyroelectric responses.

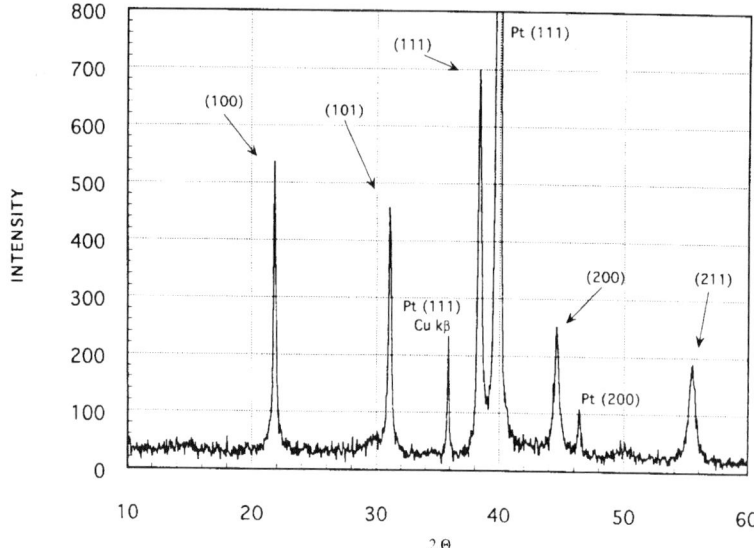

FIG. 3. X-ray diffraction spectrum for a 3600-Å-thick lead zirconium tantalate (PZT) (54/46) thin film showing mixed-phase crystallinity. X-ray diffraction spectrum is typical of most ferroelectric thin films that exhibit excellent pyroelectric properties. Pt, platinum; Cu, Copper.

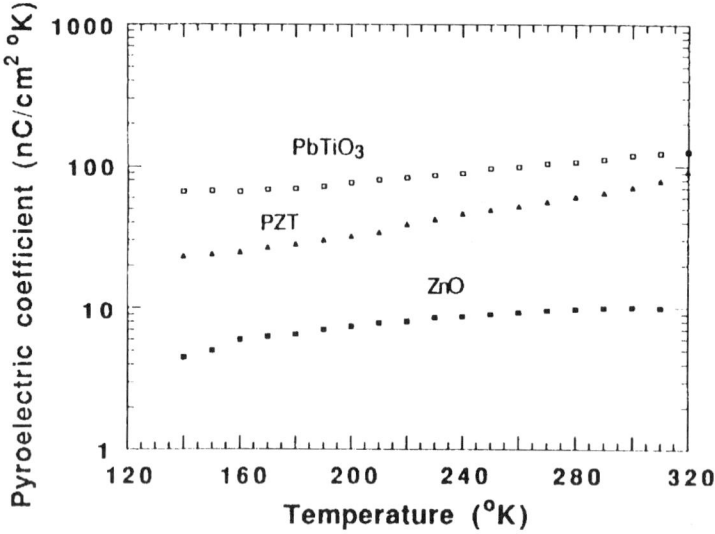

FIG. 4. Pyroelectric coefficient measured as a function of temperature for zinc oxide (ZnO), PZT (54/46), and PbTiO$_3$.

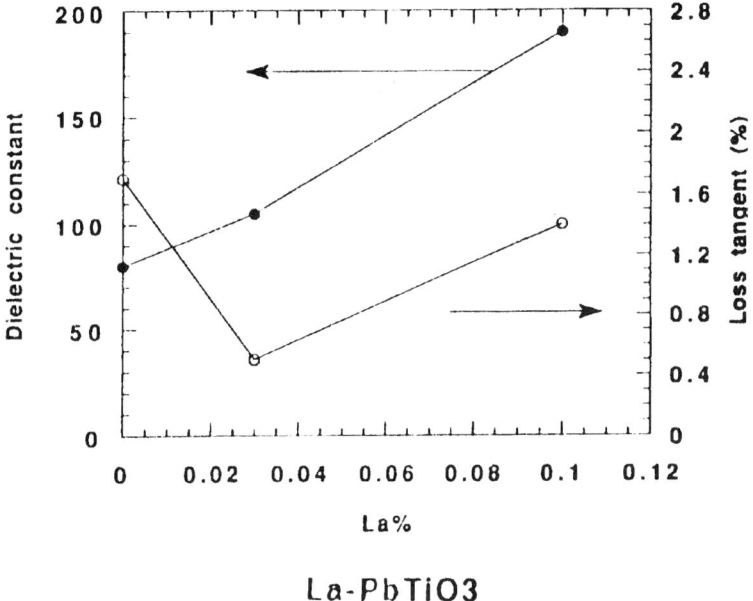

FIG. 5. Measured dielectric constant and loss tangent for 3000–4000-Å thick lead tantalate (PbTiO$_3$) thin films with intentional lanthanum doping.

The dielectric constant also determines the sensitivity of the overall pyroelectric response. Figure 5 shows the dielectric constant and loss tangent data for 3000-Å-thick PbTiO$_3$ thin films as a function of intentional lanthanum doping.

3. THERMAL ISOLATION STRUCTURES

Micromechanical supports and thermal isolation structures are formed by micromachining techniques. Micromachining therefore creates a void (usually an air gap) directly underneath the active thermal detection site. The three most common structural thin-film materials used in MEMS applications are polycrystalline Si (Fan, and Muller, 1988; Guckel *et al.*, 1988), silicon nitride (Si$_3$N$_4$) (Seikimoto *et al.*, 1982), and aluminum (Al) (Choi and Polla, 1993). These films are usually between 1.0- and 2.0-$\mu$m thick and are elevated approximately 0.5 to 1.0 $\mu$m above the surface of the Si wafer. By photolithographic patterning, heat conduction can take place only through the regions of attachment to the substrate.

For the previous three materials, polycrystalline Si has been the most extensively developed MEMS mechanical material due to its widespread availability in most integrated circuit (IC) processes. The control of the residual intrinsic stress in this material, however, places severe limitations on its use in integrated circuit processes. Thermal annealing at 1000 to 1100°C for 30 min to 1 h is often required to reduce the internal unwanted compressive stress in this material (Guckel et al., 1988). Nonstoichiometric, Si-rich $Si_3N_4$ (Pan and Berry, 1985) is another common MEMS structural material with distinct advantages for pyroelectric application due to its slightly tensile internal stress and extremely low thermal conductivity. The use of structural thin-film materials in micromachining processes can be found in Howe and Muller (1983), Howe (1988), de Rooij (1992), and Mehregany et al. (1987).

## 4. Micromachined Sensor Process Design

A representative cross-section process flow demonstrating how solid-state micromachining for the fabrication of pyroelectric IR detectors on a Si substrate is described in Fig. 6.

As shown in Fig. 6(a), a dielectric encapsulation layer is first deposited on the Si substrate. This layer is usually high-quality $Si_3N_4$ prepared by plasma-enhanced chemical vapor deposition (PECVD) techniques. Depending on whether there are transistors already fabricated on the wafer or whether discrete microsensors are to be fabricated determines the thickness of this encapsulation layer. Generally, 1.0 $\mu$m of protective $Si_3N_4$ is necessary to (1) protect the underlying circuitry in a back-end microsensor fabrication process and (2) prevent unwanted microsensor process contamination of the transistor active areas.

Figure 6(b) shows the deposition and patterning of the spacer material used in the eventual formation of an air gap for thermal isolation. This is a temporary layer (commonly referred to as a sacrificial layer) of chemical vapor–deposited (CVD) $SiO_2$. The definition of the sacrificial layer defines the air-gap region over which thermal isolation takes place. The thickness of the sacrificial layer is usually between 0.5 and 1.0 $\mu$m. By adjustment of the conditions associated with the CVD process such as through the additional *in situ* introduction of phosphorus, the resulting $SiO_2$ (commonly called phosphosilicate glass, or PSG) can be made with a low-density and high etching rate in hydrofluoric acid (HF).

The structural support membrane (Fig. 6(c)) is deposited next and is defined by either reactive ion etching (RIE) or lift-off. The mechanical structural support is used to add rigidity to an overlying pyroelectric thin

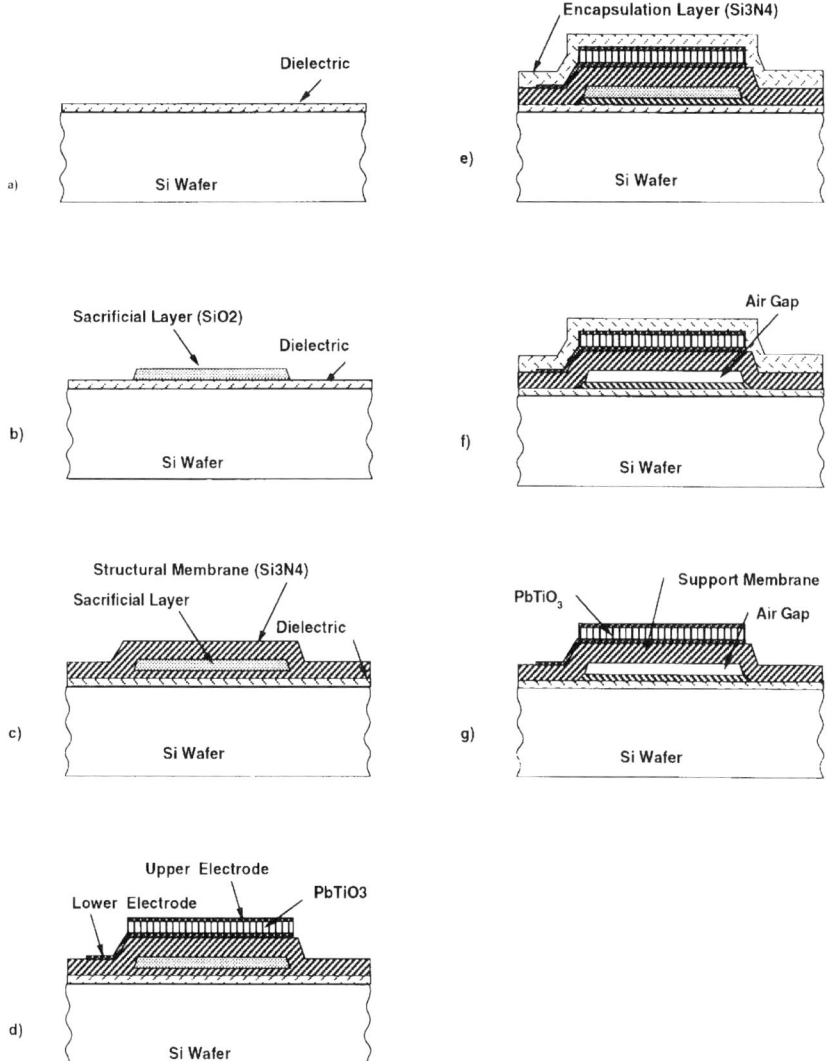

FIG. 6. Surface micromachining sequence used to fabricate ferroelectric thin-film microbridges. (a) Encapsulation of the starting wafer that might already contain transistors processed up to the point of electrical contact openings, (b) sacrificial layer deposition and patterning, (c) formation of structural membrane, (d) sol-gel deposition of ferroelectric thin films with lower and upper electrodes, (e) deposition of protective encapsulation layer, (f) lateral sacrificial etching in hydrofluoric acid, and (g) removal of encapsulation layer. $SiO_2$, silicon dioxide; $Si_3N_4$, silicon nitride; $PbTiO_2$, lead titanate.

film. Several structural support materials have been considered previously and used experimentally, including polycrystalline Si, $Si_3N_4$, and Al. Process procedures peculiar to a particular reactor are usually developed to yield materials with well-controlled internal stress characteristics (slightly tensile in nature) and good adhesion to the previously deposited materials.

Figure 6(d) shows the fabrication of the active pyroelectric thin-film sensing element. For the IR detector structures fabricated in integrated arrays, this is formed by sol-gel deposition of $PbTiO_3$. For the structure shown, titanium–platinum (Ti–Pt) is used as the lower electrode and Pt is used as the upper electrode. The $PbTiO_3$ thin film is selected to be approximately 3000 to 4000 Å thick to yield both a high responsivity and to be compatible with the topography of the wafer and its constraint on photolithography.

After definition of the $PbTiO_3$ capacitor stack, a 1.5- to 2.0-$\mu$m-thick layer of plasma-enhanced chemical vapor deposition (PECVD) $Si_3N_4$ or amorphous silicon (a-Si) is deposited at 350 to 400°C, as shown in Fig. 6(e). This encapsulation layer serves as the protective layer for the active $PbTiO_3$ thin film during subsequent solid-state micromachining. Rather than low-pressure CVD (LPCVD), PECVD is used due to the maximum processing temperature constraint of the pyroelectric thin film. Reactive ion etching is used to expose the underlying sacrificial layer. The exposed $SiO_2$ glass protrudes laterally approximately 5 $\mu$m from underneath the mechanical structural support.

Solid-state micromachining is next applied, as shown in Fig. 6(f). Buffered HF is used to selectively remove the low-density $SiO_2$ without appreciably attacking the other materials with which it comes into contact. The development of the selectivity of this lateral etch has enabled integrated microsensors to be reliably fabricated.

The encapsulation layer is then removed in selected regions to gain access to the electrical connection points needed for the microaccelerometer, as shown in Fig. 6(g).

## 5. INTEGRATED CIRCUITS

The technology for on-chip signal conditioning electronics is usually determined by the constraints set by the operative physical sensing mechanism (Muller et al., 1985). For pyroelectric detection, charge is spontaneously produced in response to a change in temperature. Field-effect transistors are usually used to detect this charge due to their high input resistance. Metal-oxide semiconductor (MOS) pre-amplifiers are therefore attractive for Si technologies (Gray and Meyer, 1993; Sansen et al., 1994) and

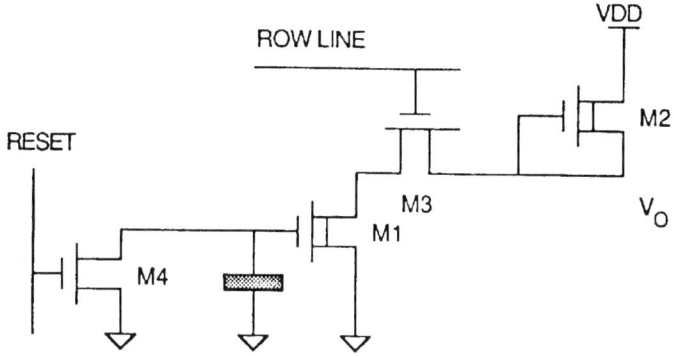

FIG. 7. Basic pyroelectric microsensor signal processing. A thin-film capacitor is directly connected to the gate of an input metal-oxide semiconductor field-effect transistor (MOSFET) with active load. A reset transistor is used to clear pyroelectric charge between readout and a transmission gate is used for column accessing.

metal-semiconductor field-effect transistor (MESFET) pre-amplifiers are used in gallium arsenide (GaAs) technologies (Harrold, 1993; Polla et al., 1990). Note that bipolar transistors usually are not used for either pyroelectric or piezoelectric amplifier applications due to their low input resistance (charge leakage path).

A simple integration approach of pyroelectric thin films with Si NMOS circuits is shown in Fig. 7. Here the pyroelectric detector element is directly connected to the gate of an input depletion-mode NMOS transistor (Polla et al., 1983). Depletion-mode transistors are used as the input due to their normally-on conducting channel with zero dc gate-source as presented by an unchanged pyroelectric capacitor. A depletion pull-up load transistor is used to complete a simple two-transistor amplified structure. For the circuit shown, two additional transistors are used as a transmission gate for row addressing and an a re-set transistor to clear residual pyroelectric charge between imaging read-out events.

## III. Process Design

In this work are three microfabrication technologies that are compatibly merged in the integrated pyroelectric IR detector. First, on-chip signal processing circuitry is needed to convert pyroelectric signal charge into an appropriate voltage signal. Therefore, a buffer charge-to-voltage NMOS amplifier is designed for each pixel element. Second, thermal isolation structures need to be fabricated by solid-state micromachining techniques. Third, highly oriented ferroelectric thin films need to be deposited and delineated over the thermal isolation structure. Therefore, Si NMOS–CMOS (complementary MOS), solid-state micromachining, and ferroelectric thin-film deposition need to be compatibly integrated in one overall process with fabrication steps merged in a manner that prevents chemical contamination, control composite thin-film stress, produce highly oriented ferroelectric thin films, and adhere to both circuit and sensor thermal processing constraints.

In the two forms of solid-state micromachining described in Fig. 1, surface micromachining is a preferred choice for pyroelectric array applications due to the possibility of individual reticulation of each heat-sensitive element. In general, bulk micromachining does not allow multiple precision geometries without substantially weakening the mechanical integrity of the semiconductor substrate. Of the two approaches, surface-micromachining is attractive due to the following reasons: (1) no two-sided photolithography is required, (2) no nonstandard processing chemicals are used, (3) precise dimensional control can be achieved, and (4) the mechanical integrity of the substrate is not weakened.

Ferroelectric fabrication begins with either MOS circuits processed up to the point of source-drain contact opening if integrated microsensor structures are to be made, or with silicon nitride–silicon dioxide ($Si_3N_4$–$SiO_2$) covered Si substrates if off-chip electronics are to be used. A 0.3 $\mu$m-thick layer of LPCVD $Si_3N_4$ and a 0.8-$\mu$m-thick layer of PSG are first deposited at 800 and 450°C, respectively. The $Si_3N_4$ layer forms an encapsulation layer to protect the almost finished CMOS circuitry from subsequent processing of the on-chip sensors and the PSG serves as the sacrificial oxide spacer used in the formation of sensor membrane structures. The PSG is patterned and chemically etched to form anchor regions for a subsequent phosphorus-doped PSG or $Si_3N_4$ microstructure membrane deposition (by LPCVD). Sensor membrane regions are defined and anisotropically patterned in a $SF_6/CCL_2F_2$ plasma by RIE. A 500- 1000-Å-thick lower Pt electrode is then sputtered over the entire wafer. The Pt serves the dual purpose of providing an adhesion–nucleation surface for the subsequently deposited PZT of $PbTiO_3$ thin film and serving as the lower electrode for pyroelectric microsensors.

Sol-gel spin-casting of ferroelectric PZT or $PbTiO_3$ is then carried out as previously described. The ferroelectric films are then patterned by either chemical etching or ion-beam sputter etching. Photolithography is then used to protect the ferroelectric thin films in carrying out a lateral sacrificial etching step of the PSG layer. This steps is commonly called surface-

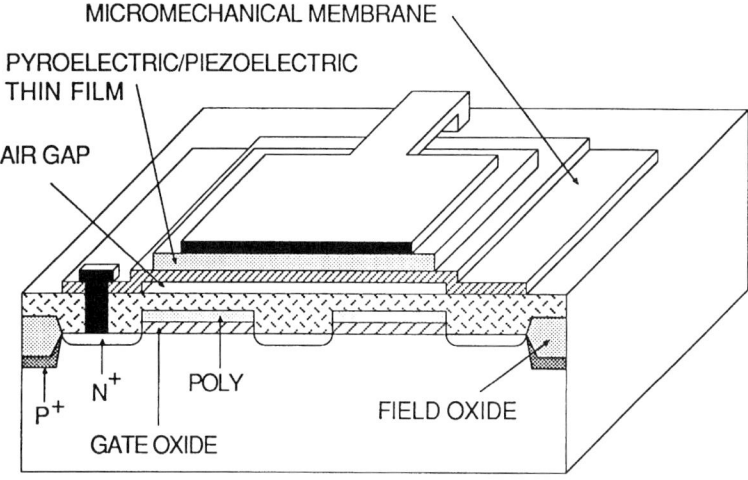

FIG. 8. Vertical stacking of pyroelectric microbridge directly above the pre-amplifier cell.

micromachining. Doubly-supported, suspended, structures are released using 48 wt% HF to undercut the PSG spacer layer. The composite membrane therefore forms a low thermal mass structure with low thermal conductivity to the underlying semiconductor substrate.

A top electrode to the ferroelectric thin films is then deposited by titanium–gold (Ti–Au) evaporation. Contact openings to both the MOS transistors and the sensor bottom polysilicon electrode are then chemically etched. This is followed by sputtering Al (2% Si) to form interconnects and bonding pads. A vertically stacked, integrated pyroelectric pixel element is shown in Fig. 8.

## IV. Silicon-Based Integrated Pyroelectric Detector Arrays

The ability to fabricate microsensors on a semiconductor substrate is attractive because of the high performance and low cost when batch manufactured. Therefore, Si is an attractive semiconductor substrate for integrated pyroelectric microsensors. Furthermore, the eventual goals of MEMS is to increase more systems functionality on a single semiconductor chip. This might include additional microsensors, such as inertial and temperature sensors; microactuators, such as flaps and valves; and additional electronics, such as mixed analog–digital circuits.

1. CELL STRUCTURE

Figure 8 shows the schematic representation of the vertical integrated microsensor concept. A scanning electron micrograph of a single integrated structure is shown in Fig. 9. In order to keep the cell size small while maintaining high sensitivity, a doubly-supported bridge structure is used. The bridge structure minimizes the thermal conductivity between the detector element and underlying substrate. Further, the doubly-supported microbridge provides rigidity to minimize unwanted piezoelectric response that may be associated with inertial movements of the free-standing pixel. A micromechanical membrane made of $Si_3N_4$ having a bridge structure is micromachined directly above the underlying transistor devices. The air gap formed by micromachining forms a useful low thermal mass and low thermal conductivity path for IR detection. A 0.3- to 0.4-thick $PbTiO_3$ pyroelectric thin-film capacitor is supported by the microbridge and directly

FIG. 9. Scanning electron micrograph of a lead tantalate ($PbTiO_3$) infrared pixel element with an underlying pre-amplifier cell. Micrograph corresponds to drawing in Fig. 8.

interconnected to an underlying NMOS transistor gate. The interconnection distance between the microsensor and the first pre-amplification stage is as short as the length of the electrical contact, and therefore the parasitic interconnection capacitance is small.

Each microsensor generates charge through pyroelectric effect, which is the result of the alteration of the internal dipole moment of the capacitor film through a change in temperature. The pyroelectric charge produced is directly coupled to the driver gate of a two-transistor depletion-mode NMOS pre-amplifier. The spontaneous charge therefore is transduced into a voltage and detected by a change in drain-source current in the pre-amplifier.

## 2. CIRCUIT OPERATION

Figure 10 shows a block diagram of a Si-based pyroelectric imaging array and associated signal processing circuitry. In this work, a diagnostic 64 × 64 element imaging array has been fabricated and tested with on-chip electronics located directly beneath each active pixel. Each detector pixel element is fabricated on a separate $Si_3N_4$ thermal isolation microbridge. Immediately below each detector, there is a simple pre-amplifier that consists of two transistors, one of which is shared by 63 other preamplifiers in the same column. There are two 6-to-64 decoders for selecting an individual detector. The row-decoder can be seen on the left, and the column-decoder can be seen at the bottom of the array. The signal processing circuit is on the bottom left of the array. An optical photograph of the finished die that measures 1.0 × 1.0 cm is shown in Fig. 11.

In the operation of the Si imaging array, a 12-bit digital word is used to drive the two decoders that select one of the pyroelectric microsensors. Each sensor detects an IR signal and generates charge through the pyroelectric effect. This charge, which is proportional to the incoming IR heat flux, is converted into a voltage and coupled directly to an underlying pre-amplifier. The pre-amplifier raises the detected signal and provides low-output impedance. Each output of a column of pre-amplifiers is connected to the source of a metal-oxide semiconductor full-effect transistor (MOSFET). The drains of the MOSFETs in each column are connected to the column output. The gates of these transistors are connected to the output of the row decoder. Only one of the outputs of the decoder is high at any time; thus only the output signal of one preamplifier is present in the column at any time. For each of the 64 columns of preamplifiers there is a column output. These columns are selected using the same selection technique with a column decoder. With this selection technique only one signal of the preamplifier is

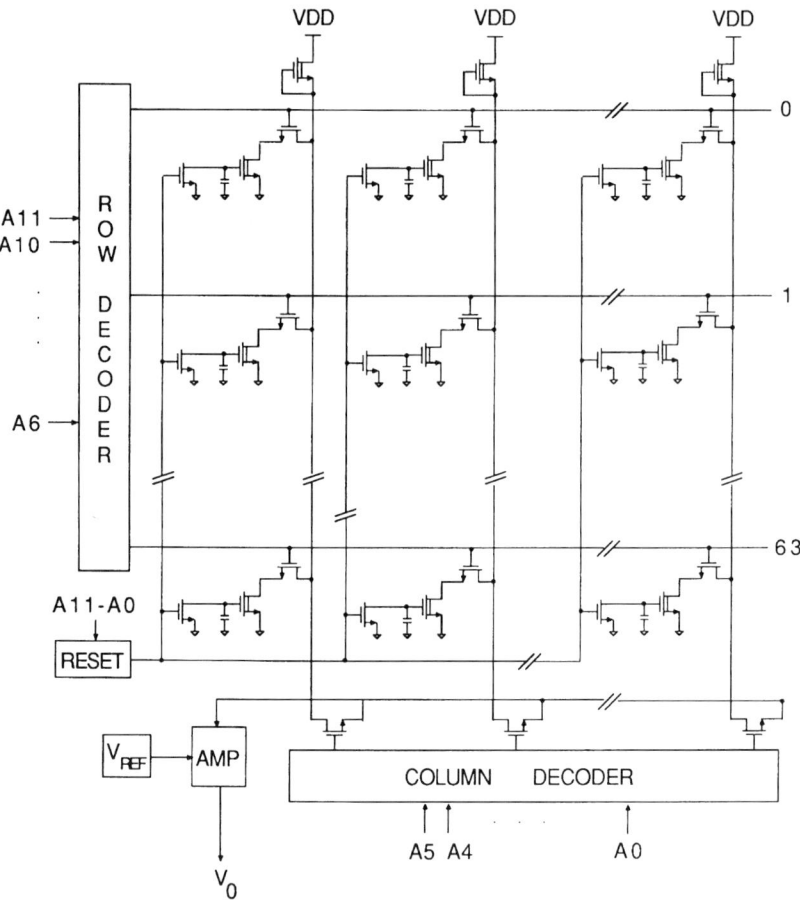

FIG. 10. Pyroelectric 64 × 64 element array architecture.

present at the output of the array at any time. The signal from the selected preamplifier and a reference signal are fed into a level shifter. The outputs of the level shifter are amplified by a differential amplifier to provide a single-ended output voltage. Figure 12 shows the three-stage signal conditioning circuit and the reference circuit. The first and third stages provide voltage gain; the second stage provides power gain. The analog output voltage is sampled and converted to a digital signal with an 8-bit A–D converter. The process is repeated until the whole array is sequentially scanned.

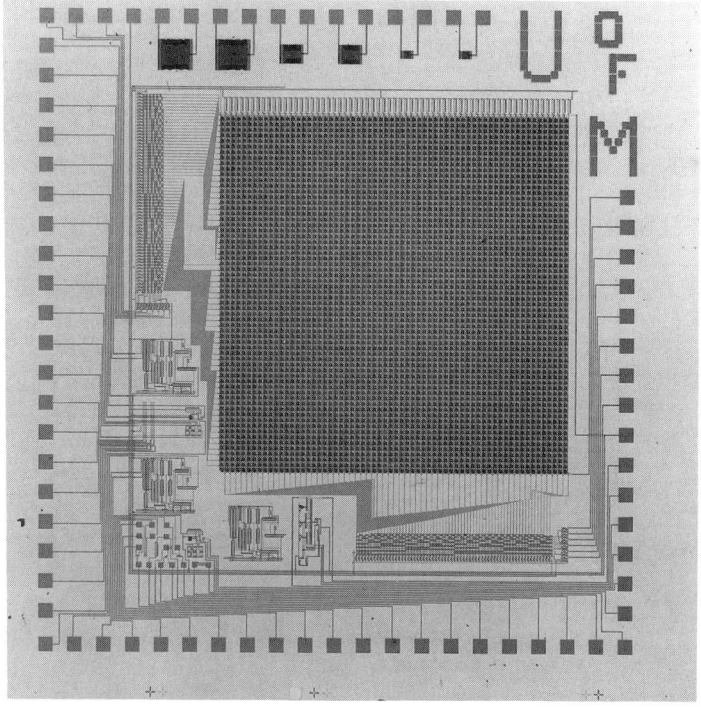

FIG. 11. Silicon-based monolithic array of 64 × 64 elements with on-chip metal-oxide semiconductor transistors. The die size shown is 1 × 1 cm.

The preamplifier has a cascode configuration. The circuit of the preamplifier is shown in Fig. 4. The load transistor (M2) is shared by 63 other preamplifiers (shown in Fig. 7) in the same column. Note that under proper operating conditions only 1 of 64 inverting transistors (M1) is connected to the load transistor at any time. This condition is exploited to reduce considerably the number of transistors in the chip. In this circuit, transistors M1 and M2 are the main components of the pre-amplifier. The circuit has a configuration similar to that of a depletion-load inverter, except that transistor M1 is also a depletion device. Being a depletion device transistor M1 enables the amplifier to operate in its linear region when the input is very small, less than 1 mV, which is typically the range of the sensor signal. Transistor M3 provides a low impedance conducting path that connects transistors M1 and M2 when it is turned on. Note also that transistor M3, connected as a common-gate amplifier, reduces the Miller effect that limits

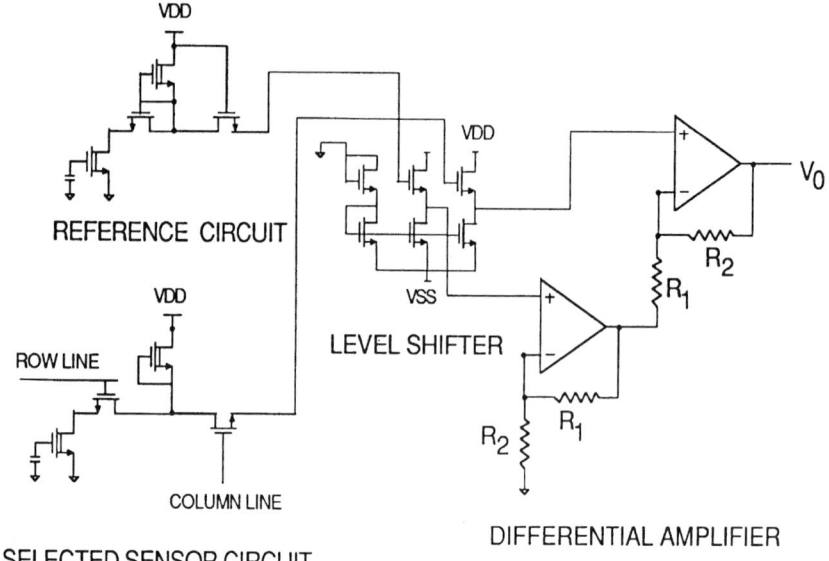

Fig. 12. Reference circuit used to cancel unwanted piezoelectric effects and ambient temperature drift.

the high-frequency response of the preamplifier. Transistor M4 is the reset transistor that is fully off until the last element of the array is accessed.

The output of the pre-amplifier is connected to a level shifter, which shifts down the dc level of the reference and pre-amplifier signals by approximately 5 V. It is used to minimize the common mode gain effect of the following differential amplifier stage. Because the input impedance of this circuit is very high, it also acts as a buffer or an isolation amplifier, coupling two stages with little loss in signal level. The level shifter, having a source-follower configuration, does not provide voltage gain but does provide large current and some power gain.

The third signal conditioning stage is a differential amplifier. This stage amplifies the difference between the reference signal and the sensor signal derived from the level shifter and provides a signal-ended output signal. The voltage gain, approximately 50 (34 dB), is independent of the operational amplifiers and is controlled only by the ratio $R_1/R_2$ in Fig. 13. The operational amplifiers consist of all enhancement NMOS transistors. This two-op amp circuit arrangement cancels the offsets of the operational amplifiers.

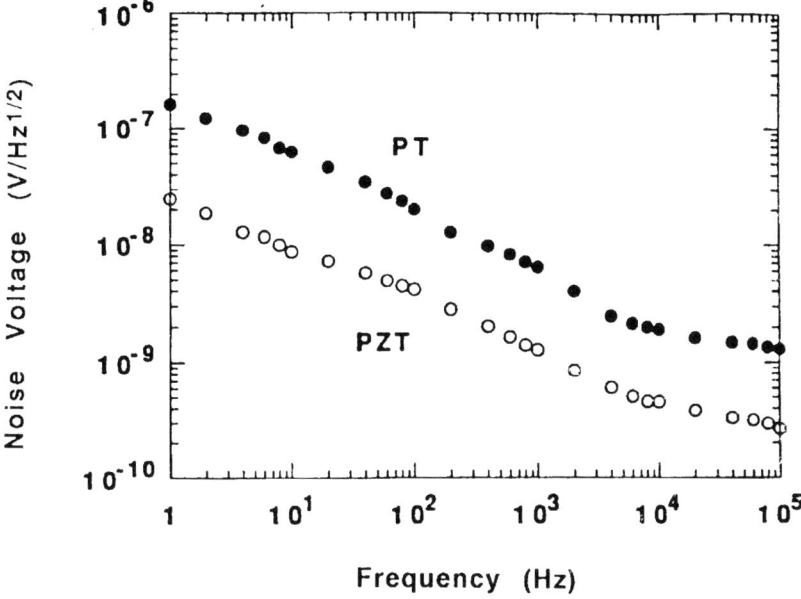

FIG. 13. Measured noise characteristics for 40 × 40 $\mu m^2$ lead zirconium tantalate (PZT) and lead tantalate ($PbTiO_3$) capacitors. Pt, platinum.

### 3. SILICON-BASED $PbTiO_3$ ARRAY PERFORMANCE

Prototype 64 × 64 element pyroelectric IR imager chips have been fabricated at the University of Minnesota as a demonstration of a three-dimensional (stacked) integrated microsensor concept. Both polycrystalline Si and $Si_3N_4$ have been used separately as the micromechanical structural membrane, with $PbTiO_3$ as the active pyroelectric material.

The basic photomask set used makes use of a merged 3-$\mu$m NMOS process with a back-end ferroelectric microsensor process. Each of the 4096 polysilicon–$Si_3N_4$ microbridges measure 50 × 50 $\mu m^2$ with 75-$\mu$m pitch and has its own simple preamplifier fabricated directly beneath an air gap. Each 0.36-$\mu$m-thick $PbTiO_3$ thin-film sensor measures 30 × 30 $\mu m^2$ and generates charge primarily through the pyroelectric effect.

Sensor and circuit characterization were carried out in atmosphere at room temperature. The circuits in the diagnostic chips were completely functional with no observable shift in threshold voltages or degradation of mobility due to the back-end micromachining and microsensor fabrication

TABLE II
SUMMARY OF INTEGRATED SYSTEM PERFORMANCE PROPERTIES FOR A $PbTiO_3$ SILICON BASED IR DETECTOR ARRAY

| | |
|---|---|
| Pixel Element Size | $30 \times 30 \, \mu m^2$ |
| Array Size | $64 \times 64$ Pixels |
| Chip Area | $1 \times 1 \, cm^2$ |
| Pyroelectric Responsivity | $90 \pm 10 \, nC/cm^2$-K |
| Blackbody Voltage Responsivity at 30 Hz | $1.2 \pm 0.2 \times 10^4 \, V/W$ |
| Detectivity D* at 297 K, 30 Hz | $2.0 \pm 0.4 \times 10^8 \, cm\text{-}Hz^{1/2}/W$ |

steps. Relevant material, circuit, and integrated system performance properties are summarized in Table II.

The pyroelectric response for the IR detectors of this work was characterized by a quasi-static method similar to that initially used by Byer and Roundy in 1972. The array was bonded in a 64-pin, dual, in-line ceramic package and was mounted on a copper block, optically shielded, and resistively heated (or coded) at a constant rate. The pyroelectric response was measured directly with a Keithley Model 617 Electrometer. A temperature change rate ranging from 1.0 to 1.5°C/min was applied using a Lake Shore Cryogenics temperature controller. A well-characterized bulk PZT sample was used as an additional reference in these measurements. The measured response for a single pyroelectric element is $90 \pm 10 \, nC/cm^2$ K. This response is due to a combined pyroelectric effect ($60 \, nC/cm^2$ K) and a possible secondary thermal stress response originating at the $PbTiO_3$– structural support interface. While several laboratories have independently characterized our samples, the detailed physics of the pyroelectric response has not been studied further.

Figures 13 and 14 show the performance characteristics of a representative single $PbTiO_3$ pyroelectric element. The measured broadband blackbody voltage responsivity for a $PbTiO_3$ pyroelectric element within the array at 297 K and a chopping frequency of 30 Hz is $1.2 \times 10^4$ V/W. This response value has been normalized with respect to the voltage gain of the on-chip pre-amplifier. The noise voltage shown in Fig. 13 was also measured for the detector elements fabricated in this work. The ratio of the measured voltage responsivity to measured noise voltage was used to calculate a normalized detectivity D*, as shown in Fig. 14. The calculated D* at 297 K and 30 Hz is $2 \times 10^8 \, cmHz^{1/2}/W$.

The IR imaging chip was operated in a serially scanned mode. All elements were functional, with 90% of the elements showing a responsivity variation within 10% of one another. No adjacent element responsivity was

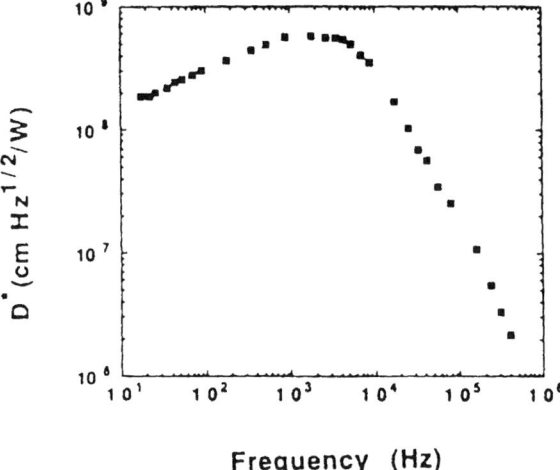

FIG. 14. Calculated D* for a representative pixel element obtain by dividing the measured voltage responsivity by the measured noise voltage. D*, Defectivity.

observed under careful single-element IR exposure. The measured voltage gain for the on-chip circuitry is approximately 48 dB and demonstrates the functionality of the imager. Further work on optical systems is required to test this array for its practicality as an IR imaging device.

The major result of this work demonstrates the feasibility of integrating MOS electronics, thermal isolation microbridges, and pyroelectric thin-film capacitors in a common IC process. This basic feasibility should allow for both single-element and scanned array imaging applications.

## V. Gallium Arsenide–Based Integrated Pyroelectric Detectors

Monolithic gallium arsenide (GaAs)-based pyroelectric microsensors previously have been shown to be feasible by Choi and Polla (1993). The approach taken in this work has been to form monolithic uncooled IR detectors in a radiation-hard and optoelectronics-compatible GaAs metal-semiconductor field-effect transistor (MESFET) technology. While the pyroelectric material used in this work is ZnO, work is also underway using $PbTiO_3$.

Figure 15 shows the integration approach using semi-insulating GaAs as a substrate and wafer for MESFET circuits, surface-micromachining of Al

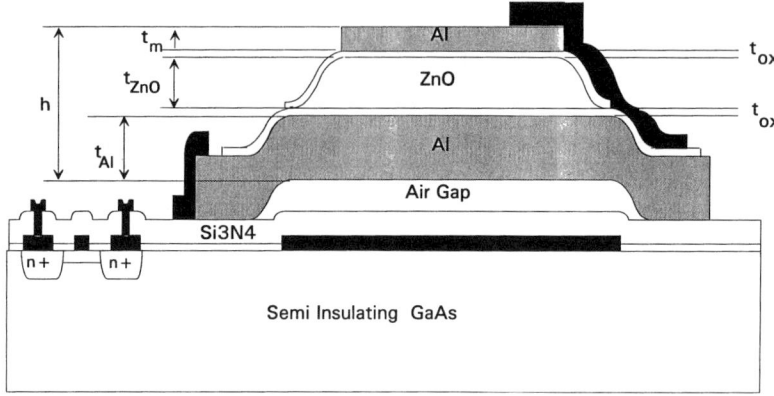

FIG. 15. Cross section of integrated gallium arsenide–based infrared detection concept with aluminum microbridge.

FIG. 16. Scanning electron micrograph of gallium arsenide–based monolithic infrared detector corresponding to Fig. 15.

membranes for creating a thermal isolation region, and ZnO as the active pyroelectric thin films. Note that ZnO has a lower pyroelectric coefficient than does $PbTiO_3$ and Al has a higher thermal conductivity than does $Si_3N_4$. While several of the approaches taken do not represent a best materials and best geometry approach, trade-offs were made in processing ease versus optimized detector performance. These trade-offs are a direct result of the thermal budget constraint set by GaAs, which is generally limited by the out-diffusion of As for processing temperatures in excess of 500°C and by the limited experience in GaAs-based micromachining.

Figure 16 shows a scanning electron micrograph of a GaAs-based pyroelectric microsensor with ZnO thin films and MESFET ICs. The active pixel element shown measures $40 \times 40 \, \mu m^2$ and is directly connected to the gate of an ion-implanted, enhancement-mode MESFET transistor connected in a differential-pair amplifier with on-chip MESFET current source. The measured normalized detectivity is approximately $3.1 \times 10^7 \, cmHz/W^{-1}$.

## VI. Summary

Pyroelectric thin-film capacitors have been monolithically integrated with on-chip signal processing electronics for uncooled IR detection applications. This has been demonstrated for both Si and GaAs technologies for several pyroelectric thin-film materials. Solid-state micromachining techniques have been applied to form low thermal conduction microbridges elevated approximately $1.0 \, \mu m$ above the semiconductor substrate. The ability to apply surface-micromachining methods allows for individual reticulation of a heat-sensitive pixel element, thereby significantly reducing element-to-element thermal cross talk in IR detector array applications.

The most advanced concept presented in this work successfully integrates the following technologies: (1) MOS signal processing electronics, (2) $Si_3N_4$ surface-micromachined thermal isolation bridges, and (3) $PbTiO_3$ pyroelectric thin films. The resulting normalized broadband blackbody detectivity D* at room temperature and 30-Hz mechanical chopping frequency is $2 \times 10^8 \, cmHz^{1/2}/W$ showing feasibility for both single-element and imaging array applications.

### Acknowledgments

The authors gratefully acknowledge the support of their colleagues in this work. Professor Lorraine F. Francis and Dr. Chen Hsueh substantially contributed to the development of the ferroelectric thin-film materials used in this work. Drs. Takashi Tamagawa, Gregory T. Cibuzar, Lai Pham, and Peter J. Schiller are responsible for the investigative development of many of the solid-state micromachining methods successfully used by the authors.

This work was supported by National Science Foundation grants (ECS-8821103, ECS-8906121, and ECS-8814651.

REFERENCES

Barth, P. W. (1995). *Rec. Int. Conf. Solid State Sens. Actuators (Transducers '95), 8th,* Stockholm, p. 276.
Budd, K. D., Dey, S. K., and Payne, D. A. (1985). *Proc. Br. Ceram. Soc.* **36,** 107.
Byer, R. L., and Roundy, C. B. (1972). *Ferroelectrics* **3,** 333.
Campbell, S. A. (1996). "The Science and Engineering of Microelectronics Fabrication." Oxford University Press, New York.
Chau, K. H.-L., et al. (1995). *Rec. Int. Conf. Solid State Sen. Actuators (Transducers '95), 8th,* Stockholm, p. 593.
Choi, J. R., and Polla, D. L. (1993). *J. Micromech. Microeng.* **3,** 60–.64.
Cooney, T. G., Glumac, D. E., Robbins, W. P., and Francis, L. F. (1994). *Mater. Res. Soc. Symp. Proc.* **360,** 401.
Fan, L. S., and Muller, R. S. (1988). *Rec. IEEE Solid State Sens. Actuators Workshop,* p. 55.
Fung, C. D., Cheung, P W., and Ko, W. H., eds. (1985). "Micromachining and Micropackaging of Transducers." Elsevier, Amsterdam.
Gardner, J. W. (1994). "Microsensors Principles and Applications." Wiley, New York.
Gray, P. G., and Meyer, R. G. (1993). "Analysis and Design of Analog Integrated Circuits." Wiley, New York.
Guckel, H. et al. (1988). *IEEE Trans. Electron Devices* **ED-35,** 800.
Harrold, S. J. (1993). "An Introduction to GaAs IC Design." Prentice Hall, New York.
Hornbeck, L. J. (1989). *Proc. Soc. Photo-Opt. Instrum. Eng.* **1150,** 86.
Howe, R. T. (1988). *J. Vac. Sci. Technol. B* [2] **6,** 1809.
Howe, R. T., and Muller, R. S. (1983). *J. Appl. Phys.* **54,** 4674.
Hsueh, C.-C., Tamagawa, T., Ye, C., Helgeson, A., and Polla, D. L. (1991). *Rec. Int. Symp. Integ. Ferroelect., 3rd.*
Hsueh, C.-C., Tamagawa, T., Ye, C., Helgeson, A., and Polla, D. L. (1993). *Integr. Ferroelectr.* **3,** 21.
Jerman, H. (1990). *Rec. Solid State Sen. Actuator Workshop,* p. 65.
Ko, W. H., Bao, M.-H., and Hong, Y.-D. (1982). *IEEE Trans. Electron Devices* **ED-29,** 48.
Lahiji, G. R., and Wise, K. D. (1982). *IEEE Trans. Electron Devices* **ED-29,** 14.
Lebbink, G. K. (1994). "Microsystems Technology Exploring Opportunities." Stichting Toekomstbeeld der Techniek, Den Haag.
Lee, Y. S., and Wise, K. D. (1982). *IEEE Trans. Electron Devices* **ED-17,** 927.
Lin, C. T., Scanlan, B. W., McNeill, Webb, J. D., Li, L., Lipeles, R. A., Adams, P. M., and Leung, M. S. (1992). *J. Mater. Res.* **7**(9), 2546.
Liu, S. T. (1976). *Ferroelectrics* **10,** 83.
Liu, S. T., Zook, J. D., and Long, D. (1975). *Ferroelectrics* **9,** 39.
Mehregany, M., Howe, R. T., and Senturia, S. D. (1987). *J. Appl. Phys.* **62,** 3579.
Middelhoek, S., and Audet, S. A. (1989). "Silicon Sensors." Academic Press, London.
Muller, R. S., Polla, D. L., and White, R. M. (1985). *IEEE ElectroTechnol. Rev.* **1,** 56–58.

Muller, R. S., Howe, R. T., Senturia, S. D., Smith, R. L., and White, R. M., eds. (1990). "Microsensors." IEEE Press, New York.
Obermeier, E., Kopystynski, P., and Niessel, R. (1986). *Rec. IEEE Solid State Sens. Workshop,* p. 40.
Pan, P., and Berry, W. (1985). *J. Electrochem. Soc.* **132,** 3001.
Petersen, K. E. (1982). *Proc. IEEE* **70,** 420–457.
Pham, L., and Polla, D. L. (1994). *IEEE Trans. Ultrason. Ferroelectr. Freq. Control* **UFFC-41,** 552.
Polla, D. L., Muller, R. S., and White, R. M. (1983). *IEEE Int. Electron Devices Meet.*
Polla, D. L., Muller, R. S., and White, R. M. (1985). *Rec. IEEE Ultrason. Symp.,* pp. 495–498.
Polla, D. L., Yoon, H., Tamagawa, T., and Voros, K. (1989). *IEEE Int. Electron Devices Meet.*
Polla, D. L., Tamagawa, T., and Choi, J. R. (1990). *Int. Symp. Circuits Syst.,* New Orleans, *1990.*
Polla, D. L., Ye, C., and Tamagawa, T. (1991). *Appl. Phys. Lett.* **59,** 3539.
Polla, D. L., Ye, C., and Pham, L. (1993). *Rec. IRIS Detect. Spec. Group Meet.*
Ristic, L. (1994). "Sensor Technology and Devices." Artech House, Norwood, MA.
Rockstad, H. K. (1993). *Rec. Int. Conf. Solid State Sens. Actuators (Transducers '93), 8th,* p. 836.
Roylance, L. M., and Angell, J. B. (1979). *IEEE Trans. Electron Devices* **ED-26,** 1911.
Sampsell, J. B. (1993). *Rec. Int. Conf. Solid State Sens. Actuators (Transducers '93), 8th,* p. 24.
Sansen, W., Huijsing, J. H., and van de Plassche, R. J. (1994). "Analog Circuit Design." Kluwer, Boston.
Seidel, H. (1987). *Rec. Int. Conf. Solid State Sens. Actuators, 4th,* pp. 120–125.
Sekimoto, M., Yoshihara, H., and Ohkubo, T. (1982). *J. Vac. Sci. Technol.* **21,** 1017.
Shorrocks, N. M., Patel, A., Walker, M. J., and Parsons, A. D. (1995). *Miocroelectron. Eng.* **29,** 59.
Sze, S. M. (1988). "VLSI Technology." McGraw-Hill, New York.
Sze, S. M., ed. (1994). "Semiconductor Sensors." Wiley, New York.
Tamagawa, T., Polla, D. L., and Hsueh, C.-C. (1990). *IEEE Int. Electron Devices Meet.,* San Francisco.
Watton, R. (1989). *Ferroelectrics* **91,** 87.
Whatmore, R. W., Patel, A., Shorrocks, N. M., and Ainger, F. W. (1990). *Ferroelectrics* **104,** 269.
Whatmore, R. W., Kirby, P., Patel, A., Shorrocks, N. M., Bland, T., and Walker, M. (1995). In "Science and Technology of Electroceramic Thin Films" (O. Auciello and R. Waser, eds.). Kluwer, The Netherlands.
Wolf, S., and Tauber, R. N. (1986). "Silicon Processing for the VLSI Era," Vols. 1–3. Lattice Press, Sunset Beach, CA.
Ye, C., and Polla, D. L. (1994). *Ferroelectrics* **157,** 347.
Ye, C., Tamagawa, T., Schiller, P., and Polla, D. L. (1991a). *Sens. Actuators, A: Phys.* **35,** 77.
Ye, C., Tamagawa, T., and Polla, D. L. (1991b). *J. Appl. Phys.* **71,** 5538.

CHAPTER 6

# Thermoelectric Uncooled Infrared Focal Plane Arrays*

*Nobukazu Teranishi*

MICROELECTRONICS RESEARCH LABORATORIES
NEC CORPORATION
SAGAMIHARA, KANAGAWA
JAPAN

I. INTRODUCTION . . . . . . . . . . . . . . . . . . . . . . . . . 203
II. THERMOPILE INFRARED DETECTOR . . . . . . . . . . . . . . . 204
   1. *Mechanism for Uncooled Infrared Detector* . . . . . . . . . . 204
   2. *Comparison Among Uncooled Infrared Detector Schemes* . . . . 205
   3. *The Seebeck Effect* . . . . . . . . . . . . . . . . . . . 206
   4. *Various Thermopile Infrared Detectors* . . . . . . . . . . . 209
III. A 128 × 128 PIXEL THERMOPILE INFRARED FOCAL PLANE ARRAY . . 210
   1. *Polysilicon Thermopile Infrared Detector* . . . . . . . . . . 210
   2. *Characteristics of a Thermopile Infrared Detector* . . . . . . 211
   3. *Signal Read-out Circuit* . . . . . . . . . . . . . . . . . 211
   4. *Charge-Coupled Device Scanner* . . . . . . . . . . . . . . 213
   5. *Package* . . . . . . . . . . . . . . . . . . . . . . . . 214
   6. *Performance* . . . . . . . . . . . . . . . . . . . . . . 215
   7. *Future Improvements* . . . . . . . . . . . . . . . . . . . 217
IV. SUMMARY . . . . . . . . . . . . . . . . . . . . . . . . . . 217
   *References* . . . . . . . . . . . . . . . . . . . . . . . 218

## I. Introduction

In this chapter the thermopile scheme for infrared (IR) focal plane arrays (FPAs) is described. The merits of this scheme are that the detectors can be fabricated in silicon integrated circuit (IC) plants because the materials and processes used are commonly used in silicon plants. Also, this scheme does

---

*Partially reprinted from Teranishi (1994). Uncooled infrared focal plane array having 128 × 128 thermopile detector elements. *Infrared Technology XX, Proc. SPIE* **2269**, 450; with permission from the author and the SPIE.

not need a chopper or a temperature stabilizer. Therefore, a low-cost, simple IR camera can be realized. The merits of thermoelectric uncooled FPAs, when compared with conventional photon detectors, are shown in Fig. 1 (see color plate section). A 128 × 128 pixel thermopile IR FPA5 is described in detail.

## II. Thermopile Infrared Detector

1. MECHANISM FOR UNCOOLED INFRARED DETECTOR

The mechanism for the uncooled IR detector is shown in Fig. 2. The uncooled IR detector basically consists of three parts: a thermal isolation structure, an IR absorber, and a thermometer. Incident IR radiation is absorbed at the IR absorber, which increases the temperature of the IR absorber. If the thermal conductance for the thermal isolation structure is small, the temperature increase becomes large. When the object temperature is smaller than the IR absorber temperature, the incident IR intensity is smaller than the emitting IR intensity and the IR absorber temperature is decreased. This temperature change is detected and converted into an electric signal by the thermometer. Therefore, the electric signal indicates the amount of incident IR intensity.

Since the IR absorptance at the IR absorber does not depend on the surrounding temperature, and since the thermometer does not need to be at a cryogenic temperature, the detector can operate at room temperature.

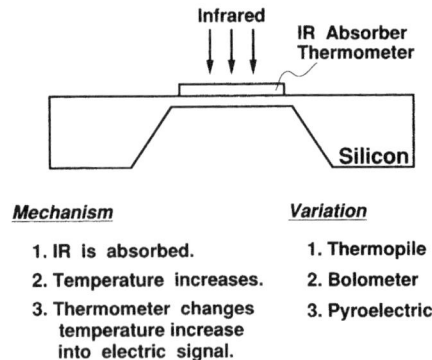

FIG. 2. Mechanism for uncooled infrared (IR) detector.

However, if the thermoelectric conversion characteristic for the thermometer is very sensitive to the operating temperature, which depends on the thermometer material, a temperature stabilizer is needed.

The spectral response for the uncooled IR detector is determined by the optics transmittance and spectral response of the IR absorber. Therefore, the spectral responsivity of the uncooled IR detector is broad and relatively flat. This is in contrast to the spectral responsivity of photon detectors, which has sharp cut-off wavelengths.

## 2. COMPARISON AMONG UNCOOLED INFRARED DETECTOR SCHEMES

The uncooled IR detectors are classified according to the thermometer scheme: thermopile scheme, bolometer scheme, and pyroelectric scheme, and others. A comparison among them is shown in Table I.

Thermopiles detect the temperature gradient between hot junctions and cold junctions by the Seebeck effect. The signal form is an electromotive voltage $\Delta V$. The bolometer detects the temperature itself by the carrier density change and the mobility change. The signal form is a resistance change $\Delta R$. The pyroelectric scheme detects the temperature change by the dielectric constant change and the spontaneous polarization change. The signal form is a polarization change $\Delta Q$.

Because thermopiles detect the temperature difference between the hot and cold junctions, and because the cold junction is located on the heat

TABLE I
COMPARISON AMONG UNCOOLED INFRARED DETECTORS

|  | Thermopile | Bolometer | Pyroelectric |
|---|---|---|---|
| Physics | Seebeck effect | Carrier density mobility | Dielectric polarization |
| Object | Temperature gradient | Temperature | Temperature change |
| Signal | Voltage $\Delta V$ | Resistance $\Delta R$ | Polarization $\Delta Q$ |
| Temperature control | Unnecessary | Necessary | Unnecessary/ necessary |
| Chopper | Unnecessary | Unnecessary | Necessary |

reservoir, the cold junction plays the important role of the temperature reference. Therefore, the thermopile does not need temperature stabilization, whereas bolometers generally do. Because the operational temperature range, for example, 0 to 50°C, is much larger than the temperature change at the IR absorber by the incident IR, and because it is difficult for the preamplifier to sense the resistance change according to the range for the whole operation temperature change, operation temperature stabilizing is needed for the bolometer. To increase the temperature dependence of the resistance and the dielectric constant and to realize large responsivity, bolometers and pyroelectric detectors often use thermoelectric material having a transition point, and the operation temperature then needs to be set near the transition temperature. In this case, operation temperature control is required.

Because the thermopile detector has a temperature reference inside, the chopper is not needed. Neither does the bolometer need the chopper, because it detects the temperature itself. Conversely, pyroelectric detectors require choppers because they respond to temperature change. The pyroelectric IR detector cannot be used under circumstances in which vibrations are large because it is adversely affected by the microphonic noise, whereas the thermopile detector and the bolometer detector do not suffer from microphonics.

The thermopile's freedom from requirements for temperature stabilization and chopping are significant advantages for realizing portable and low-cost IR cameras. Moreover, if polysilicon (PS) is chosen as the thermoelectric material, three large advantages are obtained: First, the material development process can be omitted; second, the monolithic structure is easily realized, while the pyroelectric scheme uses a hybrid structure by bump technology; third, the PS thermopile IR detector can be fabricated in a silicon IC plant. Therefore, mass-production and low cost can be realized easily.

## 3. THE SEEBECK EFFECT

There are several thermoelectric effects in the solid state: the Seebeck effect, the Peltier effect, the Thomson effect, and others. In this section, the Seebeck effect, which has an important role in thermocouples and thermopiles, is briefly reviewed.

The Seebeck effect occurs when two kinds of metals or semiconductors are electrically contacted at two points, as shown in Fig. 3. If the temperature difference, $\Delta T$ is applied between the two points, an electromotive voltage $V_S$ is generated. This phenomenon is called the Seebeck effect. The

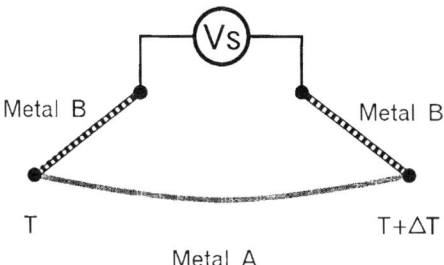

FIG. 3. Diagram of the Seebeck effect.

structure shown in Fig. 3 is called a thermocouple, and a series of thermocouples is called a thermopile. The electromotive voltage $V_S$ is denoted by the Seebeck coefficient (the thermoelectric power) $\alpha$, as $V_S = \alpha \Delta T$. The Seebeck coefficient is originally defined by a combination of two kinds of metals and semiconductors. If an ideal metal, having zero Seebeck coefficient, is assumed, then an absolute Seebeck coefficient can be defined for each metal or semiconductor. The Seebeck coefficients of semiconductors are generally 1.5 decades larger than those of metals. The Seebeck effect of semiconductors is used for uncooled IR detectors and is explained subsequently.

As shown in Fig. 4, an N-type semiconductor is set, having two metal electrodes that are electrically connected at the two ends of the semiconductor. The metals used for the electrodes are of the same kind. The temperature difference $\Delta T$ is applied between the two ends. The hotter end is called a hot junction; the colder end is called a cold junction. There are two mechanisms to generate the Seebeck effect in semiconductors:

1. If the electron density of the semiconductor is controlled by donors, namely, under a freeze-out region, then the electron density is exponentially increased by an increasing temperature. In this case, the electron density at a hotter area becomes larger than that at a colder area, and electrons diffuse from the hotter side to the colder side. Excess electrons are then accumulated at the colder side, which generates an electric field. The hot junction becomes positive, and the cold junction becomes negative. A balance between the drift current generated by this electric field and the diffusion current generated by the electron density difference determines the electron density profile and the electric field size. This electric field contributes to the Seebeck effect.

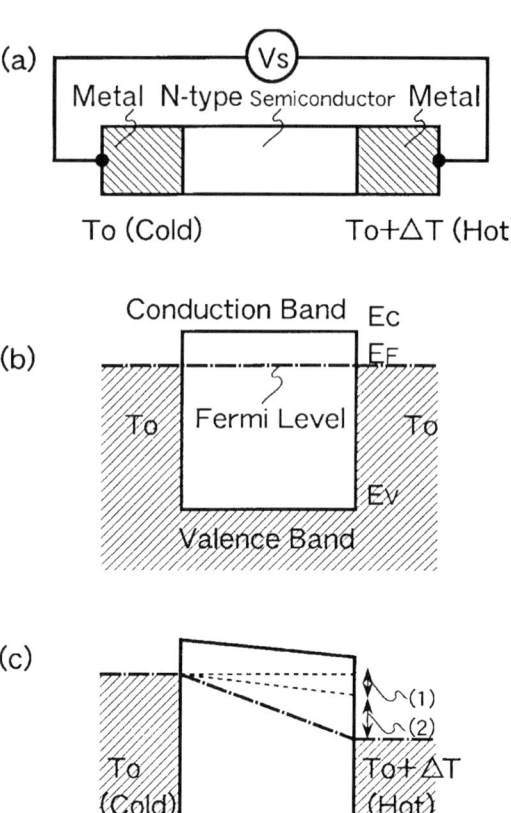

FIG. 4. The Seebeck effect at an N-type semiconductor.

2. The Fermi level $E_f$, measured from the conduction band bottom edge $E_c$, is increased by a temperature increase. The Fermi level at the hot junction becomes larger than that at the cold junction, and this Fermi level difference also contributes to the Seebeck effect.

In the case of a P-type semiconductor, majority carriers are holes and therefore the thermoelectric field becomes opposite. In other words, the sign of the Seebeck coefficient for an N-type semiconductor is positive, whereas that for a P-type semiconductor is negative. If a semiconductor is under an intrinsic region, both electron density and hole density are increased. Then, both electrons and holes diffuse from the hotter area to the colder area, and only the mobility difference between electron and hole contributes to the Seebeck effect, and it is much smaller when compared with the freeze-out region.

If there is a temperature difference, as shown in Fig. 3, the phonon density is increased at the hotter area, and the phonons also diffuse from the hotter to the colder area. Due to collisions between electrons and phonons, electrons receive momenta from the phonon diffusion current. This effect, called the phonon drag, accelerates the Seebeck effect.

## 4. VARIOUS THERMOPILE INFRARED DETECTORS

There have been many reports on thermopiles. One example is a 32-pixel linear thermopile IR detector array, which has been reported by the University of Michigan (Choi and Wise, 1986; Bear et al., 1991). Each pixel has 40 thermocouples of p-type PS and gold (Au) that provide a responsivity of 64 volts per watt (V/W). Its detector array period is 500 $\mu$m, and the detector area is 400 × 700 $\mu$m. Another example reported by Xensor Integration is a 10-pixel linear thermopile IR detector array, which uses a P-type diffusion region in a silicon (Si) epitaxial layer and aluminum (Al) for thermoelectric materials. Sensitivities of 10's of volts per watt have been demonstrated (Sarro et al., 1988; van Henvaarden, 1989). Physikalisch-Technisches Institut (Jena, Germany) has developed a single thermopile, made of P-type $(Bi_{1-x}Sb_x)_2Te_3$ and N-type $Bi_{1-x}Sb_x$ films. It has 72 junctions beneath a 0.2-mm$^2$ sensitive area. A sensitivity of 500 V/W is obtained (Volkein et al., 1991). ETH has fabricated a single thermopile IR detector that uses N-type and P-type PS for thermoelectric materials. It has an operational amplifier on chip, integrated by complementary metal-oxide semiconductor (CMOS) technology. Its sensitivity is 150 V/W (Elbel et al., 1992; Lenggenhager and Baltes, 1993). Honeywell Inc. (Plymouth, Minnesota) has developed a 120-pixel linear thermopile IR image sensor (Wood et al., 1995), which was incorporated in an imaging radiometer by Infrared Solutions, Inc., as described in Chapter 10. The Japan Defence Agency and the NEC Corp. fabricated a 128 × 128 pixel thermopile IR FPA that uses N-type and P-type PS for thermoelectric materials. A sensitivity of 1.550 V/W is realized (Kanno et al., 1994).

The thermopile IR detectors mentioned above are array sensors, or experiments with array sensors. Among them, JDA/NEC's 128 × 128 FPA is the most large scale, most sensitive, and most practicable. Therefore, the details of this sensor are described in the next section.

## III. A 128 × 128 Pixel Thermopile Infrared Focal Plane Array

1. POLYSILICON THERMOPILE INFRARED DETECTOR

To realize the large-scale 128 × 128 pixel FPA, the pixel size is reduced to 100 × 100 μm. A sketch is shown in Fig. 5 (see color plate section). This device has a monolithically integrated structure to increase the fill factor. The charge-coupled device (CCD) for signal charge accumulation and signal charge read-out are fabricated on the Si surface. Over the CCD, silicon dioxide ($SiO_2$) diaphragms for a thermal isolation structure are made by using micromachining technology. The diaphragm size is 80 × 84 μm, and the fill factor is as large as 67%.

The diaphragm is made of 450-nm-thick $SiO_2$ film. Under the diaphragm there is a hollow to isolate the diaphragm thermally. The hollow is formed by etching a 1-μm-thick sacrificial PS layer. On each diaphragm, 32 pairs of P-type boron-doped PS and N-type phosphorous-doped PS thermopiles are formed. The PS electrode is 70 nm thick and 0.6 μm wide. Hot junctions are located at the central part of the diaphragm; cold junctions are located on the outside edge of the diaphragm, where the heat conductance is very large. The temperature at the cold junction is always the same as the temperature for the substrate, which is regarded as a heat reservoir. The hot and cold junctions are backed by Al layers to reduce contact resistance.

The measured Seebeck coefficient for PS is shown in Fig. 6. The Seebeck coefficients for P-type and N-type PS are almost the same, but the signs are

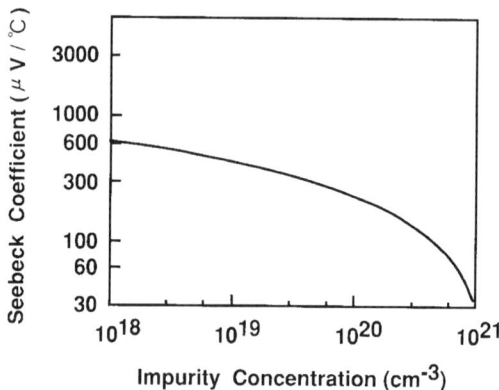

FIG. 6. The Seebeck coefficient for polysilicon.

opposite. The PS Seebeck coefficient greatly depends on impurity concentration. The measured PS Seebeck coefficient is in good agreement with the calculated single-crystal Si coefficient. The impurity concentration used here is about $10^{19}$ to $10^{20}\,\mathrm{cm}^{-3}$. The Seebeck coefficient then becomes 300 to $400\,\mu\mathrm{V}/^\circ\mathrm{C}$. This value is large compared with the Seebeck coefficient for metals. The Seebeck coefficient of platinum is $4.4\,\mu\mathrm{V}/^\circ\mathrm{C}$ at $0^\circ\mathrm{C}$, of iron $15.0\,\mu\mathrm{V}/^\circ\mathrm{C}$, of bismuth $110\,\mu\mathrm{V}/^\circ\mathrm{C}$, and of antimony $47\,\mu\mathrm{V}/^\circ\mathrm{C}$.

## 2. Characteristics of a Thermopile Infrared Detector

Scanning electron microscope (SEM) photographs for the 32-pair thermopile IR detector array are shown in Fig. 7. It is seen that the diaphragm is suspended in air. Because the $SiO_2$ film has a residual compression stress, the diaphragm is slightly curved. Trianglar patterns located in the corner of the diaphragm and small rectangles located around the diaphragm periphery are etching holes for the sacrificial layer etching. Square projected patterns located in the central part and just outside the diaphragm are the hot and cold junctions, respectively. Thin striped patterns on the diaphragm are the PS electrodes. Under the hollow, there is a CCD for signal charge accumulation and transfer. There is an $SiO_2$ layer over the CCD for surface smoothing.

The IR responsivity for the 32-pair thermopile was measured, as shown in Fig. 8. The conditions during the measurement are as follows. The device has a gold black top layer that serves as the IR absorber. The absorptance for the gold black layer is more than 90%. The device is put in a vacuum to reduce the thermal conductance. The temperature for the blackbody, which is the IR source, is set at 500 K. The responsivity in the low chopper frequency region was found to be 1550 V/W, a value much larger than previously reported values, such as 56 V/W (Choi and Wise, 1986). The cut-off frequency is 130 Hz when it is defined as $-3\,\mathrm{dB}$. It can be said that the cut-off frequency is sufficiently large to image a moving object with a 30 or 60 frames per sec frame rate.

## 3. Signal Read-out Circuit

The signal read-out circuit, the equivalent circuit for one pixel, is shown in Fig. 9. The electromotive voltage from the thermopile is applied to the gate of the read-out metal-oxide semiconductor (MOS) transistor, which controls the drain current. The source is grounded, while the drain is

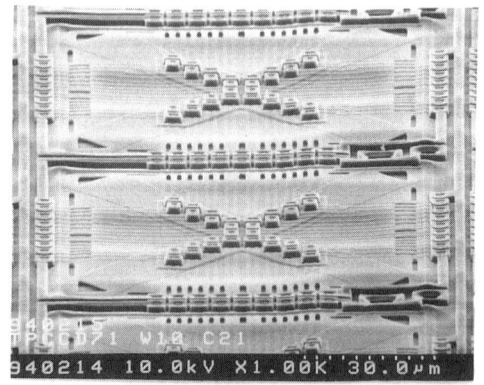

(a) **Diagonal View**

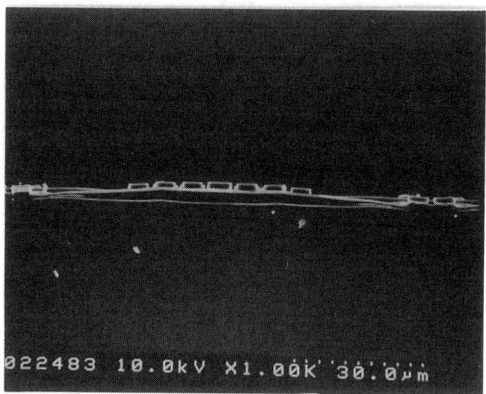

(b) **Cross Sectional View**

FIG. 7. SEM photographs showing thermopile infrared focal plane array pixel. (a) Diagonal view. (b) Cross-sectional view.

connected to an accumulation capacitor, which is the CCD itself. The handling capability is $2 \times 10^7$ electrons. The other end of the thermopile is connected with a bias voltage source.

The bias voltage determines the operating point at the read-out transistor, to control the proper charge flow into the accumulation capacitor. The read-out transistor is usually operated in a weak inversion region. If the bias voltage is turned off, the drain current becomes zero; therefore, the accumulation period is equal to the bias voltage application time.

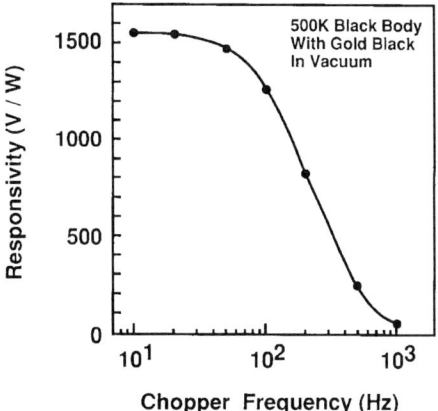

FIG. 8. Chopper frequency dependence of responsivity.

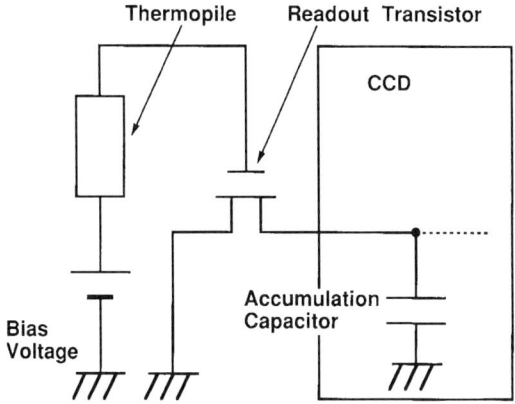

FIG. 9. Signal read-out circuit. CCD, charge-coupled device.

4. CHARGE-COUPLED DEVICE SCANNER

The scanner part consists of a vertical CCD for the image area, a vertical CCD for the storage area, and a horizontal CCD, with the floating diffusion scheme output having a source follower amplifier. The frame rate for this device is 120 Hz, and clock frequencies for the vertical CCD and the

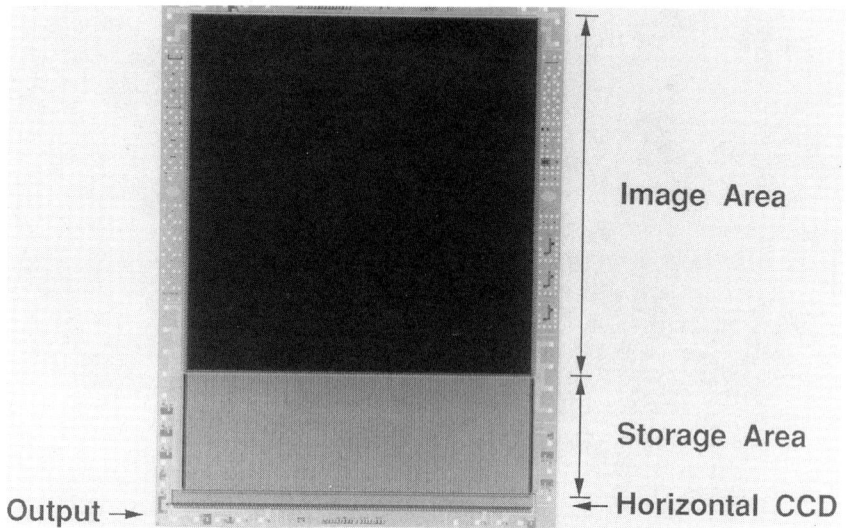

FIG. 10. Chip micrograph for the 128 × 128 pixel thermopile infrared focal plane array. CCD, charge-coupled device.

horizontal CCD are 17 kHz and 2.6 MHz, respectively. Figure 10 shows a chip micrograph for the 128 × 128 pixel thermopile IR FPA. The chip size is 19.5 mm in the vertical direction, and 15.2 mm in the horizontal direction. Because the CCD scanner and the floating diffusion output have little noise, the dominant noise source is the shot noise from the read-out transistor.

The CCD scanner is fabricated on a (100) P-type Si substrate. Both the vertical CCD and the horizontal CCD are designed as the buried type. They have overlapping double-layer PS electrodes. The vertical CCDs for both the image and storage areas are driven by four-phase clock pulses; the horizontal CCD is driven by a two-phase clock pulse.

5. PACKAGE

Figure 11 shows the vacuum package containing the 128 × 128 pixel thermopile IR FPA. The package size is 41 mm in diameter and 14 mm in height. The package can be designed to be much more compact.

Since the absorptance for the gold black, which is the IR absorber, was measured as more than 90% from 2- to 15-$\mu$m wavelengths, the spectral

6 THERMOELECTRIC UNCOOLED INFRARED FOCAL PLANE ARRAYS 215

FIG. 11. Thermopile infrared focal plane array package.

response is determined by the transmittance for the lens and the package window. The window is made of germanium having an antireflective coating for 10-$\mu$m wavelength IR. The antireflective coating for the lens is also designed for around 10-$\mu$m wavelength.

6. PERFORMANCE

The IR camera, fabricated for device evaluation, does not use the chopper or the operation temperature stabilizer. It has a coarse analog offset correction circuit, a fine digital offset correction circuit, and a digital gain correction circuit. It also has frame memory for signal accumulation, which converts the frame rate from 120 to 30 Hz to display on the standard monitor, and to improve the signal-to-noise ratio.

Figure 12 shows an example of the reproduced IR image, where the hot area is the white area. The output offset distribution is 22%, which is corrected by both the analog and digital offset correction circuits, while no gain correction was carried out. The array-averaged noise equivalent temperature difference (NETD) was found to be 0.5°C with the f/1.0 lens. The resolution is limited mainly by the charge transfer inefficiency at both

Fig. 12. Reproduced infrared image, taken by 128 × 128 pixel thermopile infrared focal plane array.

TABLE II
DEVICE SPECIFICATIONS AND PERFORMANCES

| | |
|---|---|
| Image area | 12.8 × 12.8 mm (V) |
| Pixel number | 128 (H) × 128 (V) |
| Pixel size | 100 μm (H) × 100 μm (V) |
| Fill factor | 67% |
| Sensor scheme | Thermopile |
| Sensor material | Polysilicon |
| Scanner | Charge-coupled device |
| Noise equivalent temperature difference | 0.5 K (at $f = 1$) |

the vertical CCD and the horizontal CCD. There are 162 defect pixels remaining among 16,384 pixels in the device that was used to take Fig. 12. Table II shows the device specifications and performance. It can image moving objects; therefore, it can be said that this 128 × 128 pixel PS thermopile IR FPA has demonstrated the feasibility of the thermopile technique for uncooled IR imaging.

## 7. FUTURE IMPROVEMENTS

Because the performance shown was obtained by the device from the first fabrication lot, there are many ways it can be improved. To improve the sensitivity, there are several approaches:

1. The CCD is used for a scanner because it has low noise; however, it is much more important to increase the charge handling capability for IR FPA, because the sensitivity generally increases with the square root of the charge handling capability. If a complementary metal-oxide semiconductor (CMOS) scanner were applied, it would be expected that the charge handling capability would be 10 times or more that of the CCD.
2. The thermal conductance for the diaphragm should be reduced, even if the number of thermocouple pairs is decreased. It is estimated that, if the number of thermocouple pairs is decreased from 32 to 1, the thermal conductance can be reduced to 1/100. The sensitivity would then be expected to be 3 times larger.
3. If the number of thermocouple pairs is decreased from 32 to 1, the electric conductivity for PS can be decreased, and the Seebeck coefficient can be increased.

To increase the resolution, the transfer efficiency for the CCD should be improved. The pixel pitch of 100 $\mu$m is very large when compared with those for visible-light CCD image sensors of 5 to 15 $\mu$m. If the charge transfer length per individual transfer stage is large, the fringing field, which helps the charge transfer motion, becomes smaller. Therefore, by optimizing the impurity profile at the CCD channel, the fringing field could be increased.

## IV. Summary

Thermopile scheme IR sensors are promising for low-cost, portable, quick-start, maintenance-free, IR video camera applications. This scheme has advantages when compared with other uncooled schemes, such as the bolometer and pyroelectric approaches: (1) If PS is used for the thermoelectric material, the materials used are the same as those for silicon ICs, and the whole fabrication process can be carried out at a silicon IC plant. Therefore, it can be said that a low-cost, mass-producible, uncooled IR image sensor is realized by this technology. (2) Since the cold junctions play the role of the temperature reference, the thermopile scheme does not need a chopper or a temperature stabilizer.

The 128 × 128 element thermopile IR image sensor presented in this chapter has a monolithically integrated structure to increase the fill factor. The CCD for signal charge accumulation and signal charge read-out is fabricated on the Si surface. Over the CCD, $SiO_2$ diaphragms for thermal isolation are made by using micromachining technology. On each diaphragm, 32 pairs of P-type and N-type PS thermopiles are formed. The NETD obtained by the device is 0.5°C, with an f/1 lens. This value can be improved by increasing the charge handling capability at the scanner and reducing the thermal conductance of the diaphragm.

### REFERENCES

Bear, W. G., Hull, T., Wise, K. D., Najafi, K., and Wise, K. D. (1991). *Transducers '91 Dig. Tech. Pap.*, pp. 631–634.

Choi, I. H., and Wise, K. D. (1986). *IEEE Trans. Electron Devices* **ED-33,** 72–79.

Elbel, T., Lenggenhager, R., and Baltes, H. (1992). *Sens. Actuators A: Phys.* **35,** 101–106.

Kanno, T., Saga, M., Matsumoto, S., Uchida, M., Tsukamoto, N., Tanaka, A., Itoh, S., Nakazato, A., Endoh, T., Tohyama, S., Yamamoto, Y., Murashima, S., Fukimoto, N., and Teranishi, N. (1994). *Proc. SPIE* **2269,** 450–459.

Lenggenhager, R., and Baltes, H. (1993). *Transducers '93 Dig. Tech. Pap.*, pp. 1008–1011.

Sarro, P. M., Yashiro, H., van Herwaarden, A. W., and Middelhoek, S. (1988). *Sens. Actuators* **14,** 191–201.

van Herwaarden, A. W. (1989). *Sens. Actuators, A: Phys.* **21–23,** 621–630.

Volkein, F., Wiegand, A., and Baier, V. (1991). *Sens. Actuators A: Phys.* **29,** 87–91.

Wood, R. A., Rezachek, T. M., Kruse, P. W., and Schmidt, R. N. (1995). *Proc. SPIE* **2552.**

CHAPTER 7

# Pyroelectric Vidicon

*Michael F. Tompsett*

US ARMY RESEARCH LABORATORY
FORT MONMOUTH, NEW JERSEY

I. HISTORY . . . . . . . . . . . . . . . . . . . . . . . . . . . . . 219
II. PERFORMANCE ANALYSIS . . . . . . . . . . . . . . . . . . . . 223
   *References* . . . . . . . . . . . . . . . . . . . . . . . . . . . 225

## I. History

The history of ferroelectric camera tubes began in 1963 when Hadni at the University of Nancy in France proposed (Hadni, 1963a, b; Hadni *et al.*, 1965), using triglycine sulfate (TGS) in a camera tube operating in dielectric mode. For good sensitivity the crystal had to be maintained close to its Curie temperature of 49.2°C where the change in dielectric constant as a function of temperature can be very large, with the dielectric constant varying from 16,200 at the Curie temperature to approximately 15,001°C on either side of it. By biasing the crystal at voltage $V_0$ and then electrically isolating it while the target temperature changes, the charge displacement as the dielectric constant $\varepsilon_T$ changes with temperature causes a voltage change $\Delta V$ that can be used as a measure of the temperature change and that is given by

$$\Delta V = \frac{V_0}{\varepsilon_T}\left(\frac{\partial \varepsilon_T}{\partial T}\right)\Delta T$$

Le Carvennec adopted this concept when researching the use of a TGS target for an electron-beam scanned thermal-imaging camera tube; his research was published in 1969. Le Carvennec established a potential across the TGS target using a metal electrode on one side and an electron beam on the other. However, a camera tube using this dielectric mode of operation was found to be very unsatisfactory for two reasons. Making large crystal slices having a uniform Curie temperature and polarization curve and maintaining a uniform elevated temperature over the target area were both nontrivial tasks. Another potential but fundamental problem of operating too close to the Curie temperature is random switching of electric domains (depoling). A temperature resolution of about 5°C was demonstrated in a sealed-off tube in 1972 (Charles and Le Carvennec, 1972). Another proposal based on the dielectric mode was made in 1968 (Pel'ta et al., 1968), with the TGS crystal mounted external to the camera tube on a metal plug faceplate. However, because changes in charge across the TGS are small, the interplug sidewall capacitance would have prevented any viable spatial resolution from ever being obtained.

The first concept for a pyroelectric tube was patented in 1963 by Astheimer and was based on a complicated mechanism using a return-beam orthicon mode; however, it never was implemented. In 1968, Tompsett (1969a, b) proposed a simple thermal-imaging camera tube that would operate in true pyroelectric mode and made the first estimates of the potential performance of this pyroelectric vidicon (Tompsett, 1971). The proposed camera tube used a thin pyroelectric crystal slice suspended in vacuum as the target of a vidicon camera tube (Weimer et al., 1950), as shown in Fig. 1. The front surface of the crystal facing the scene is coated with a conducting layer, which is biased at a fixed potential close to that of the electron-gun cathode and connected to the input of a video amplifier. The target is first exposed to radiation from a uniform shutter. It is then scanned with the electron beam, and, by the process of cathode stabilization, deposits electrons on the back surface of the target until it assumes a uniform electrical potential close to that of the cathode. An infrared (IR) thermal image is then projected onto it, which causes local changes in temperature and corresponding changes in electrical polarization; that is, a charge pattern is formed on the back surface of the crystal that corresponds to the local temperature of the scene relative to that of the shutter. If the orientation of the crystal is such as to establish a positive charge pattern on this back surface, then as the electron beam rescans this surface, cathode stabilization deposits electrons onto it. The image charge appears on the front electrode and is read out as a video signal. The complement of the charge pattern can be recreated by reimposing the shutter in front of the target. However, now the back surface is negative and the previously landed electron charge must be removed. In addition, in cathode potential stabilization, the thermal spread of the electrons leads to the target becoming negatively biased relative to the cathode by about 200 mV (Holeman and Wreathall, 1971). Once this happens the number of electrons that can land on the target and read out a signal becomes very small and picture lag results from incomplete read-out. The first tube was made and described in

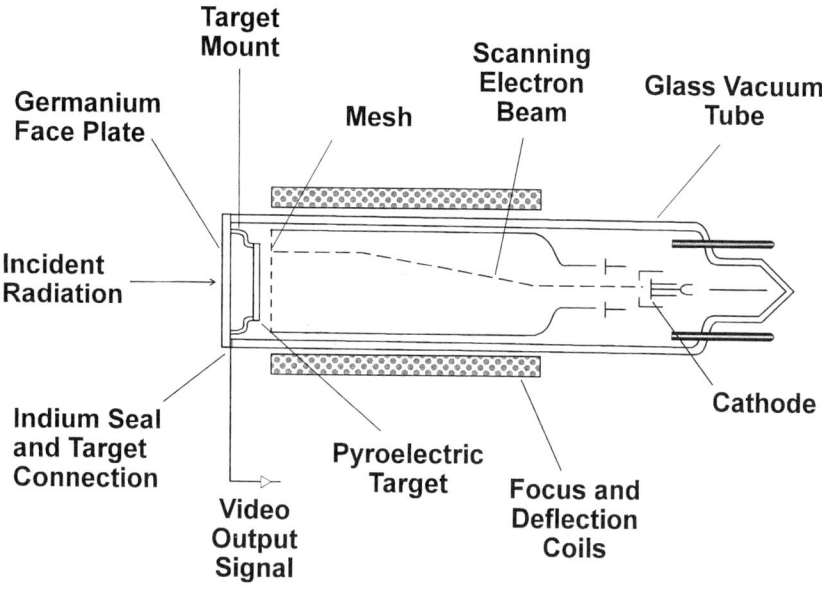

FIG. 1. Sectional schematic of a pyroelectric vidicon thermal imaging camera tube as drawn by Holeman and Wreathall. (Reprinted from Holeman and Wreathall (1971). Thermal imaging camera tubes with pyroelectric targets. *J. Phys. D. Appl. Phys.* **4**, 1898–1909; with permission from the Institute of Physics.)

1971 (Holeman and Wreathall, 1971) and subsequently improved (Putley *et al.*, 1972; Singer and Lalak, 1976; Watton, 1976; Watton *et al.*, 1974, 1976). Operation of the tube required that a positive "pedestal" (Holeman and Wreathall, 1971; Logan and Watton, 1972) be established on the surface of the target. There are several possible implementations of this (Watton, 1976). European developers used the positive ions produced between the grid and the target from residual gas in the tube, when the electron beam scans across the target, but biased so as not to land on it (Holeman and Wreathall, 1971). Anode potential stabilization was adopted in the United States, and is normally done during the flyback time (Singer *et al.*, 1974). A way to avoid the need for a pedestal and still be able to read the charge pattern is to use return-beam or electron-beam mirroring (Tompsett, 1969c, 1971); however, neither of these were found viable (Boot and Castledine, 1974; Goss, 1987). The above operation has been described in terms of a chopped mode, but the chopper can be dispensed with and the temperature resolution doubled by operating in a panned mode, where the camera is kept in continuous motion to maintain temperature difference across the target.

FIG. 2. Photograph of fire fighters in action using an Argus Fourth Generation Portable Thermal Imaging Camera containing a pyroelectric vidicon camera tube manufactured by EEV Ltd. (Courtesy of EEV Ltd., Chelmsford, England; with permission.)

The pyroelectric vidicons in manufacture today are of two types: gas tubes, which have excellent uniformity but a reduced life; and hard tubes, which have the complementary characteristics. The best performance of these tubes is quoted (Goss, 1987) as 400 lines with a minimum resolvable temperature of 0.2°C. However, commercial systems such as the EEV Argus Fourth Generation Portable Thermal Imaging Camera (EEV Ltd., Chelmsford, England) shown in Fig. 2 are specified with 2°C in chopped mode and 1°C in panned mode between pixels.

## II. Performance Analysis

The performance analysis of the pyroelectric vidicon is reviewed briefly here. The basic pyroelectric effect is characterized by the pyroelectric coefficient $p$, which is the change in electrical polarization $P_s$ of an unbiased crystal with temperature, $T$, that is, $p = \partial P_s/\partial T$. For an element of area $A$, the charge $Q$ generated for a temperature change $\Delta T$ is given by $Q = Ap\Delta T$. The corresponding open circuit voltage $\Delta V$ for a target of thickness $d$ and dielectric constant $\varepsilon_T$ is given by $\Delta V = dp\Delta T/\varepsilon_0\varepsilon_T$. It is the net difference $\Delta W$ of the incoming radiation power over the power radiated from the target that determines the rate of change of target voltage with time, that is,

$$\frac{\partial V}{\partial t} = \frac{dp}{\varepsilon_0\varepsilon}\left(\frac{\partial T}{\partial t}\right) = \frac{dp}{\varepsilon_0\varepsilon}\frac{A\Delta W}{Ads} = \frac{p}{\varepsilon_0\varepsilon s}\Delta W$$

where $s$ is the specific heat.

As an example, consider the change of radiant emittance from a temperature step of 1°C in a scene focused by an $f/1$ lens. The scene delta emittance is approximately 6 Wm$^{-2}$, which after focusing by the lens and by the atmospheric and lens attenuations, is reduced to $\Delta W = 1$ Wm$^{-2}$. Substituting into the previous expression using the parameters for TGS at 25°C— namely, $p = 2.8 \times 10^{-4}$ Cm$^{-2}$°C$^{-1}$, $\varepsilon_T = 42$ and $s = 2.55 \times 10^6$ Jm$^{-3}$C$^{-1}$— gives $\partial V/\partial t \approx 300$ mVs$^{-1}$. Now for a target suspended in a vacuum, heat loss is primarily by radiation with a time constant $\tau$ given by $\tau = ds/4\sigma(\alpha_f + \alpha_b)K^3$, where $\sigma$ is Stefan's constant, $\alpha_f$ and $\alpha_b$ are the emissivities of the front and back surfaces respectively, and $K$ is the absolute temperature of the target. At 300 K for a single crystal slice of TGS thinned to 20 μm, $\tau \approx 5$ sec. Thus the thermal time constant is typically much longer than the shutter period, which is generally chosen to correspond to the television frame period of 16.6 or 20 msec, depending on the country of use, in order to view moving images and be compatible with normal television displays.

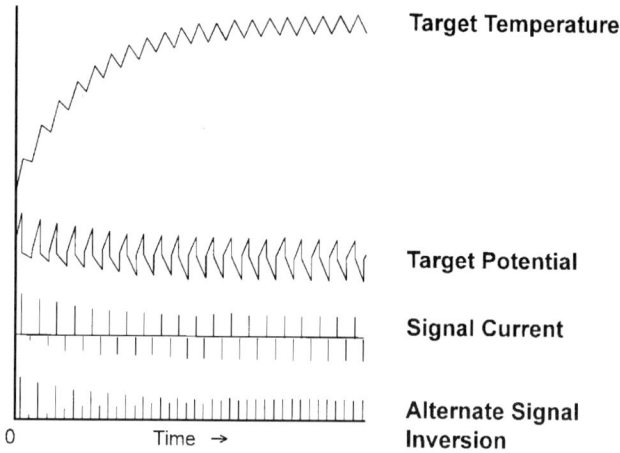

FIG. 3. Target temperature, target potential, signal current, and alternate field-inverted signal current from a pyroelectric vidicon target with chopped illumination following an instantaneous change in scene temperature at time zero, as described by Holeman and Wreathall. (Reprinted from Holeman and Wreathall (1971). Thermal imaging camera tubes with pyroelectric targets. *J. Phys. D. Appl. Phys.* **4**, 1898–1909; with permission from the Institute of Physics.)

The resulting instantaneous target temperatures, voltages, and read-out currents both with and without alternate signal inversion (Holeman and Wreathall, 1971) are shown in Fig. 3. There are several advantages to using signal inversion, the most important being that the influence of lateral heat spread is limited to that which occurs during a chopper period. The eye scans two fields, so that the effective response time for transients is only a single chopper period, despite the 5 sec target time constant and the fact that the spatial noise on the target not modulated by the shutter is eliminated.

A final conclusion can be drawn from the previous equation, which shows that the figure of merit for pyroelectric-mode operation of a material used in this application, where the target capacitance dominates the amplifier capacitance, is $p/\varepsilon\varepsilon_0 s$, that is: (Pyroelectric coefficient)/(Dielectric constant) × (Specific heat). A comprehensive discussion of dielectric materials and their physical properties and operation in the pyroelectric vidicon is given by Whatmore (1986).

A very important aspect of television images is resolution. This is a particular challenge in bolometric devices, where, if a simple thermal sensitive property of a laterally continuous target is used, thermal conduction would cause the resolution to be very poor. However, in the case of the pyroelectric vidicon, using the shuttered mode leads to detection of only those temperature changes between scans, that is, the effective time for thermal diffusion is reduced to that of the shutter period (Tompsett, 1971). In the region of a hot image there will be temperature oscillations at the shutter frequency that will decrease exponentially with distance from the image with a half-width on the order of $k(\pi f \rho s)^{-1}$, where $k$ is the thermal conductivity, $\rho$ is the density, $f$ is the shutter frequency, and $s$ is the specific heat. For TGS with a shutter frequency of 33 Hz, the half-width is approximately 33 μm, which corresponds to a resolution on the order of 300 lines in a nominal 1-inch vidicon. Full time-dependent analyses have been done (Holeman and Wreathall, 1971). These analyses are verified by experimental results, which show that a simple 15 × 15 mm target can produce a resolution of 300 lines. This number can be improved to 400 lines by a process of reticulation, in which grooves are cut or etched into the target to separate the pixels (Thomson-CSF, 1975; Singer and Lalak, 1976; Watton, 1976; Goss, 1987).

## References

Astheimer, R. W. (1963). U.S. Pat. 3,256,435.
Boot, H., and Castledine, J. (1974). *Electron. Lett.* **10,** 452.
Charles, D. R., and LeCarvennec, F. (1972). *Adv. Electron. Electron Phys.* **33A,** 279–284.
Goss, A. J. (1987). *Proc. SPIE* **807** (Passive Infrared Syst. Technol.), 25–32.
Hadni, A. (1963a). *J. Phys. (Paris)* **24,** 694.
Hadni, A. (1963b). Fr. Pat. 945,749.
Hadni, A., Henniger, Y., Thomas, R., Vergnat, P., and Wynche, B. (1965). *J. Phys. (Paris)* **26,** 345–360.
Hadni, A., Thomas, R., Magnin, J., and Bagard, M. (1978). *Infrared Phys.* **18,** 663–668.
Holeman, B. R., and Wreathall, W. M. (1971). *J. Phys. D* **4,** 1898–1909.
LeCarvennec, F. (1969). *Adv. Electron. Electron Phys.* **28A,** 265–272.
Logan, R. M., and Watton, R. (1972). *Infrared Phys.* **12,** 17–28.
Pel'ta, S. D., Dun, L. M., Novik, V. K., and Rez, I. S. (1968). *Radio Eng. Electron. Phys.* **13,** 157–158.
Putley, R., Watton, R., Wreathall, W. M., and Savage, S. D. (1972). *Adv. Electron. Electron Phys.* **33A,** 285–291.
Singer, B., and Lalak, J. (1976). *Ferroelectrics* **10,** 103–107.
Singer, B., Crowell, M. H., and Conklin, T. (1974). *IEEE Trans. Electron. Devices* **ED-21,** 744.
Thomson-CSF. (1975). Br. Pat. 1,395,341, filed in France 1971.
Tompsett, M. F. (1969a). U.S. Pat. 3,646,267.
Tompsett M. F. (1969b). U.K. Pat. 1,239,243.
Tompsett, M. F. (1969c). U.K. Pat. 1,263,325.
Tompsett, M. F. (1971). *IEEE Trans. Electron. Devices* **ED-18,** 1070–1074.
Watton, R. (1976). *Ferroelectrics* **10,** 91–98.
Watton, R., Smith, C., Harper, B., and Wreathall, W. M. (1974). *IEEE Trans. Electron. Devices* **ED-21,** 462–469.
Watton, R., Jones, G. R., and Smith, C. (1976). *Adv. Electron. Electron Phy.* **40A,** 301–312.
Weimer, P. K., Forgue, S. V., and Goodrich, R. R. (1950). *Electronics* **23,** 70–73.
Whatmore, R. W. (1986). *Rep. Progr. Phys.* **49,** 1335–1386.

CHAPTER 8

# Tunneling Infrared Sensors

*T. W. Kenny*

DEPARTMENT OF MECHANICAL ENGINEERING
STANFORD UNIVERSITY
STANFORD, CALIFORNIA

I. INTRODUCTION . . . . . . . . . . . . . . . . . . . . . . . . . . . 227
II. SENSOR MODELING . . . . . . . . . . . . . . . . . . . . . . . . 229
   1. *Sensor Thermal Model* . . . . . . . . . . . . . . . . . . . . . 229
   2. *Sensor Mechanical and Electrical Model* . . . . . . . . . . . . 232
   3. *Noise Model and Considerations* . . . . . . . . . . . . . . . . 236
III. TUNNELING TRANSDUCER BACKGROUND . . . . . . . . . . . . . . . 239
   1. *Comparison of Tunneling and Capacitive Transducers* . . . . . . 241
   2. *Tunneling Transducer Design Considerations* . . . . . . . . . . 243
IV. TUNNELING INFRARED SENSOR DESIGN AND FABRICATION . . . . . . 245
V. TUNNELING TRANSDUCER OPERATION . . . . . . . . . . . . . . . 253
VI. INFRARED SENSOR OPERATION AND TESTING . . . . . . . . . . . 259
VII. FUTURE PROSPECTS FOR THE TUNNELING INFRARED SENSOR . . . . . 264
VIII. CONCLUSION . . . . . . . . . . . . . . . . . . . . . . . . . . 266
   *References* . . . . . . . . . . . . . . . . . . . . . . . . . . 266

## I. Introduction

Uncooled infrared (IR) sensors for detecting mid-to-long wavelength ($3\,\mu\mathrm{m} < \lambda < 100\,\mu\mathrm{m}$) IR radiation are generally based on thermal detection principles. In thermal detectors, IR radiation is captured by an absorber, converted into heat, and detected by a thermometer (Jones, 1953). By measuring the temperature of the thermometer, it is possible to determine the flux of incoming IR radiation.

In all thermal detectors, the thermometer is connected to a temperature reference by a finite thermal conductance. The thermometer also has a finite heat capacity. In the design of thermal IR detectors, sensitivity is optimized by increasing the absorber efficiency and thermometer sensitivity and by minimizing the heat capacity of all elements. The optimal value of the thermal conductance is dependent on the level of ambient background radiation and the response function of the thermometer (Mather, 1984).

The availability of good thermal detectors is generally limited by the availability of sensitive thermometers. Elsewhere in this book, IR detectors based on the best miniature thermometers are described. In every case in which nature provides a good solid-state thermometer, man has developed an IR detector.

In addition to these many solid-state electronic effects, thermal expansion may be used for IR detection. Most solid materials have a thermal expansion coefficient of between 1 and 100 ppm/°C, and therefore thermal expansion can result in a very small displacement. Of all choices, the thermal expansion of gas offers the best characteristics, because the expansion coefficients are large (0.3%/K at room temperature), and the heat capacities are very small ($10^{-4}$ J/Kmm$^3$ at standard temperature and pressure). The thermal expansion of gas was originally used in the Golay cell as an efficient thermometer more than 50 years ago, (Golay, 1947a, 1949; Hadni, 1967). Among room temperature IR sensors with moderate-sized apertures ($\sim$ mm), the Golay cell still offers the best resolution.

As originally designed, the Golay cell required use of extremely fragile membranes and a complicated read-out scheme based on deflection of an optical beam (Golay, 1947b). The membranes were made by dropping colloid compounds on the surface of water, where they spread to a thickness on the order of 1000 Å, and could be lifted by passing a wire beneath the surface of the water. These membranes were then coated with very thin metal films, which can act as efficient IR absorbers, and suspended in the middle of a gas cell, one end of which is sealed by a flexible mirror, also made of a colloidon membrane. This membrane can be either a quarter wavelength thick (100 Å), or it can support a reflective metal film. In either case, an optical read-out is accomplished by focusing a grid pattern on the flexible mirror, and re-focusing on a photodetector. Because the flexible mirror is deformed, the light reaching the photosensor is increased, with a linear relationship between deflection and intensity.

This system is a very complicated instrument, and relies on precise optical alignment and very fragile membranes. Mechanical shock or abrupt changes in illumination can cause damage. More recent commercial implementations of this IR sensor rely on a simplified optical system or on a capacitive displacement transducer. In either case, it has been necessary to retain the use of very thin and fragile membranes in order to produce a detectable deflection.

Because the Golay cell is still commercially viable and its performance still competitive, regular attempts have been made at miniaturization and improvement. Generally, the Golay cell has proven difficult to modify without sacrificing performance. Because much simpler devices such as pyroelectric detectors offer good performance, modern Golay cells that sacrifice performance have not been commercially viable.

Golay cells have been designed that use capacitance displacement transducers to detect the membrane deflection. Miniaturization of such detectors has been limited by the performance and cost of miniature capacitive displacement transducers. A recent report by Chevrier describes work on a micromachined Golay cell with a capacitive detection scheme (Chevrier *et al.*, 1995). This device is made by bulk micromachining techniques similar to those used for the research described later in this paper. Although these authors have presented some preliminary sensitivity results, it seems the noise in their device is not as low as necessary to be competitive with other uncooled sensors.

In this chapter, we describe a new miniature Golay cell IR detector that is fabricated by silicon (Si) micromachining techniques, allowing miniaturization and integration with electronics as needed for arrays. The deflection of the membrane is detected by a tunneling displacement transducer, which measures variations in a quantum mechanical tunneling current between the moving membrane and a fixed tip. The tunneling transducer offers the required sensitivity with very simple circuitry and enables the development of a miniature, robust, sensor assembly. While the use of a tunneling displacement transducer may seem to be unnecessarily complicated, its implementation in this device is quite straightforward.

## II. Sensor Modeling

In the following sections, the design and operation of the tunneling IR sensor are presented in detail. Before describing the tunneling transducer, we analyze its thermal and mechanical properties.

### 1. SENSOR THERMAL MODEL

To understand the design of the detector described herein, we need to consider the transport of heat through the device. The thermodynamic characteristics of the tunneling IR sensor may be accurately represented by a simple model. An illustration of the sensor geometry is shown in Fig. 1. The sensor consists of a volume of trapped gas bounded by two 0.7-$\mu$m-thick silicon nitride ($Si_3N_4$) membranes and 4 solid walls. The dimensions of the gas volume are approximately 1.7 × 1.7 × 0.4 mm. Midway between the membranes in the middle of the trapped gas is an IR absorber, also supported by a $Si_3N_4$ membrane. The bottom membrane is coated on the

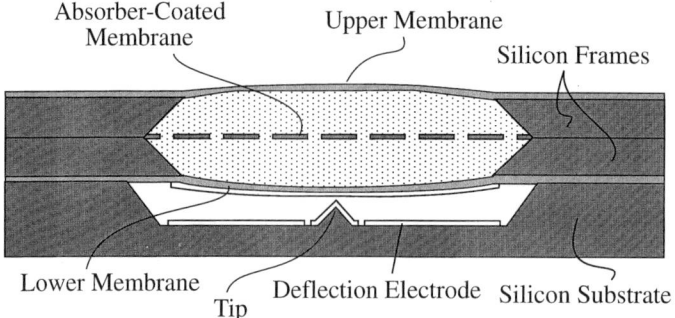

FIG. 1. Drawing of the architecture for the tunneling infrared sensor. Infrared radiation is absorbed by a membrane suspended in the middle of a cell filled with trapped gas. The membrane is warmed, causing the trapped air to expand, which causes a downward force on the lower membrane. The tunneling transducer detects the deflection of the membrane, and causes the feedback system to exert a balancing force on the membrane, which keeps it in place. The thermal energy is transported through the trapped air to the membranes, where it escapes the device.

outside surface by a 2000-Å gold (Au) film, which serves as the reference electrode for the tunneling transducer.

We begin by accounting for the thermal conductance from the absorber membrane to the surrounding room: Thermal conductance from the absorber membrane to either boundary membrane is due to convective heat transport through the trapped air, and is approximately $4 \times 10^{-4}$ W/K. The thermal conductance between the upper membrane and the surrounding room air is $3 \times 10^{-5}$ W/K; that along the lower membrane to the Si frame is approximately $8 \times 10^{-4}$ W/K; and that between the lower membrane and the sensor substrate is $1.4 \times 10^{-3}$ W/K. The dominant path for heat transport is from the absorber through the lower half of the cell, and through the very thin layer of air trapped between the lower membrane and the sensor substrate.

The heat capacity of the upper and absorber membranes is simply that of the nitride, $3 \times 10^{-6}$ J/K. Because the lower membrane also has a 2000-Å Au film, its heat capacity is $4.5 \times 10^{-6}$ J/K. The heat capacity of the trapped air is $1.5 \times 10^{-7}$ J/K.

The thermal model of the IR sensor may be simplified by accepting the following approximations:

- Conduction from the upper membrane to the room due to convection may be ignored, because it is so much less conductive than the path through the lower membrane.

- Heat capacity of the trapped air may be ignored, because it is so much less than that of the membranes.
- Heat conduction from the lower membrane to the substrate is so much larger than through any of the other parts of the path that the lower membrane can be treated as a heat sink.

With these assumptions, the main heat path is through the lower half of the gas cell to the lower membrane, which is held at room temperature. The gas in the upper half of the cell and the upper membrane follow the temperature of the absorber membrane, with some phase lag at frequencies above the G/C cut-off associated with those elements. These assumptions are illustrated in the simple thermal model shown in Fig. 2. At very low frequencies, the capacitances do not matter, and we have an effective G of $3.8 \times 10^{-4}$ W/K. At very high frequencies, the capacitance of the absorber dominates (since the upper membrane is separated by a finite conductance, and cannot keep up with the heat oscillation), and therefore the effective capacitance is about $3.1 \times 10^{-6}$ J/K. Within these assumptions, this detector thermal model is very nearly equivalent to a simple thermal time constant of 8 ms, with corresponding 3 dB point at 20 Hz.

Overall, the heat capacity of this device is dominated by that of the nitride membranes, and thermal conductance is limited by the conduction through the air in the cell from the absorber to the lower membrane. Therefore, if

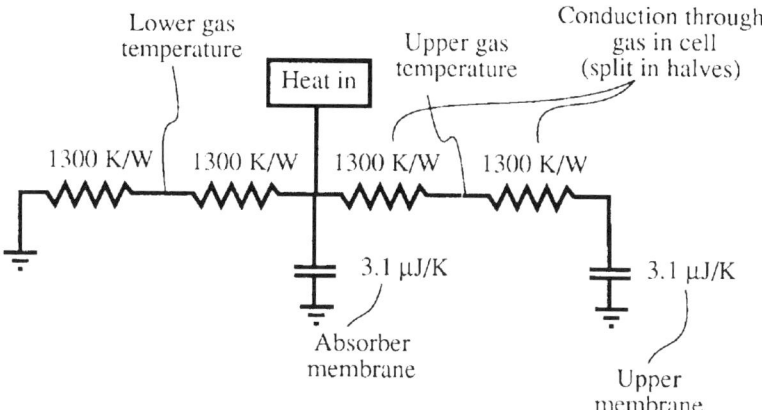

Fig. 2. Simplified thermal model of the infrared sensor. Heat is transported from the absorber membrane, through the gas in the cell, to the membranes. Following the approximations in the text, we consider the lower membrane to be efficiently heat sunk, and we ignore the heat capacity of the gas.

the cell is miniaturized while preserving the aspect ratios, the heat capacity will decrease as $1/L^2$ and the thermal conductance will decrease as $1/L$. As we shall see later, this implies that the sensitivity and the cut-off frequency increase linearly as the dimensions are reduced.

## 2. Sensor Mechanical and Electrical Model

To convert these thermal expansions to a measurable temperature change, we need to account for the sensitivity of the displacement transducer. We define the performance of the trapped gas and the displacement transducer using figures of merit for thermometers. In general, a thermometer produces an electrical signal (a voltage or a current) that is related to the temperature by

$$I(T) = I(T_0)(1 + \alpha_{T_0}(T - T_0))  \quad (1)$$

The temperature coefficient of this thermometer, $\alpha_{T_0}$, is given by

$$\alpha_{T_0} = \frac{1}{I(T_0)} \frac{\partial I(T)}{\partial T}\bigg|_{T_0} \quad (2)$$

The temperature coefficient is an important parameter in the determination of the performance of a thermal IR detector. We calculate the temperature coefficient of our system as follows. For this sensor, temperature changes are detected by measurement of thermal expansion of the trapped gas by the displacement transducer. The resulting temperature coefficient is a product of the sensitivity of the tunneling displacement transducer and the thermal expansion coefficient of the trapped gas:

$$\alpha_T = \frac{1}{I} \frac{\partial I}{\partial z} \frac{\partial z}{\partial T} \quad (3)$$

Now, the tunneling displacement transducer operates by the detection of current, which tunnels between a pair of very closely spaced electrodes. To obtain a nonequilibrium current, a small voltage bias is applied between the electrodes, and the resulting current is proportional to the bias voltage and exponentially dependent on the separation between these electrodes. For the tunneling displacement transducer, the current is related to the operating conditions (Binnig and Rohrer, 1986):

$$I \propto V \exp(-\alpha_I \sqrt{\Phi} z) \quad (4)$$

where $\Phi$ is the effective height of the tunneling barrier, $V$ is the bias voltage (small compared with $\Phi$), and $\alpha_I = 1.025$ ($\text{Å}^{-1}\,\text{eV}^{-1/2}$). Typical values of $I$, $\Phi$, and $z$ are 1.5 nA, 0.2 eV, and 10 Å, respectively. Although the current varies exponentially with separation, we can use a linear approximation for small temperature changes, and we obtain

$$\frac{1}{I}\frac{\partial I}{\partial z} = \frac{1}{I}(-\alpha_I\sqrt{\Phi}I) = -\alpha_I\sqrt{\Phi} = 0.4\,\text{Å}^{-1} \tag{5}$$

Therefore, a change in separation between the electrodes of 1 Å results in a 40% change in the tunneling current! We begin to see why this transducer represents a good choice for this application.

The thermal expansion of the cell may be calculated. This is somewhat complicated by the fact that the thermal expansion is resisted by the stiffness of the membranes. For now, we make a simplifying assumption that will be justified later. If we assume that the thermal expansion is not limited by membrane stiffness, then only the volume is a function of the gas temperature. In this case, the ideal gas law may be reduced to

$$\frac{\partial T}{T} = \frac{\partial V}{V} = \frac{\partial h}{h} \tag{6}$$

where $h$ is the separation between the upper and lower membranes. Simply put, this states that fractional changes in cell temperature are equal to fractional changes in cell height. Since $T = 300\,\text{K}$ and $h = 0.4\,\text{mm}$,

$$\frac{\partial h}{\partial T} = \frac{h}{T_0} = \frac{4\times 10^6\,\text{Å}}{300\,\text{K}} = 1.3 \times 10^4\,\text{Å/K} \tag{7}$$

Then the thermal response coefficient for this "thermometer" is

$$\alpha_T = \frac{1}{I}\frac{\partial I}{\partial z}\frac{\partial z}{\partial T} = (0.4\,\text{Å}^{-1})(1.3 \times 10^4\,\text{Å/K}) = 5 \times 10^3\,\text{K}^{-1} \tag{8}$$

The low-frequency responsivity of any thermal IR detector is given by

$$S = \frac{\partial I}{\partial P} = I\alpha_T\frac{\partial T}{\partial P} = \frac{I\alpha_T}{(G^2 + (\omega C)^2)^{1/2}} \tag{9}$$

Therefore, the responsivity of our model sensor is

$$S = \frac{I\alpha_T}{(G^2 + (\omega C)^2)^{1/2}} = \frac{(1.5 \times 10^{-9}\,\text{A})(5 \times 10^3\,\text{K}^{-1})}{(4 \times 10^{-4}\,\text{W/K})} = 0.02\,\text{A/W} \quad (10)$$

What can we make of these numbers? Well, ordinary high-input impedance amplifiers have equivalent current noise on the order of $10^{-13}\,\text{A}/\sqrt{\text{Hz}}$. Therefore, it is not difficult to measure current fluctuations that are consistent with IR power fluctuations as small as $10^{-11}\,\text{W}/\sqrt{\text{Hz}}$. The state of the art for uncooled IR sensors of this size is about 20 times larger than this, and therefore this combination of structure, transducer, and amplifier should not be limited by the ability to measure the current.

What about the performance of the displacement transducer? If we wish to be competitive with state-of-the-art uncooled IR sensors, we must be able to detect fluctuations in IR power in the range of $10^{-10}\,\text{W}/\sqrt{\text{Hz}}$. For this physical configuration, a power fluctuation of $10^{-10}\,\text{W}$ gives rise to a temperature change of

$$\Delta T = \frac{\Delta P}{G} = \frac{10^{-10}\,\text{W}}{3.8 \times 10^{-4}\,\text{W/K}} = 2.6 \times 10^{-7}\,\text{K} \quad (11)$$

From Eq. (7) above, we have

$$\frac{\partial h}{\partial T} = 1.3 \times 10^4\,\text{Å/K}$$

Therefore, we need to detect deflections as small as 0.003 Å. The implications of this requirement for tunneling or capacitive transducers will be discussed below.

This model for the tunneling IR sensor includes several approximations. The actual cell geometry includes tilted internal sidewalls. Because of these tilted sidewalls, the volume of the cell has probably been underestimated by 10% to 20%. However, we have also neglected the thermal conductivity from the air in the cell to these tilted sidewalls, which is expected to result in a 10% to 20% underestimation of thermal conductance. These geometric approximations probably result in an error in the time constant of the device on the order of 10%. At this point in our calculations, we are mainly interested in determining if this device will have useful performance, and therefore such errors are not of concern.

We also assumed that the ideal gas in the cell expands freely on heating. This assumption ignores the stiffness of the membranes. To see if this is a reasonable approximation, let us calculate the stiffness of the membranes and see if a 1% pressure change produces adequate force to cause a 1% volume change. The membrane actually used in this device features a circular corrugation with a 1-mm inside diameter. The effect of the currugations is to extend the linear range of deflection of the membrane. Ordinarily, a flat membrane begins to exhibit nonlinearities in load-deflection when the deflection becomes comparable to the thickness. By including the currugations, the behavior should be linear for deflections up to 10 times the thickness.

The region outside this corrugation is under tensile stress, and can be assumed not to deflect as much as the unstressed region in the center. We will assume that all of the deflection occurs within the corrugations, resulting in a slight overestimate of the actual diaphragm stiffness. The stiffness of an unstressed circular diaphragm is (Young, 1989)

$$k = \frac{64\pi}{12} \frac{Et^3}{r^2} \text{ (N/m)} \tag{12}$$

where $E$ is Young's modulus for $Si_3N_4$, $t$ is the thickness of the diaphragm, and $r$ is the radius. Young's modulus for $Si_3N_4$ (Petersen, 1982) is $1.9 \times 10^{11} \text{ N/m}^2$, and therefore the stiffness of this 1-mm-diameter diaphragm is 4.4 N/m. A 1% increase in the atmospheric pressure on this diaphragm introduces a force of about $8 \times 10^{-4}$ N, which would lead to a deflection of 180 μm. This much expansion would represent a 50% volume increase. A 1% change in temperature will cause the gas to expand until either the pressure or the volume changes by 1%. Since the 1% volume change only requires a deflection of 4 μm, whereas the 1% pressure change would induce a deflection of 180 μm, the expansion is clearly limited by volume change. Therefore, it is a good approximation to ignore the diaphragm stiffness in favor of the gas expansion in the previous calculation.

These thermodynamic calculations show that a micromachined Golay cell can be expected to have a simple thermal response with very high sensitivity. The use of Si micromachined materials, such as $Si_3N_4$, enables large-scale fabrication of this sensor within ordinary Si micromachining foundries.

The mechanical characteristics of this sensor can be expected to be greatly superior to that of the traditional Golay cell. Silicon nitride membranes of this size have been shown to withstand very large accelerations and pressure differences as large as 1 atm. Therefore, this device can be expected to be very robust. In contrast, conventional Golay cells are very fragile and can be broken by dropping onto almost any surface and by exposure to bright light.

## 3. NOISE MODEL AND CONSIDERATIONS

Previously, we have shown that noise associated with use of an operational amplifier to measure tunneling current would not prevent this device from meeting the state of the art. There are other noise sources in thermal IR detectors that need to be evaluated before an accurate performance prediction can be made. The fundamental sources of noise in the tunneling IR sensor are similar to those for all thermal IR sensors: expected fundamental noise contributions from the transducer and circuit, thermal noise from temperature fluctuations in the gas, and fluctuations in the absorption and emission of photons from and to the background. Each of these noise sources contributes fluctuations in the output signal of this sensor, and the best way to express them is as equivalent noise in IR power. Therefore, we express these independent noise sources as noise equivalent power (NEP) levels. The NEP due to a given noise source is simply the IR signal power, which would be detectable in the presence of that noise source with a unity signal-to-noise ratio (SNR) within a 1-Hz bandwidth. The total NEP is then the square root of the sum of the squares of the individual NEPs.

In tunneling, electrons traverse a quantum mechanical barrier as individual quanta. Therefore, there will be noise in the current associated with the statistical nature of the quantized transport. This noise source is commonly referred to as shot noise, and it appears as a significant noise source if the sensitivity of the detector is small. For reference, fluctuations in a current of 1.5 nA are

$$I_n = \sqrt{2eI} = \sqrt{2(1.6 \times 10^{-19})(1.5 \times 10^{-9})} = 2 \times 10^{-14} \, A/\sqrt{Hz} \quad (13)$$

which correspond to power fluctuations on the order of $10^{-12}$ W. This turns out to be a very small contribution, primarily because the sensitivity (0.02 A/W) is very high. As an aside, the tunneling junction represents a "resistance," and therefore the contribution due to Johnson noise in this system is worth mention. If the device were unbiased, there would be some quantity of current propagating in both directions through this device. On average, these currents cancel out. However, in any finite time interval, there will be statistical fluctuations in both of these currents that do not cancel out, giving rise to Johnson noise. In our device, we impose a bias on the situation, which gives rise to a large imbalance in the currents. The statistical fluctuations in this imbalance also give rise to noise, expressed as shot noise, above. Which is larger? If we were to treat these as independent phenomena, the Johnson noise would be given by

$$V_n = \sqrt{4kBTR\,\Delta F} = \sqrt{4(1.38 \times 10^{-23})(300)(10^8)\Delta F}$$
$$= 1.3 \, \mu V/\sqrt{Hz} \quad (14)$$

Taken across the 100-MΩ resistor, this corresponds to a noise current of about $10^{-14}$ A/$\sqrt{\text{Hz}}$, which is two times smaller than the shot noise current we calculated previously. It turns out that as long as the bias voltage across the tunnel junction is larger than 25 mV, the statistical fluctuations in the imbalance are larger than the statistical fluctuations in the balanced currents. Now, in fact, these are really related phenomena. One way to think of all of this is that the Johnson noise is nothing more than the "zero-bias shot noise." A proper accounting would analyze the fluctuations in the current flow in both directions as a function of the bias, and regard the statistical fluctuations accordingly. In practice, it is usually very accurate to consider only the larger of the shot and Johnson noise contributions as expressed previously. For the case of the tunneling transducer, shot noise is the larger of the two and therefore is the term we consider.

Another important noise source in thermal IR detectors arises in the form of temperature fluctuations. Heat transport through the thermal resistance is a dissipative (nonreversible) process. All dissipative processes introduce fluctuations. This situation is exactly analogous to the generation of voltage noise (Johnson noise) in a resistor. For thermal IR sensors, this noise source generates an error in IR power measurement given by (Jones, 1953; Hadni, 1967)

$$P_{\text{thermal}} = \sqrt{4k_B T^2 G} \tag{15}$$

The similarity to the expression for Johnson noise in a resistor is no coincidence, since this noise source may be attributed to statistical imbalances in the transport of heat-carrying quanta (phonons) across the thermal resistance. It is interesting to note that this noise source leads to an error in measured power that is frequency-independent. This is because the noise-induced temperature fluctuations are subject to the same thermal response as are the signal-induced temperature fluctuations.

Finally, the incoming photons are also subject to statistics. This noise source is generally referred to as *background noise* or *photon noise*, and there is nothing the sensor designer can do about it because it is mixed with the incoming signal. If the device is limited by photon noise, then it represents the final state of the art.

Altogether, the theoretical NEP for this IR sensor may be expressed as the square root of the sum of the squares of these independent noise sources

$$(\text{NEP})^2 = 4k_B T^2 G + 16A\sigma k_B T^5 + \frac{2eI(G^2 + (\omega C)^2)}{I^2 \alpha^2} \tag{16}$$

where $k_B$ is Boltzmann's constant ($1.38 \times 10^{-23}$ J/K), and $s$ is the Stephan-Boltzmann constant ($5.67 \times 10^{-8}$ W/m²K⁴). For the design parameters

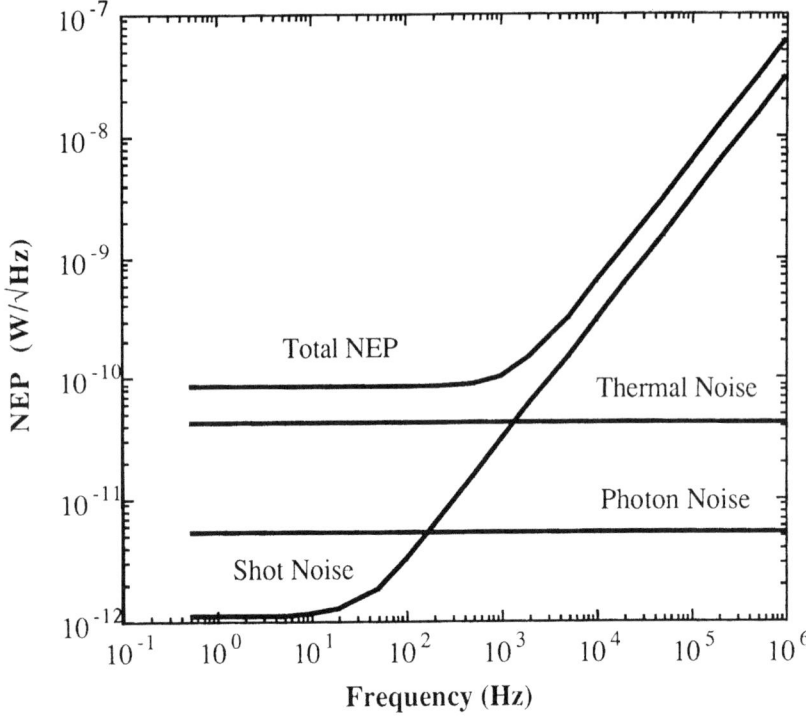

FIG. 3. Plot of the expected contributions to the noise equivalent power (NEP) as a function of frequency for the tunneling infrared (IR) sensor described in the model. For comparison, the measured NEP for a comparably sized (2 mm) pyroelectric IR sensor is shown.

described previously, the NEP of this sensor is expected to be dominated by thermal noise at frequencies up to about 1 kHz. When additional responsivity reductions due to the actual IR absorber efficiency of 60% are taken into account, the NEP for signals below 1 kHz is equal to about $7 \times 10^{-11}$ W/$\sqrt{\text{Hz}}$. A plot of these noise sources as a function of signal frequency for this sensor design is shown in Fig. 3.

It is interesting to note that the predicted NEP is not expected to degrade at frequencies immediately above the reciprocal of the thermal time constant at 20 Hz. The dominant noise both below and above the thermal roll-off frequency is due to temperature fluctuations. As mentioned previously, this particular noise source has a frequency distribution that is affected by the thermal roll-off in the same manner as the responsivity. Therefore, the SNR in the presence of this noise source is expected to remain frequency-independent above the thermal roll-off. Since thermal noise dominates the

NEP at low frequencies, the total NEP of this sensor is expected to remain frequency-independent until noise from some other source, such as the amplifier or the transducer exceeds the temperature fluctuation noise.

In practice, an additional, nonfundamental noise source appears in the operation of this device. This noise has a $1/f$ character and represents a larger error than do all of the previously calculated noise sources by a factor of from two to five depending on operational conditions. We have observed this noise source in all of our tunneling displacement transducers, and attribute it to fluctuations in the tunnel current because of migration of Au atoms or adsorbates on the tunnel electrodes during operation. This noise source is currently being studied in an attempt to understand how it may be reduced or eliminated. As will be shown later, the operation of the tunneling IR sensor presently is limited by this nonfundamental transducer-related noise source. The preceding discussion still represents the fundamental limitations to the performance of this device, and it is useful in calculating the performance of this sensor as a function of the device geometry.

### III. Tunneling Transducer Background

To convert the physical displacement of the membrane to an electrical signal, we use a tunneling displacement transducer. In this section, we describe the characteristics of the tunneling transducer and provide some justification for its use in this device.

Electron tunneling through vacuum or air barriers has been the subject of theoretical description since the beginning of quantum mechanics. Calculations of transmission probabilities through one-dimensional barriers have been standard fare in textbooks on quantum mechanics for more than 50 years. The development of a physical device that allows verification of simple vacuum tunneling theory took many years because tunneling only occurs under circumstances that are very difficult to obtain and maintain.

In principle, tunneling can occur between any pair of metal electrodes. A simple switch, for example, offers a suitable geometry: a pair of metal electrodes that can be brought into contact, and then separated. If an accurate electrical system was constructed to measure the resistance of the switch while it is being closed, the classic prediction is that the resistance would be infinite until contact, and then it would be finite (and small if the switch were a good one).

From a quantum mechanical perspective, the resistance between two metal contacts is never infinite. It is extremely large when the contacts are separated by a macroscopic distance. In this case, a macroscopic distance is any distance greater than a few nanometers! Then, when the distance becomes as small as 1 nanometer, the resistance drops to less than 1 GΩ, and decreases very rapidly as the separation is reduced further. Therefore, we may understand tunneling as the nonclassical behavior of switch contacts when separated by distances on the order of 1 nm.

In practice, the arrangement of a separation of 1 nm is very difficult and can only be accomplished through the use of precise actuators and feedback control. The first successful experimental demonstration of such a system was at IBM Zurich in 1982 (Binnig and Rohrer, 1986). In the scanning tunneling microscope (STM), piezoelectric actuators were used to position a sharp metallic tip about 10 Å above a metallic surface. This tip was then raster-scanned across the surface using the same piezoelectric actuators. When a 100-mV dc voltage bias is applied between tip and surface, a tunneling current of about 1 nA can be measured. As the tip scans across the surface, the surface topography causes the separation between the tip and surface to change. Variations in the tip–surface separation appear as variations in tunneling current, which are recorded to produce high-resolution images of the surface.

This STM tunnel junction may be evaluated as a new displacement transducer. For this device, an electronic circuit capable of detecting 1% variations in a 1-nA current from 100-MΩ source would be good enough to detect deflections as small as 0.003 Å. If the detection were limited by shot noise in the tunnel current, the minimum detectable deflection would be $2 \times 10^{-5} \text{ Å}/\sqrt{\text{Hz}}$. This capability for the detection of sub-nm deflections has previously been restricted to large, sophisticated measuring instruments.

It is especially important to note that the displacement sensitivity of the tunneling transducer is not directly dependent on the overall dimensions of the transducer. In this device, the only important effect is that, somewhere, a pair of atoms on opposite sides of the contacts are within 1 nm of each other. Because of this, it is *not* necessary to tailor the shape of the tip of a tunneling sensor. No matter what the actual shape is, somewhere, a pair of atoms will be closer than the others, and that pair will dominate the system.

Because the important effect happens between a single pair of atoms, the sensitivity of the tunneling transducer is independent of the size of the mechanical structure used to support the electrodes. As a result, the tunneling transducer may be miniaturized without direct loss of displacement resolution. This represents an important opportunity for miniaturization of a broad class of sensors. A position sensor based on electron tunneling has been incorporated into the design for an accelerometer (Waltman and Kaiser, 1989; Kenny *et al.*, 1991a; Baski *et al.*, 1988), an IR sensor (the subject of this chapter) (Kenny *et al.*, 1991b, 1992, 1996), and a magnetometer (Wandass *et al.*, 1989). Several other applications also are being considered. In addition, MacDonald *et al.* have developed single crystal-based fabrication techniques for miniaturized tunneling structures (MacDonald, 1992; Yao *et al.*, 1992; Zhang and MacDonald, 1993), and Fujita *et al.* have used surface micromachining to construct an operational lateral tunneling sensor (Kobayashi *et al.*, 1992). Recent work at the University of Michigan has included demonstration of a pressure sensor that has been operated in the field emission regime, and may soon be extended to the tunneling regime (Yeh and Najafi, 1994). Theoretical issues surrounding the fundamental limits to displacement detection with tunneling transducers also have been studied (Presila *et al.*, 1992; Bocko *et al.*, 1988; Gabrielson, 1993; Yurke and Kochanski, 1990).

1. COMPARISON OF TUNNELING AND CAPACITIVE TRANSDUCERS

A comparison between tunneling transducers and capacitive transducers for measurement and control of small displacements may be carried out. If one were to construct a capacitive transducer for this IR sensor application, the sense electrodes would likely have dimensions no larger than $0.5 \times 0.5$ mm, and a separation between electrodes of no less than $10 \mu$m. The capacitance for this arrangement of electrodes then is 0.2 pF. In order to measure the smallest IR power variations that are allowed by thermodynamics in this device, $7 \times 10^{-11}$ W/$\sqrt{\text{Hz}}$, it is necessary to detect temperature variations as small as

$$\Delta T = \frac{\Delta P}{G} = \frac{7 \times 10^{-11} \text{ W}}{4.1 \times 10^{-4} \text{ W/K}} = 1.7 \times 10^{-7} \text{ K}/\sqrt{\text{Hz}} \qquad (17)$$

which corresponds to less than a 1 part in $10^9$ change in temperature/$\sqrt{\text{Hz}}$. Since the gas cell is 400-$\mu$m thick, the transducer must be capable of measuring a diaphragm deflection of 0.003 Å in a 1-Hz bandwidth.

For a 0.2-pF capacitor with an electrode separation of $10 \mu$m, this corresponds to a capacitance change of about $10^{-8}$ pF. Therefore, the required SNR for the capacitance measurement must be at least $2 \times 10^7/\sqrt{\text{Hz}}$.

There are many approaches to the measurement of capacitance, each of which offers various performance advantages and imposes various system requirements. Among the easiest to describe, study, and implement are those based on relaxation oscillators — a measurement technique based on the measurement of the decay time of an RC filter that uses a reference resistor and the transducer capacitance as the elements. Capacitance measuring systems based on relaxation oscillators are ultimately limited by Johnson noise due to the resistance of the capacitor discharge circuit (Spencer et al., 1988). In this case, the voltage on the capacitor during the discharge fluctuates as

$$V^2 = 4k_B TR\Delta F. \tag{18}$$

Since the low-pass filter of the discharge resistance and the sense capacitance have a finite bandwidth ($2\pi\Delta F = \pi/2RC$), the voltage noise is

$$V^2 = \frac{k_B T}{C} = \frac{1.38 \times 10^{-23} 300}{2 \times 10^{-13}} = (450\,\mu V)^2 \tag{19}$$

which is spread across the frequency spectrum from dc to the frequency of the relaxation oscillator. Assuming that the electronics implementation limits the amplitude of the bias oscillation voltage to 5 V, we find an SNR of $(5\,V/450\,\mu V) = 1.1 \times 10^4$, within the full bandwidth of the oscillator. From the earlier calculation of required SNR in a 1-Hz bandwidth, we may calculate the required frequency of the relaxation oscillator:

$$f = \left[\frac{2 \times 10^7}{1.1 \times 10^4}\right]^2 = 3.3\,\text{MHz} \tag{20}$$

Therefore, thermodynamic limits to the minimum capacitance measurement by a relaxation oscillator require that the oscillator run at frequencies above a few MHz. Given the ability to perform capacitance measurements in this frequency range, performance comparable to the tunneling transducer indeed is feasible.

However, the characteristics of the oscillator and the amplification circuitry for this capacitance measuring system must include stability at the level of 1 part in $2 \times 10^7$ within the measurement bandwidth. This accuracy imposes impractical cost requirements for what otherwise is expected to be an inexpensive device.

The tunneling transducer avoids this practical limitation on electronics stability for two reasons. The transducer is operated with a position-control feedback system that attempts to maintain a working separation of 10 Å that is at least three orders of magnitude smaller than could be expected from a nonfeedback-controlled micromechanical structure. Even with feedback control of a capacitive system, electrode roughness and dielectric breakdown introduce a minimum electrode separation near 1 $\mu$m.

In addition, there is built-in gain in the tunneling transducer that results from the nonlinearity of the tunneling process, which quantify as follows. We have shown that a 1% change in tunneling current results from a 3-mÅ change in electrode separation. The 3-mÅ separation change represents a 0.03% (0.003 Å/10Å) change in electrode separation. Since a 0.03% separation change produces a 1% change in signal, the transducer itself offers a "gain" on the order of 33. The combination of measuring changes in a 1000 times smaller average electrode separation and adding a gain on the order of 33 times from the transducer produces an advantage of 33,000 times over a linear capacitor with respect to electronics stability requirements. Therefore, to meet the resolution requirements of this application, the tunneling transducer requires stability of only 1 part in 1000, which is achieved easily in inexpensive electrical circuits. This large advantage is the main reason tunneling transducers enable accurate measurements of sub-Å deflections with very inexpensive circuitry.

## 2. Tunneling Transducer Design Considerations

This section outlines some of the constraints and considerations imposed on the design of the IR sensor due to use of a tunneling displacement transducer.

Since tunneling takes place between the atoms on the surfaces of the tunneling electrodes, it is critically important that the electrode material be clean and metallic. In all of our successful experiments, we have used Au films as tunneling electrodes. Gold is a nearly ideal metal for this purpose because it does not undergo chemical reactions to form insulating surface layers when exposed to air.

Because tunneling only occurs when the tip is nearly 10 Å from the counterelectrode, the gap between the electrodes must be controlled by feedback during operation. This is usually accomplished by measuring the tunneling current, comparing it with a reference value, and applying correction signals to an electromechanical actuator. This actuator functions by applying forces to the moving sensor element to keep the tip-element gap constant (force-rebalance design). If the gain of the transducer, circuit, and actuator is sufficient, the tunneling current will be maintained in the presence of external disturbances. By monitoring the feedback signals produced by the control circuit, the signal forces applied to the sensor element can be detected.

Through use of the tunneling transducer, the extreme sensitivity to displacement allows the amplification and measurement of small deflections with relatively simple circuitry. As a result, the characteristics of the electromechanical actuator used to control the separation between the tunneling electrodes often impose the dominant limitations to tunneling system performance. We have developed actuators that use electrostatic deflection of micromechanical components to control their position.

It is important for any transducer to be insensitive to environmental sources of noise. As is well known, early STMs were extraordinarily sensitive to vibration and required the construction of large, complex vibration isolation systems for their use. Since the tunneling transducer is fundamentally a mechanical structure, the sensitivity to vibration is best eliminated through careful mechanical design. When a mechanical element is subjected to an acceleration at frequencies below its resonance, the amplitude of deflection is inversely proportional to the square of the mechanical resonant frequency. Therefore, sensitivity to vibration is best reduced by increasing the mechanical resonant frequency of all elements of the transducer. The micromachined elements used in this device accomplish this goal easily. However, these actuators have small range of deflection ($<5\,\mu$m), which precludes their use for coarse approach between tunneling electrodes. By using Si micromachining techniques to better define the initial separation between the tip and membrane, it is possible to accomplish the needed coarse approach during assembly.

The active element for the tunneling IR sensor consists of a thin $Si_3N_4$ diaphragm. For the IR sensor, the dimensions of the diaphragm are $2\,\text{mm} \times 2\,\text{mm} \times 0.5\,\mu\text{m}$. A series of concentric circular corrugations with an inside radius of 0.5 mm are used to release the remaining tensile stress. It is expected that the dynamic characteristics of the diaphragm will be dominated by the unstressed circular region inside the corrugations. The resonant frequency and stiffness for an unstressed circular diaphragm of these dimensions are calculated to be about 18 kHz and 4 N/m, respectively.[28]*

Another important advantage to the use of the membrane as a tunneling contact is that this element is not rigid and does not carry much momentum.

*The resonant frequency is approximately given by

$$f = \frac{10.2}{2\pi}\sqrt{\frac{Et^3 g}{12\,wd^4}}$$

(from Young, 1989, Table 36, p. 716) where $E = 1.9 \times 10^{11}\,\text{N/m}^2$, $w$ is an average of the densities for silicon nitride and gold. The stiffness is given by

$$k = \frac{64\pi}{12}\frac{Et^3}{r^2}$$

(derived from Young, 1989, Table 24, p. 429).

Therefore, the amount of force between the tunneling electrodes during a "crash" is limited. For use of tunneling as a transducer in any practical application, we strongly recommend that one of the contacts be supported by a compliant element. Early experiments on devices that did not feature a compliant electrode always resulted in device failure due to mechanical destruction of the tip or counterelectrode during a crash.

For the purposes of this sensor development, it was necessary to design and fabricate the components with a geometry that allows easy assembly and operation. To meet these requirements, a simple bulk micromachined approach to fabrication of separate sensor elements was chosen.

## IV. Tunneling Infrared Sensor Design and Fabrication

The basic design of the tunneling IR sensor is shown in Fig. 4. The sensor consists of three parts, the two halves of the gas cell and the substrate. A thin metal film that acts an IR absorber is suspended on a membrane in the middle of the cell. The upper membrane is perforated with a small pinhole that eliminates the dc response of the sensor. The lower membrane is fabricated by a process that introduces corrugations in the membrane, extending the linear range of the deflection of the membrane. The lower membrane is also coated with a Au film that acts as the ground contact for both the tunneling and deflection electrodes. The substrate features a fixed

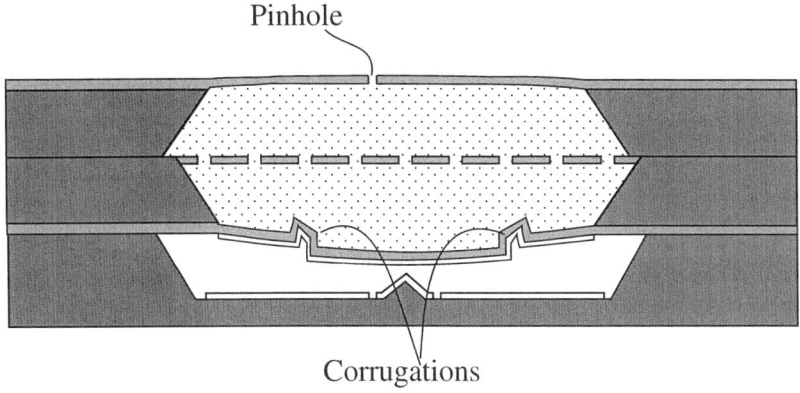

FIG. 4. This drawing shows the design of the tunneling sensor modified to include the corrugated lower diaphragm, which allows deflection through larger range, and the pinhole in the upper membrane, which allows the operating point to be independent of sensor temperature.

Si tip coated with a Au film and a deflection electrode that is placed in a recess surrounding the tip.

The IR sensor is operated in the following manner. An increasing voltage is applied by the feedback circuit to the deflection electrode. When this voltage is large enough to draw the lower membrane to within 10 Å of the fixed tip, a tunneling current appears and is sensed by the feedback circuit, which halts the increase in deflection voltage and applies variations only as needed to keep the current constant. If an oscillating IR signal is brought to bear on the sensor, it is absorbed by the thin metal film on the intermediate electrode, which heats the gas and causes an oscillatory thermal expansion. This expansion applies a force to the lower membrane, which is sensed by the feedback circuit. The feedback circuit reacts by adding an oscillation to the deflection voltage, so that the sum of the pressure force and the electrostatic force on the lower membrane is constant, and the position of the membrane is unchanged. By monitoring the amplitude of the voltage oscillation applied to the deflection electrode, we measure an electrical quantity that is proportional to the IR power oscillation.

*Nitride Membranes.* Double-polished Si wafers (3-inch diameter, $\langle 100 \rangle$ orientation, lightly doped (phosphorus), 200-$\mu$m thick) are coated on both sides with 0.7 $\mu$m of low-pressure chemical vapor deposited (LPCVD) $Si_3N_4$. This $Si_3N_4$ is doped with extra Si to enable better matching of the thermal expansion coefficient of the film with the underlying substrate. The resulting film have tensile stress less than 100 MPa, which is more than 10 times lower than stoichiometric LPCVD nitride films would have.

*Pinholes.* One side of these wafers is coated with a 200-Å Aluminium (Al) film. A layer of e-beam resist is patterned with individual 0.5-$\mu$m-diameter pinholes, which are transferred to the Al film by etching in a dilute potassium hydroxide (KOH) solution. The wafer is then etched in a carbon tetrafluoride ($CF_4$) plasma, which etches the pinholes into the nitride film.

*Membrane Release.* A standard photoresist–plasma etch process is then used to etch a series of diagonal openings filling a 2-mm square area on the opposite surface of the wafer. Finally, the wafer is submerged in a 40% KOH, 80°C solution, which etches Si through the pattern of diagonal lines at a rate of about 70 $\mu$m/h. Because these thin openings in the mask are diagonal to the $\langle 100 \rangle$ crystal planes, the etch undercuts the lines, forming a square pit that surrounds the openings, and leaving the freestanding ribbons of nitride between the lines behind. After 3 to 4 h, the etch stops when it reaches the nitride film with the pinhole on the opposite side of the wafer.

*Corrugation Process.* The lower half of the cell is formed in a manner similar to the upper membrane. First, the wafers are passed through a phororesist process that exposes a set of 20 concentric circular openings. The diameter of these openings ranges from 1.0 to 1.4 mm, and are each 5-$\mu$m wide with 5-$\mu$m spacings. These wafers are then etched in a $SF_6$ plasma, removing about 1 $\mu$m of Si from the exposed circular openings. After stripping of the photoresist, the wafers are coated with low-stress nitride, as described previously. A standard photoresist–plasma etch process is used to open 2-mm square windows on the side of the wafer opposite the original trench etching. The wafers are wet-etched in KOH as above until the nitride membranes are released. Because these membranes were grown on a surface with etched circular trenches, the released membranes are corrugated.

*Tip Formation Process.* The substrate wafers are also standard Si wafers (3-inch diameter, $\langle 100 \rangle$ orientation, lightly doped, 400-$\mu$m thick), which are coated by silicon dioxide ($SiO_2$), grown by wet oxide diffusion. These wafers are patterned in a standard photoresist process, etched in $CF_4$, and wet-etched in KOH or ethylene diamine pyrocatechol (EDP) to form the 50-$\mu$m deep recess that surrounds the tips. The tips are formed during this process by undercutting a 40-$\mu$m square of mask material during the wet recess etch. Because of the selectivity of the wet etchant to certain crystal planes of Si, this process produces pyramid-shaped tips, 50-$\mu$m tall, with ends that are either sharp with radii of curvature on the order of 1 $\mu$m, or flat-ended with mesas of diameter 2 to 5 $\mu$m. After etching, these surfaces are passivated with 1 $\mu$m of thermally grown $SiO_2$, and a Au electrode pattern is created by a modified lift-off technique. A scanning electron micrograph of a typical tip and electrode pattern that results from this process is shown in Fig. 5. It is important to note that we do not employ any tip sharpening techniques. We have found that, for the purposes of a *z*-component transducer, controlling the "sharpness" of the tip is unimportant for sensor performance. We have used tips with a 1- to 5-$\mu$m radius of curvature such as that in Fig. 5, or we have used tips that have a 5-$\mu$m flat mesa on the end. Sensors with "sharp" or "flat" tips have had indistinguisable behavior and performance. The exponential nature of the tunneling process virtually guarantees that the observed current will be dominated by individual atoms on the tunneling electrodes, whatever the shape of the electrode surfaces.

*Substrate Metal Process.* The substrates are metallized by lift-off of electronic-beam-evaporated Au films. In lift-off, photoresist is patterned, metal is deposited, and the photoresist is dissolved, "lifting" the parts of the pattern that were deposited on photoresist. Since lift-off requires strong adhesion between the Au and the $SiO_2$ coating on the tip and surrounding

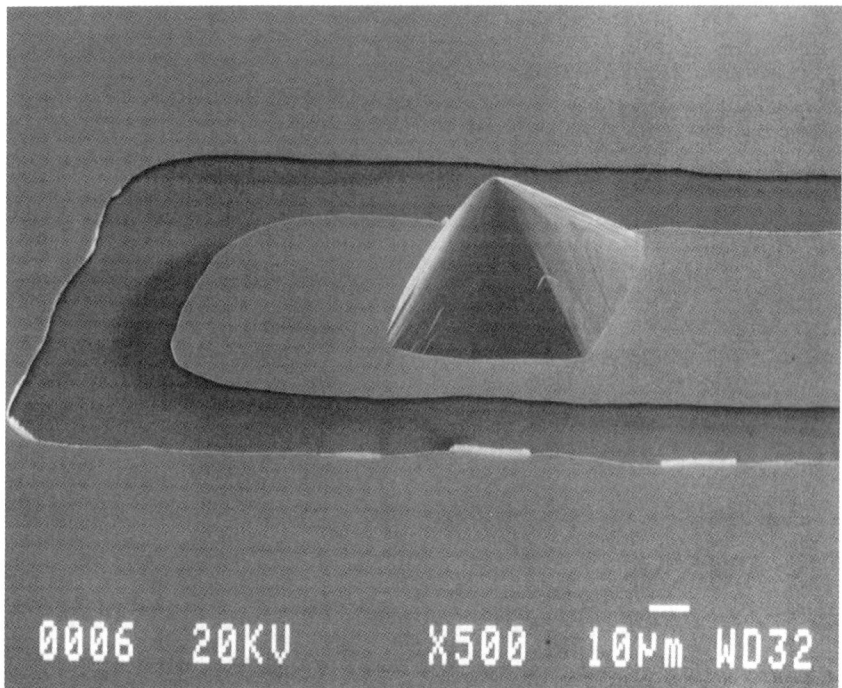

FIG. 5. This scanning electron microscope (SEM) photograph shows the shape and size of one tip that was fabricated for the tunneling infrared sensor. Also showing is the gold electrode that coats the tip and the deflection electrode that surrounds the tip. These two electrodes are separated by a 20-30-$\mu$m-wide region of bare silicon dioxide ($SiO_2$). This particular tip is rather attractive looking; generally, tips that feature flatter ends are also used with no quantitative difference in sensor performance.

surfaces, it is common to use an intermediate adhesion layer. We have found that deposition of a Titanium (Ti) adhesion layer, followed by a Platinum (Pt) diffusion barrier layer, and then the Au layer results in clean tunneling electrodes. In this case, the Pt layer is thermodynamically stable and prevents migration of Ti through to the Au surface. Tunneling sensors based on this electrode combination have already been operated for as long as 60 months in air without evidence of electrode degradation. Other combinations of adhesion layers and diffusion barriers can be made to work; however, it is *extremely* important to guard against the possibility of contaminant diffusion to the Au surface.

*Infrared Absorber Process.* The IR absorber is deposited on the layer of nitride that was released by wet etching through the thin diagonal openings.

This nitride layer resides in the middle of the gas cell after assembly. An optimal IR absorber would have no heat capacity, no thermal conductivity, and 100% efficiency. A fair approximation to these ideal characteristics is available through use of a very thin metal film. As the thickness of metal films varies from 0 to macroscopic dimensions, the films evolve from mostly transparent to mostly reflecting. In the middle, where the transmittance is equal to the reflectance, a metallic film offers some absorption of radiation due to resistive dissipation of electric field-induced motion of electrons. This method for absorption of IR radiation is widely used in thermal IR sensors, and features as much as 50% absorber efficiency, which is approximately independent of radiation wavelength over a broad range in the IR spectrum (Hadni, 1967).

This absorber mechanism is optimized for thin metals films with sheet resistances equal to $188\,\Omega$/square. It can be difficult to produce metal films with stable resistance in this regime. We have finally settled on use of Pt films as the IR absorber. Platinum offers relatively high conductivity, meaning that appropriate resistive films will need to be in the 100-Å thickness range. However, Pt offers the advantages that it may be controllably evaporated at low rates and is extremely stable after evaporation. To determine the appropriate thickness for the Pt films to be used as absorber, we have measured the IR transmission for a series of Pt film samples. In these experiments, 2 to 20 microns broadband IR radiation from thermal IR sources near room temperature is incident on a bare nitride film or a nitride film with a sample thickness of Pt. The relative transmission is recorded as a function of Pt film thickness and plotted in Fig. 6. This data show a smooth variation of transmission as a function of film thickness. Values of transmission ranging from 60% to 65% were achieved for film thicknesses measured at 35 to 40 Å; this range of values of transmission is desired because radiation transmitted through the membrane is reflected by the tunneling electrode on the bottom of the cell and is offered another chance at being captured by the absorber.

*Assembly Process.* After the fabrication of the individual sensor components is complete, the sensors are assembled. The substrate elements are mounted in standard IC-style chip carriers with rapid-curing epoxy, and connections between the electrodes are made to the pads of the carrier by Au wire bond. The two halves of the gas cell are prepared by coating the surfaces of the components that are to be joined with a slow-drying adhesive, such as ordinary photoresist, using a fine paintbrush. We began using photoresist because it was readily available and have found that it offers several convenient characteristics. After the bonding surfaces are coated with a bead of photoresist, the elements are aligned using a Zeiss contact aligner, and pressed together.

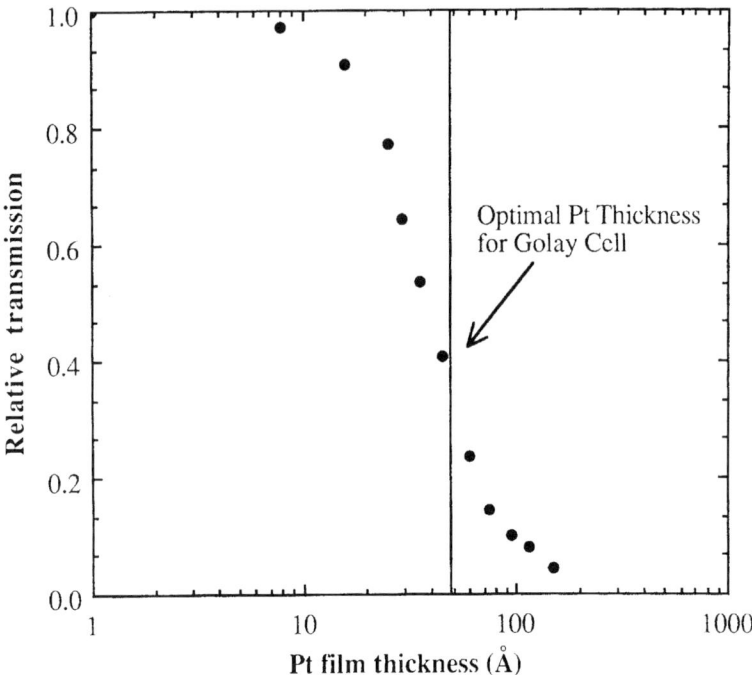

FIG. 6. This plot shows a measurement of the infrared (IR) transmission relative to a bare nitride membrane as a function of the deposited platinum (Pt) film thickness. These measurements used broadband radiation from 2 to 20 microns in wavelength. This plot shows a smooth, repeatable transition from transparent to reflecting. Optimal IR absorption on a single pass occurs when the transmittance is 25%. Since this sensor benefits from a second chance to absorb transmitted light, the optimal thickness occurs for a transmittance closer to 40%, which corresponds to a Pt "thickness" of about 40 Å. The actual thickness of these films is not well known, because the thickness monitor has not been calibrated. The relationship between thickness and IR properties is critically dependent on film morphology, and therefore another calibration would be necessary if a different deposition system were used. These data are useful because they indicate the likely range for useful Pt thicknesses, and because they illustrate the repeatability of the IR properties of these films on nitride membranes.

After mounting of the substrate and bonding of the gas cell, the bonded cell is placed on the substrate and clamped down. The geometry of the sensor and the location and arrangement of electrical connections are shown in Fig. 7. Voltages are applied to the deflection electrodes until tunneling is observed. Then, the lateral position of the cell on the substrate and the

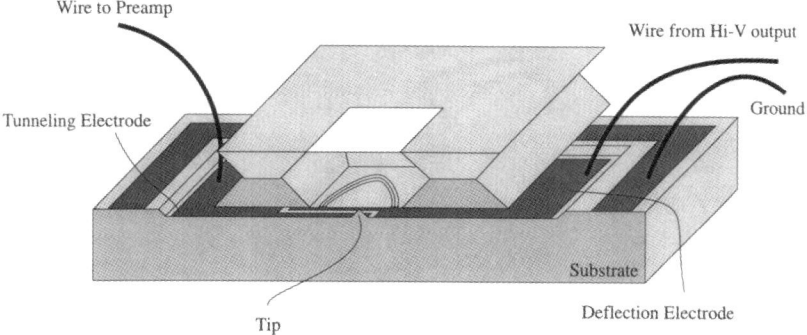

FIG. 7. This drawing shows the design of the assembled sensor. An electrical contact is established between the metal films on the lower surface of the membrane and the upper surface of the substrate. This arrangement allows all the wire bonds to the sensor to be made on the upper surface of the substrate. Contacts to the tip and deflection electrodes are made on the substrate as shown.

position and force of the clamps are adjusted to minimize the deflection voltage required to maintain tunneling. For a typical sensor, this process takes 15 min and reduces the required deflection voltage from over 300 V to less than 200 V. Once a satisfactory (low) operating voltage is achieved, additional quick-curing epoxy is applied to the perimeter of the gas cell, bonding it to the substrate. Once this epoxy is cured, the clamps are removed and the sensor is ready for testing. A drawing of the completely assembled sensor mounted in a chip carrier is shown in Fig. 8.

It is important to note that the fabrication and assembly techniques described previously can be accomplished on low-grade fabrication equipment available throughout industry and academia. The only precision fabrication step is the electron-beam patterning of the pinhole, which may be abandoned altogether in early experiments. The remainder of the lithographic patterns were drawn with 25 $\mu$m features. This entire set of masks for these experiments was printed from postscript files using a commercially available linotronic printer, which may be found at many commercial printing–copying establishments. Typically, a 40 × 40 inch pattern printed at high resolution (allowing feature sizes as small as 25 $\mu$m) can be printed in less than an hour for around $20. For these experiments, the linotronic printer produced patterns on optically transparent plastic films. These films offered sufficient contrast for near-uv radiation to allow contact printing onto ordinary mask blanks or directly onto wafers. It was usually

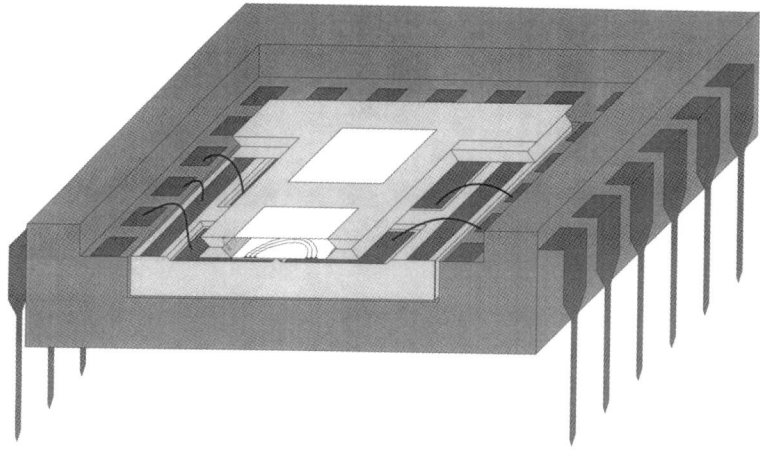

FIG. 8. This illustration shows the mounting of the tunneling infrared sensor prototypes in standard integrated circuit carriers. After assembly is complete, and the sensors are made operational, the upper element is "potted" into the carrier by application of quick-curing epoxy around the perimeter of the device.

necessary to increase exposure time to overcome the uv absorption in the film, as well as to saturate the bubblelike defects that appear throughout the "clear" regions of the pattern. The $Si_3N_4$ membranes used in these experiments are grown by a specialized LPCVD system, optimized for low-stress nitride growth. Such systems are becoming common in industry and academia. Initially, the films used in these experiments were grown at University of California at Berkeley (Berkeley, California); later these films were grown at Stanford University (Stanford, California) and at the Jet Propulsion Laboratory of the California Institute of Technology (Pasadena, California). These institutions are presently providing nitride depositions as a service to outside parties. All other lithographic, ething, and deposition processes were carried out on ordinary, widely available processing equipment.

We encourage all interested parties to attempt fabrication of similarly simple tunnel sensor elements. It is our opinion that tunneling transducers can be produced throughout industry and academia, and that more attempts at production will lead to more rapid development of production-compatible techniques and earlier introduction of tunneling transducers as a commercially viable technology.

In our evidence, the important aspects to the design and fabrication of a tunnel sensor are reliance on a simple mechanical design and attention to the correct materials for the tunneling electrodes. In addition to this, we highly recommend the development of a simple test structure to begin experimentation with materials and control systems in parallel with the development of the final mechanical structure. Without this approach, it is likely that the "optimal," mechanical design will impose materials or other constraints on the fabrication process that will make tunneling difficult or impossible to achieve.

## V. Tunneling Transducer Operation

Conventional scanning tunneling microscope (STM) feedback loops must control complex electromechanical structures with low-frequency resonances. The design and operation of STM feedback loops are complicated by the presence of low-frequency resonances. In this design, the lowest mechanical resonant frequency is above 10 kHz. Therefore, the gain and bandwidth of the electrical circuit used to control the sensor may be substantially larger than that used in typical STMs or in previous tunneling sensors. Because of this, the feedback circuitry for tunneling sensors may be very simple.

A typical feedback circuit used to control the tunneling IR sensor is shown in Fig. 9. This circuit uses a single operational amplifier to produce correction voltages that are added to a high voltage offset and applied to the deflection electrode. This simple version of the circuit is operated in the following manner. A 163-mV tunneling bias is applied to the electrode on the actuator, and the electrode on the tip is grounded through a 10-M$\Omega$ resistor. A large voltage is applied to the deflection electrodes, electrostatically attracting the membrane down toward the tip. When the membrane is within 10 Å of the tip, a tunnel current of 1.5 nA appears. A voltage decrease of 15 mV across a 10-M$\Omega$ resistor in series with the tip always occurs with this amount of tunneling. A single high-input impedance, moderate noise op-amp is then used to compare the tip voltage with a set-point and generate an error signal. This low-voltage, wide-bandwidth error signal is then added to a high-voltage, narrow-bandwidth offset to produce the voltage that is applied to the deflection electrodes. In this simplest circuit, the high voltage signal may be generated by a power supply that is initially adjusted to set the error signal near zero.

A CA3140 amplifier is used for the error amplifier in the circuits studied in this work. Other field-effect transistor (FET) input amplifier also may be

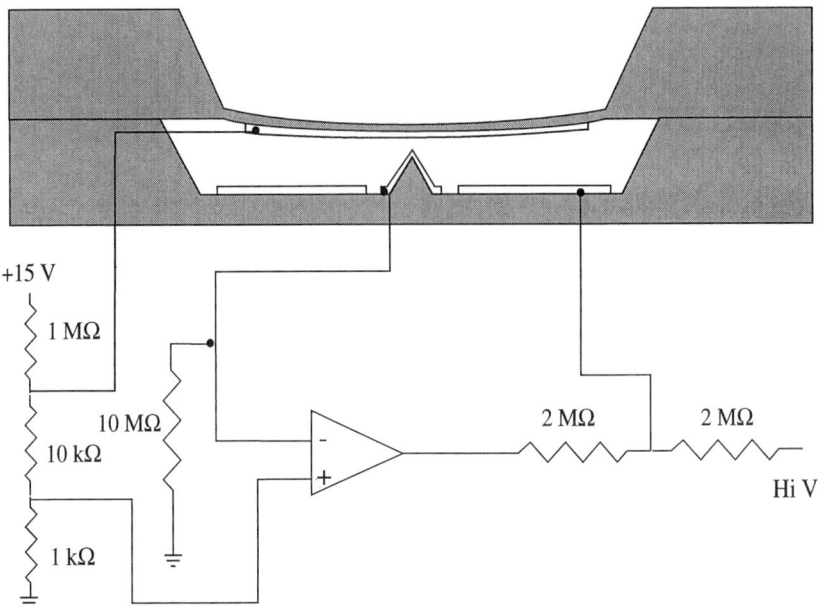

FIG. 9. This drawing shows the design of the simple tunneling transducer feedback circuit. The membrane is biased at a voltage of 163 mV, and the tip is grounded through a 10-MΩ resistor. The setting on a high-voltage power supply is manually increased until some current begins to flow through the tunneling junction. When this current exceeds the set-point of 1.5 nA, the op-amp begins to lower its output, so that the sum of the op-amp output and the power supply is exactly as needed to keep the tunneling current equal to 1.5 nA. If infrared (IR) radiation changes the pressure in the cell, the op-amp reacts by changing toe force on the membrane to keep the current constant. By monitoring the op-amp output, the IR signal may be recorded. This circuit requires occasional manual adjustment of the power supply setting to keep the op-amp output in the middle of its control range.

used. The input impedance of the first amplifier should be substantially higher than is the 100-MΩ source impedance of the transducer. Also, the input current noise of the amplifier should be low enough that it does not dominate the final noise of the sensor. The CA3140 has input noise of $2 \times 10^{-13}\,\text{A}/\sqrt{\text{Hz}}$, which is lower than other noise sources we have observed in operation of these sensors. Since the noise and bandwidth in these systems have not been limited by preamplifier characteristics, we generally have not been concerned with optimization of these elements.

The electrostatic force between a pair of electrodes varies as $(V/d)^2$. This apparent nonlinearity is remedied by the fact that the correction voltage generated by the feedback circuit is added to a large offset voltage. Therefore, the electostatic force is proportional to

$$F \alpha (V_o + V_s)^2 = V_o^2 + 2V_o V_s + V_s^2 \qquad (21)$$

where $V_o$ is the offset voltage ($\sim 100$ V), and $V_s$ is the signal voltage (about 100 mV). Since $V_s \ll V_o$, the last term may be ignored. The middle term is then linear in $V_s$, and is the term that provides the rebalance force for the transducer. For this situation, the linearity is accurate to within 1% as long as $V_s < 1$ V. Since the normal noise levels in this transducer are below 1 mV in a 1-Hz bandwidth, the dynamic range of this sensor is well in excess of 60 dB. This dynamic range is very good compared with that normally expected from open-loop capacitors or strain gauges, and is a result of the use of a force-feedback approach to signal measurement.

These sensors were made operational, and routine characterization were carried out on the transducer alone, as well as on the complete IR sensors. For all the measurements described herein, the devices were operated in laboratory air at atmospheric pressure. Stable tunneling was achieved in a typical device with an average deflection voltage of 120 V. Figure 10 shows a measurement of the voltage noise from the feedback circuit measured at the output of the error amplifier.

This noise spectrum exhibits a typical $1/f$ character, in agreement with tunneling noise spectra seen elsewhere (Moller *et al.*, 1989; Tiedje *et al.*, 1988; Kleint, 1971). The source of the $1/f$ noise in these devices has not been determined experimentally. Migration of individual Au atoms at room temperature has been observed in STM experiments. Others have attributed noise in tunneling in air to migration of adsorbed water molecules through the tunneling region. The adsorbed water can mediate forces between tip and counterelectrode through the formation of a meniscus (Weisenhorn *et al.*, 1989). Either of these mechanisms can be expected to produce randomly timed steps in the feedback signal, which could be expected to produce the shape of the observed spectrum. Also, relaxation of the package that holds the two elements of the structure also can be expected to introduce low-frequency noise. Finally, we have observed that building pressure fluctuations from the air handlers in the adjacent clean room can be a source of low-frequency noise in our measurements. Packaging the sensors in air-filled, stiff-walled containers generally results in reduction in the observed noise spectrum by a factor of five for all frequencies below 1 kHz.

Since the dominant long-term drift effects may be attributed to characteristics of the rapidly curing epoxy, we have begun evaluations of other epoxies that are more widely used in commercial packaging and of those used in the aerospace industry. We certainly expect that better long-term

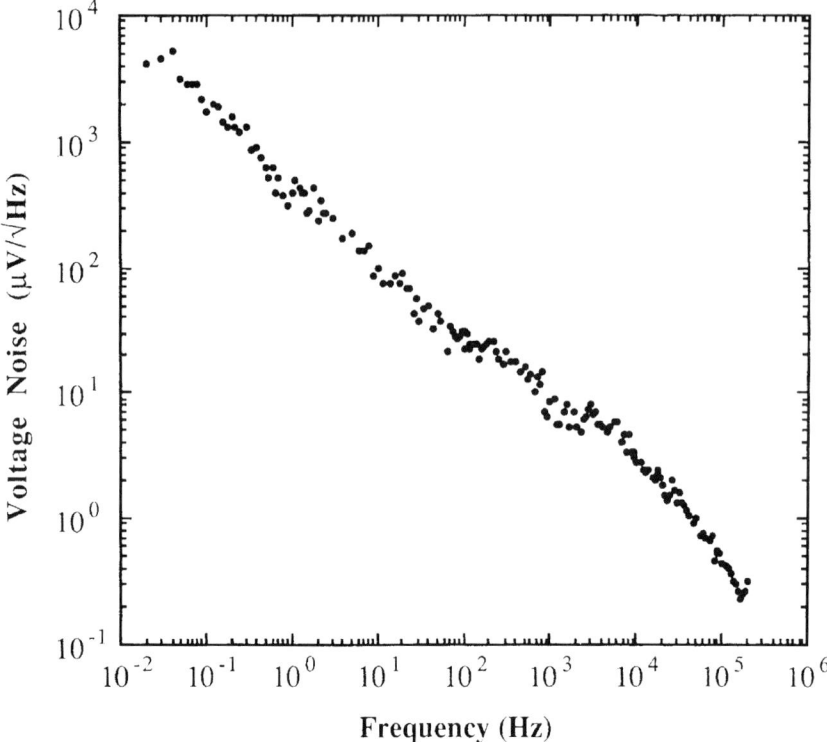

FIG. 10. This plot shows the measured voltage noise at the output of the feedback loop during operation of a tunneling transducer. This noise spectrum is entirely due to nonfundamental processes at the tip or in the mechanical package that holds the structure together. Noise due to electronics or thermal fluctuations would be substantially below that measured here.

drift performance can be obtained. Ultimately, a monolithic device that does not include epoxy can certainly be expected to be much more stable.

It is necessary to perform an independent measurement of transducer stiffness to convert noise in the feedback signal to errors in the position measurement. We used a laser interferometer (Polytec Laser Vibrometer OVF-501-0, Polytec P.I. Inc., 3152 Redhill Suite 110, Costa Mesa, CA 92626) to measure electrostatically induced deflections of the membrane near the operating voltage. A typical sensor was brought to tunneling at an average deflection voltage of 120 V. The deflection voltage was reduced to 110 V, and a 5-V, 40-Hz sinusoidal oscillation was added to the 110-V offset. The resulting oscillation in the deflection of the membrane was measured with the laser interferometer, which recorded an amplitude of 400 Å. Therefore, the transducer responsivity is given by

$$\frac{\partial s}{\partial V} = \frac{400\,\text{Å}}{5\,\text{V}} = 80\,\text{Å/V} \tag{22}$$

With the measured voltage noise of the transducer operating in air in the laboratory, as shown in Fig. 10, the displacement resolution may be calculated. The measured displacement resolution of this transducer is then $0.007\,\text{Å}/\sqrt{\text{Hz}}$ at 10 Hz and less than $0.0001\,\text{Å}/\sqrt{\text{Hz}}$ at 10 kHz. The interferometer measurement is independent of oscillation frequency to beyond 2000 Hz, indicating that squeeze film damping does not affect the dynamics of the diaphragm at these frequencies and amplitudes.

The measured deflection of 80 Å/V corresponds to a membrane stiffness of about 30 N/m, which is substantially larger than the calculated value of 4 N/m. The discrepancy is likely due to remaining tensile stress in the diaphragm, added stiffness of the Au film, and contributions from nonbulk effects. There is no reason to suspect measurement error of this size, because the optical interferometer-based deflection measurement should be very accurate. Therefore, we attribute the discrepancy to real effects not included in the calculation.

Measurements of the effective tunnel barrier height $\phi$ are important indications of the cleanliness of the tunneling electrodes. For tunneling between clean Au electrodes in air, values for $\phi$ of between 0.05 and 0.5 eV are to be expected. The tunneling barrier height was measured in the following manner: A 1-mV, 10-Hz oscillation voltage was added to the reference input of the feedback loop, causing the feedback loop to modify deflection voltages to produce a 0.1-nA oscillation in the tunneling current. To produce this oscillation, the feedback loop adds a small voltage oscillation to the deflection voltage, which was measured as 2 mV in amplitude at the output of the op-amp. From Eq. (4), the height of the tunnel barrier $\phi$ is given by

$$\sqrt{\phi} = -\frac{1}{\alpha I}\frac{\partial I}{\partial s} = -\frac{1}{\alpha I}\frac{\partial I}{\partial V}\frac{\partial V}{\partial s} = \frac{1}{1.025 \times 1.5\,\text{nA}}\frac{0.1\,\text{nA}}{0.002\,\text{V}}\frac{1}{80\,\text{Å/V}} \tag{23}$$

$$\phi = 0.17\,\text{eV}$$

At different times and with different sensors, this measured barrier height may vary by as much as a factor of five.

The effect of variations in barrier height is explained as follows. It is important to remember that the feedback loop applies rebalance forces to maintain the *position* of the membrane. In this case, the system will provide feedback signals as needed to keep the position of the membrane very nearly constant in the presence of external, IR radiation-induced forces. Since the feedback force necessary to achieve balance is always very nearly equal to the signal force, regardless of the actual value of the barrier height, the sensitivity (Volts/Watt) should stay approximately constant. For example, if the gain in the system is 100 at the signal frequency, a reduction of two times in the barrier height produces a change in responsivity $R$ given by

$$R = R_0 \left(1 - \frac{\sqrt{2}}{100}\right) = (0.986)R_0 \qquad (24)$$

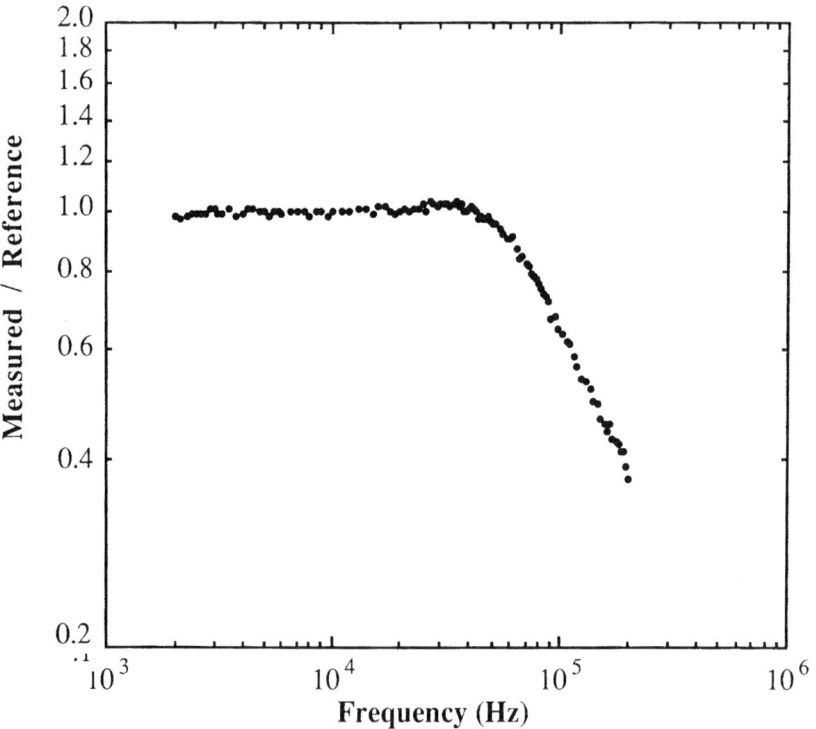

FIG. 11. This plot shows the ratio of measured modulation to input white-noise modulation in the tunneling current as a function of frequency for frequencies between 2 and 200 kHz. The feedback control system is able to reproduce the input white-noise modulation at all frequencies up to 50 kHz.

or less than a 2% reduction. This is a general property of force-rebalanced sensors and explains why force-rebalanced designs are widely used for precision measurements of physical signals. The SNR for the tunnel transducer system is affected by variations in the barrier height, because the noise is proportional to the reciprocal of the square root of the barrier height. For this reason, it is important to use electrodes that offer reasonably high barrier height under operating conditions. For tunneling in air, Au films are the best available choice.

Figure 11 shows a measurement of the bandwidth of the transducer. For this measurement, a white noise voltage modulation is added to the reference input of the error amplifier in the feedback loop. The feedback loop responds by generating an amplified modulation signal at the deflection electrode, which produces modulations in the position of the membrane and modulations in the tunneling current. Within the bandwidth of the transducer system, the measured modulation matches the amplitude and phase of the reference modulation. Figure 11 shows the ratio of the current modulations measured after the buffer to the reference modulation as a function of frequency, recorded by a standard spectrum analyzer. At all frequencies up to 50 kHz, the transducer is able to accurately reproduce the reference modulations. Above 50 kHz, this response begins to roll off because of the limited bandwidth of the preamplifier.

## VI. Infrared Sensor Operation and Testing

After these experiments on the characteristics of the tunneling transducer, compete IR sensors were assembled and made operational. A simple experiment uses a room temperature chopper positioned above the sensor, and a slightly warmer human hand. The sensor is illuminated with an oscillating IR power whose amplitude of oscillation is near $20\,\mu W$. Figure 12 shows an oscilloscope trace obtained during such an experiment, clearly illustrating good response and good SNR. In this experiment, the chopper frequency is low enough to illustrate the thermal equilibrium that is obtained in the trapped gas within a few 10's of ms, yet is fast enough to be unaffected by the acoustic time constant of the pinhole in the gas cell.

Figure 13 shows a measurement of the responsivity of the tunneling IR sensor as a function of chopping frequency. These data show a thermal cut-off frequency of 20 Hz, which is in excellent agreement with that predicted. Above this frequency, the response falls smoothly as $1/f$, indicating that the thermal behavior of this sensor is in good agreement with the simple model at frequencies up to several hundred Hz.

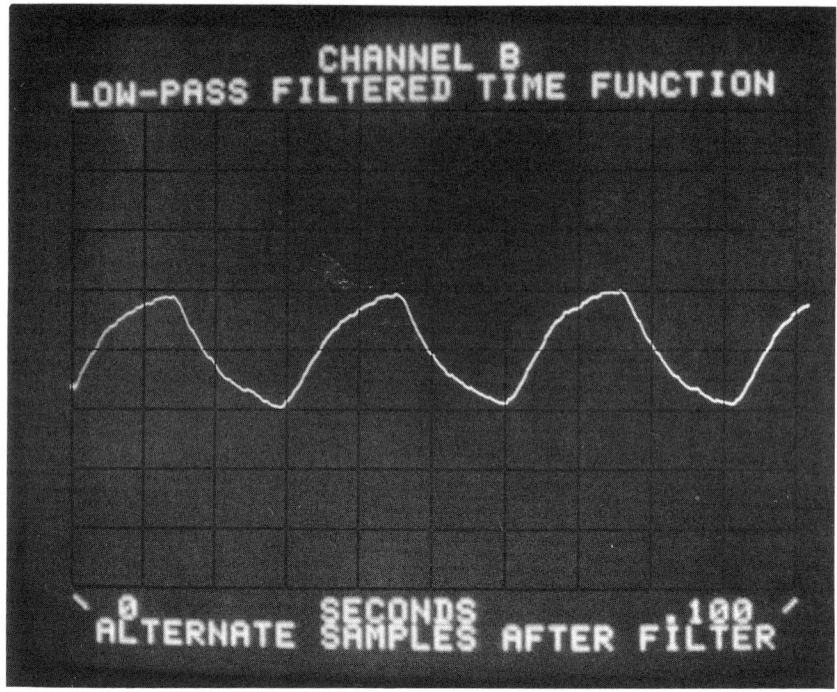

FIG. 12. This photograph shows a time record of the output of a sensor exposed to a room temperature chopper that is illuminated with a human hand, offering a 15-Hz chopped signal with amplitude on the order of 20 μW. The vertical scale is 0.5 V/division, and the horizontal scale is 10 mV/division. The low-pass filter used to acquire these data has a cut-off frequency near 2.5 kHz.

A calibrated responsivity measurement was made on this sensor using a NIST-traceable calibrated blackbody IR source heated to 800°C and a narrow-band IR filter (center wavelength = 4.2 μm). This source–filter combination had an effective area of 5.7 mm$^2$, and illuminated the sensor with 9.0 nV of IR power. The sensor output amplitude (3.3 mV) and noise (120 μV/$\sqrt{Hz}$) were measured with a HP3582B Spectrum analyzer giving a SNR of 27/$\sqrt{Hz}$ and a sensor NEP of $3 \times 10^{-10}$ W/$\sqrt{Hz}$ at 47 Hz. Noise was recorded as a function of frequency for this sensor, compared with the earlier responsivity measurements, and plotted as NEP as a function of frequency in Fig. 14, along with data from a comparably sized, commercially available pyroelectric IR sensor and the calculated performance of the thermal noise-limited tunneling IR sensor. While the performance of this

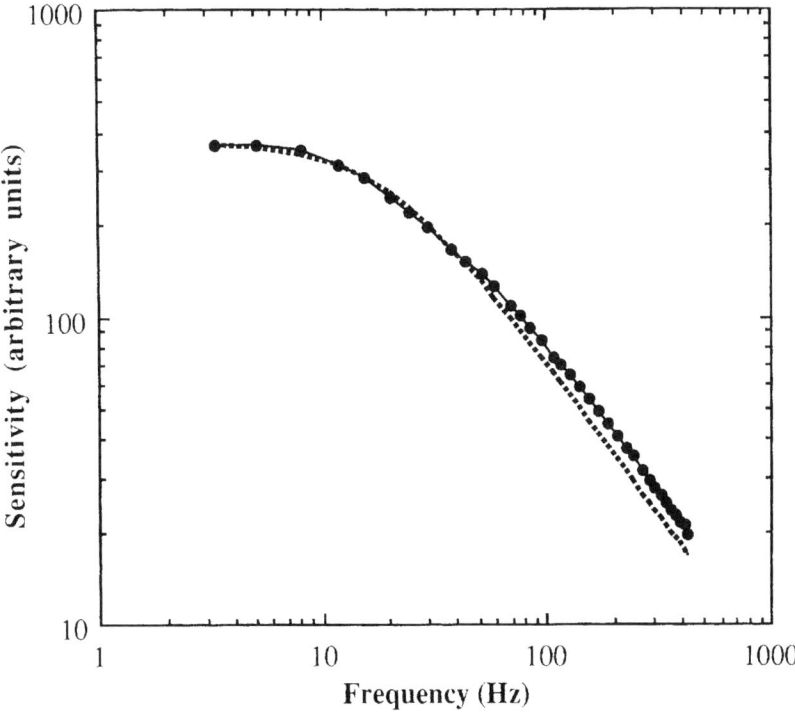

FIG. 13. This plot shows a measurement of the responsivity of the sensor as a function of chopper frequency. These data show that the response is closely fit by a simple single-pole thermal response with cut-off frequency at 20 Hz, which is shown as a dotted line.

device does not fully meet expectations, it is somewhat better than that of a good pyroelectric sensor.

The noise in this sensor is dominated by noise in the tunneling transducer. As stated earlier, the noise sources for this device, as well as for all STMs, are not fully understood. We believe that some of this noise is the result of relaxation of the stress in the epoxy used in the packaging of these sensors. The remainder of this noise may arise from migration of impurities adsorbed on the Au surfaces, and may be improved by packaging of the sensor electrodes in clean, dry environments. Noise due to migration of the Au atoms requires development of different electrode materials for improvement. We have already done some experiments on sensors with Pt electrodes and found that tunneling in air is inhibited by formation of oxide or carbon layers on the Pt surface. By deliberately crashing the membrane into the tip, tunneling can be achieved and maintained indefinitely. Once the tunneling

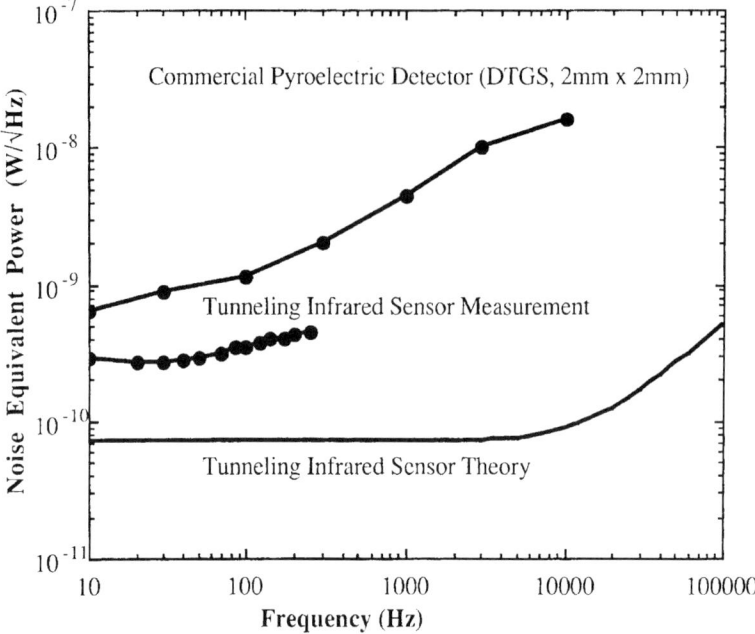

FIG. 14. This plot shows the measured noise equivalent power (NEP) of the tunneling infrared sensor as a function of frequency. For comparison, this plot also shows the predicted NEP for this sensor as well as the NEP for a comparably sized pyroelectric sensor.

is interrupted by the operator, it was our experience that additional crashes were required to recover tunneling. Experiments on other electrode materials, as well as in drier or cleaner environments have been considered, and are now underway.

Despite these issues, the sensors already constructed demonstrate performance competitive with state-of-the-art, mm-sized, uncooled IR sensors. The potential exists for development of inexpensive single-element IR sensors, as well as arrays. Applications for a new uncooled IR sensor that offers state-of-the-art performance are obviously widespread.

Other performance characteristics of this sensor are noteworthy. Because this sensor uses feedback forces to measure the thermally induced expansion of a trapped gas, its performance varies only slowly over a very broad temperature range. To first order, the responsivity scales with the reciprocal of the square of temperature, because the heat capacity decreases linearly, and the fractional responsivity to temperature changes increases as $1/T$. Thermal noise scales as $\sqrt{T}$; however, this sensor is not limited by thermal noise. For comparison, most other decent uncooled IR sensor technologies rely on a thermometer effect that is near a phase transition of some sort. As a result, the responsivity of most uncooled IR sensors must be optimized for each particular application, or the average temperature of the sensor must be actively regulated to maintain decent performance.

In addition, the tunneling IR sensor is expected to be extremely insensitive to ionizing radiation. The sensor element consists entirely of metals and insulators and can only be damaged by formation of shorts through insulators, open circuits across metals, or actual structural damage to the sensor.

These two features indicate that the tunneling IR sensor might be of particular interest for space-based measurement of long-wave thermal IR radiation from planetary surfaces. The ability to operate with good efficiency at whatever temperature the focal plane happens to be at eliminates the power requirements associated with active temperature regulation. The pixel dimensions presently being studied are far too large for imaging in the optical regime, but imaging in the long-wave IR is limited by diffraction. Pixels with dimensions near 1 mm are appropriate for many possible optical instruments intended for long-wave IR imaging.

Some informal life-testing experiments have been carried out on these transducers. For membrane transducers fabricated as described above, an individual device has been in continuous operation for more than 48 months with no observable degradation in performance. An IR sensor based on this device has been in continuous, unprotected operation in the lobby of the Jet Propulsion Laboratory MicroDevices Laboratory since December 1992, where it has been subjected to many uncontrolled experiments (very bright IR signals, mechanical abuse, power-supply failures, earthquakes). All failures to date for other devices of this type in our laboratory have resulted from operator error (mechanical rupture of membrane, application of voltage to the wrong leads). Examples of experiments that have not destroyed the sensor include illumination with soldering irons at a distance of less than 2 cm, occasional illumination with sunlight, and dropping on the floor. These sensor also have been operated *during* and after severe mechanical abuse. For example, a sensor was operated while the circuitboard it was mounted in was manually pounded on a granite workbench, which generated acceleration signals well over 1000 g and a bandwidth of many kHz. These disturbances were reflected in the signal output of the sensor, but did not result in any observable sensor degradation. This apparent immunity to mechanical abuse is an obvious result of the mechanical design of the sensor, whose only moving part is a high resonant frequency membrane with a very low mass. It is important to note that any of these experiments would certainly have destroyed a commercially available Golay cell.

We have delivered devices (by hand and via United Parcel Service shipment) to collaborators at NASA–Goddard Space Flight Center, Rose-Hillman Institute of Technology, Naval Air Warfare Center (Warminster, Pennsylvania), University of California (Los Angeles), California Institute of Technology (Pasadena), the University of Michigan (Ann Arbor) and some industrial collaborators, all of whom report successful operation. In all of these experiences, device failure was limited to cases of operator error as described previously. This collection of experience supports the conclusion that reliable, easily operated, tunneling transducers can be built and packaged for real applications.

In an experiment underway at Stanford University, several of these IR sensors are being packaged for a space flight on a student-built spacecraft. These sensors will be used to detect the earth horizon signals due to a 1-rpm spacecraft rotation, and provide information to the photography and communication systems. We look forward to this opportunity to demonstrate flight survivability, radiation immunity, long-term operation, stability, and utility for space applications.

## VII. Future Prospects for the Tunneling Infrared Sensor

The hope for future development and application of this sensor relies on successful resolution of a number of important issues.

The present method for fabrication of these sensors is unacceptable for volume production. It will be important to develop a technique that would allow parallel fabrication of large numbers of sensors all the way through the assembly stage. Initially, a technique based on the present bulk micromachining of three separate sensor components that are then bonded before dicing of the individual sensors would be acceptable. Such a process could be expected to yield individual uncooled IR sensors at a cost comparable to bulk-micromachined pressure sensors, which are packaged for the automotive industry at near $10 per sensor. Eventually, it will become desirable to develop a fabrication process that is carried out entirely on the surface of a single substrate.

Commonly known as surface micromachining, this process has been used for fabrication of the widely publicized rotary micromotors, as well as a number of other devices, including accelerometers for automotive air-bag applications. To adapt such a process for this tunneling IR sensor, it will be necessary to develop several specialized processing techniques to overcome the present incompatibility of Au electrodes with LPCVD dielectric deposition. A series of experiments to evaluate different electrode materials that are LPCVD-compatible, as well as lower-temperature dielectric deposition techniques that are Au-compatible, are underway.

As stated earlier, we believe that the fabrication techniques described herein are well within the capabilities of laboratories throughout industry and academia. We welcome and encourage all interested parties to try their hand at tunneling transducer fabrication and are happy to offer guidance and advice to all who ask. Greater participation in the development of production compatible fabrication techniques is certain to reduce time needed.

Noise and reliability issues need to be addressed more comprehensively for the tunneling IR sensor to meet widespread commercial acceptance. Studies that will lead to a better understanding of the sources of noise in these sensors are already underway. Simply replacing the quick-curing epoxy with a more mechanically stable bonding agent may eliminate much of the behavior that presently dominates low-frequency noise in these sensors. Eventually, it will become desirable to eliminate the epoxy altogether in favor of a Si–Si bond of the type already in use in commercial pressure sensors and accelerometers, or the polysilicon–Si bond that would be used in a surface micromachined structure. Recent experiments by others done elsewhere on Si-bonded or surface micromachined tunneling transducers have been complicated by electrode cleanliness issues, and have not helped resolve this issue.

For the IR sensor, it will be necessary to carry out more thorough modeling of responsivity and noise to predict the scaling of the sensor performance with size. It is our expectation that sensor NEP will decrease with size as predicted by the simple model presented earlier in this manuscript down to dimensions on the order of 0.3 mm. At this point, the present sensor architecture becomes limited by the availability of suitably thin Si wafers, and a surface micromachined approach will need to be developed. At 0.05-mm pixel sizes, micromachined bolometers recently developed at Honeywell Inc. (Plymouth, Minnesota) represent an impressive accomplishment (Wood et al., 1992). The success of the Honeywell detector makes it very difficult to justify extending the tunneling IR sensor to 0.05-mm pixels. For pixel sizes between 0.05 mm and 0.3 mm, a crossover regime between the Honeywell bolometers and these tunneling sensors is expected to occur. Better modeling and development of suitable fabrication techniques for both sensors are necessary before the best performing technology can be identified at any particular pixel size.

## VIII. Conclusion

This paper describes the design, development, and characterization of a novel uncooled IR sensor that uses a tunneling displacement transducer to measure the thermal expansion of a small volume of trapped gas. This sensor is constructed using simple Si micromachining techniques that can be duplicated at many laboratories throughout industry and academia. The tunneling transducers developed in the course of this work may be operated with simple feedback circuitry, and exhibit remarkable sensitivity and bandwidth. There remain several unanswered questions regarding the limiting noise and drift in these transducers, which are the subject of further study. Infrared sensors built in this manner were characterized and found to offer performance competitive with the best available mm-scale uncooled IR sensors. In addition, these sensors are easily operated and extremely robust. Commercial and NASA space applications exist that can be expected to benefit from the use of this new sensor technology.

### Acknowledgments

The authors acknowledge close collaboration with W. Kaiser, K. Reynolds, J. Podosek, E. Vote, S. Waltman, L. Miller, and T. George, as well as a number of helpful conversations with T. VanZandt, S. Manion, L. Bell, A. Milliken, P. Smith, S. Martin, R. Vasquez, F. Grunthaner, M. Roukes, A. Cleland, D. Marcus, T. Gabrielson, R. Muller, P. Maker, T. George, A. Partridge, T. Stowe, K. Stattenfield, and M. Hecht.

The work described herein was performed in collaboration with the Center for Space Microelectronics Technology, Jet Propulsion Laboratory, California Institute of Technology and was jointly sponsored by the Ballistic Missile Defense Organization/Innovative Science and Technology Office, the Defense Advanced Research Projects Agency, the Naval Air Warfare Center, the Army Space Technology Resrearch Office, and the National Aeronautics and Space Administration, Office of Aeronautics, Exploration, and Technology.

### References

Baski, A. A., Albrecht, T. R., and Quate, C. F. (1988). *J. Microsc.* (Oxford) **152**, 73.
Binnig, G., and Rohrer, H. (1986). *IBM J. Res. Dev.* **30**, 355.
Bocko, M. F., Stephenson, K. A., and Koch, R. H. (1988). *Phys. Rev. Lett.* **61**, 726.
Chevrier, J. B., Baert, K., Slater, T., and Verbist, A. (1995). *Microsyst. Technol.* **1**, 71.

Gabrielson, T. B. (1993). *IEEE Electron. Devices* **40**, 903.
Golay, M. J. E. (1947a). *Rev. Sci. Instrum.* **18**, 347.
Golay, M. J. E. (1947b). *Rev. Sci. Instrum.* **18**, 357.
Golay, M. J. E. (1949). *Rev. Sci. Instrum.* **20**, 816.
Hadni, A. (1967). "Essentials of Modern Physics Applied to the Study of the Infrared," pp. 269-283. Pergamon, Oxford.
Jones, R. C. (1953). *J. Opt. Soc. Am.* **43**, 1.
Kenny, T. W., Waltman, S. B., Reynolds, J. K., and Kaiser, W. J. (1991a). *Appl. Phys. Lett.* **58**, 100.
Kenny, T. W., Kaiser, W. J., Waltman, S. B., and Reynolds, J. K. (1991b). *Appl. Phys. Lett.* **59**,
Kenny, T. W., Kaiser, W. J., Podosek, J. A., Rockstad, H. K., and Reynolds, J. K. (1992). *Proc. IEEE Solid State Sen. Actuator Workshop, 1992,* p. 174.
Kenny, T. W., Kaiser, W. J., Podosek, J. A., Rockstad, H. K., and Reynolds, J. K. (1996). *Rev. Sci. Instrum.* (to be published).
Kleint, C. (1971). *Surf. Sci.* **25**, 394.
Kobayashi, D., Hirano, T., Furuhata, T., and Fujita, H. (1992). *Proc. IEEE Microelectromech. Syst. Workshop, 1992,* p. 214.
MacDonald, N. C. (1992) *Proc. IEEE Solid State Sens. Actuator Workshop, 1992,* p. 1.
Mather, J.-C. (1984). *Appl. Opt.* **23**, 584.
Moller, R., Esslinger, A., and Koslowski, B. (1989). *Appl. Phys. Lett.* **55**, 2360.
Petersen, K. E. (1982). *Proc. IEEE* **70**, 420.
Presilla, C., Onofrio, R., and Bocko, M. F. (1992). *Phys. Rev. B* **45**, 3735.
Putley, E. H. (1977). "Optical Infrared Detectors" (R. J.Keyes, ed.), pp. 71-98. Springer-Verlag, Berlin.
Spencer, R. R., Fleischer, B. M., Barth, P. W., and Angell, J. B. (1988). *IEEE Trans. Electron Devices* **35**, 1289.
Tiedje, T., Varon, J., Deckman, H., and Stokes, J. (1988). *J. Vac. Sci. Technol. A* [2] **6**, 372.
Waltman S. B., and Kaiser, W. J. (1989). *Sens. Actuators* **19**, 201.
Wandass, J. H., Murday, J. S., and Colton, R. J. (1989). *Sens. Actuators* **19**, 211.
Weisenhorn, A. L., Hansma, P. K., Albrecht, T. R., and Quate, C. F. (1989). *Appl. Phys. Lett.* **54**, 2651.
Wood, R. A., Han, C. J., and Kruse, P. W. (1992). *Solid State Sens. Actuator Workshop, 1992,* p. 132.
Yao, J. J., Arney, S. C., and MacDonald, N. C. (1992). *J. MicroElectroMech. Syst.* **1**, 14.
Yeh, C., and Najafi, K. (1994). *Proc. IEEE Solid State Sens. Actuator Workshop, 1994,* p. 123.
Young, W. C. (1989). "Roark's Formulas for Stress and Strain," 6th ed. McGraw-Hill, New York (derived from Table 24, p. 429).
Yurke, B., and Kochanski, G. (1990). *Phys. Rev. B* **41**, 8184.
Zhang, Z. L., and MacDonald, N. C. (1993). *J. MicroElectroMech. Syst.* **2**, 66.

CHAPTER 9

# Application of Quartz Microresonators to Uncooled Infrared Imaging Arrays

*John R. Vig and Raymond L. Filler*

US ARMY COMMUNICATIONS AND ELECTRONICS COMMAND
FORT MONMOUTH, NEW JERSEY

*Yoonkee Kim*

US ARMY RESEARCH LABORATORY
FORT MONMOUTH, NEW JERSEY

| | |
|---|---|
| I. INTRODUCTION | 269 |
| II. QUARTZ MICRORESONATORS AS INFRARED SENSORS | 271 |
| III. QUARTZ THERMOMETERS AND THEIR TEMPERATURE COEFFICIENTS | 272 |
| IV. OSCILLATOR NOISE | 273 |
| V. FREQUENCY MEASUREMENT | 274 |
| VI. THERMAL ISOLATION | 275 |
| VII. INFRARED ABSORPTION OF MICRORESONATORS | 277 |
| VIII. PREDICTED PERFORMANCE OF MICRORESONATOR ARRAYS | 279 |
| IX. PRODUCIBILITY AND OTHER CHALLENGES | 281 |
| X. SUMMARY AND CONCLUSIONS | 283 |
| APPENDIX. PERFORMANCE CALCULATIONS | 284 |
| References | 294 |

## I. Introduction

Frequency and time (which are closely related) are the physical quantities that can be determined with the highest accuracy. Most high-precision frequency control and timing devices are based on quartz resonators. A quartz crystal can act as a stable mechanical resonator, which, by its piezoelectric behavior and high $Q$, determines the frequency generated in an oscillator circuit. About $2 \times 10^9$ quartz resonators are manufactured annually for oscillator, clock, and filter applications. (An oscillator consists of a resonator plus feedback circuitry that can sustain the vibration of the resonator at a resonance frequency.)

An important reason that quartz resonators are useful for frequency control applications is that these resonators are among the lowest noise devices known. The frequencies of the lowest noise resonators can be measured with a precision of 14 significant figures (i.e., the noise is a few parts in $10^{14}$ at the optimum measurement times (Walls and Vig, 1995)).*

The fundamental mode resonance frequency of a quartz crystal vibrating in a thickness mode (Bottom, 1982; Gerber and Ballato, 1985) can be expressed as

$$f = \frac{v}{2d} = \frac{1}{2d}\sqrt{\frac{c_{ij}}{\rho}} \tag{1}$$

where $v$ is the velocity with which an elastic wave is propagated in the quartz plate, $d$ is the thickness of the plate, $c_{ij}$ is the elastic modulus (i.e., the ratio of the stress to the strain) associated with the elastic wave being propagated, and $\rho$ is the density of quartz.

Every term in Eq. (1) is temperature-dependent. Moreover, $c_{ij}$ and its temperature coefficients are highly dependent on the angles of cut of the plate with respect to the crystallographic axes of quartz. The thermal expansion of quartz is also highly dependent on direction with respect to the crystallographic axes. The temperature coefficients of $c_{ij}$ range from positive to negative values. One of the reasons that quartz resonators are useful for frequency control and timing applications is that, at certain angles of cut, the temperature coefficient of $c_{ij}$ cancels the temperature coefficients of the other terms in Eq. (1), that is, angles of cut exist in quartz that can provide zero temperature coefficient of frequency. At other angles of cut, it is possible to obtain steep and monotonic frequency versus temperature characteristics that can be useful for thermometers and infrared (IR) detectors. The commonly used and well characterized cuts of quartz have names such as the AT-cut, BT-cut, SC-cut, AC-cut, and so on (Bottom, 1982; Gerber and Ballato, 1985).

In the manufacture of quartz resonators, plates are cut from a quartz crystal along precisely controlled directions with respect to the crystallographic axes. After shaping to required dimensions, metal electrodes are applied to the quartz plate, which is mounted in a holder structure. The assembly is sealed hermetically in an enclosure (because contamination leads to frequency instabilities).

Whereas in frequency control and timing applications of quartz-crystal devices the resonators are designed to be as insensitive to the environment as possible, quartz resonators can also be designed to be highly sensitive to

---

*Oscillator stabilities are expressed as normalized frequency changes, $\Delta f/f$. The units of stability are, therefore, dimensionless.

environmental parameters such as temperature, pressure, force, and acceleration. Quartz-crystal sensors can exhibit unsurpassed resolution and dynamic range (EerNisse et al., 1988). Quartz thermometers can provide millidegrees of absolute accuracy and microdegrees of resolution over wide temperature ranges. Quartz sorption detectors can detect a change in mass of $10^{-12}$ g. Quartz accelerometers and force sensors are capable of resolving $10^{-7}$ to $10^{-8}$ of full scale.

The highest precision resonators are made of circular quartz plates the typical dimensions of which are 15 mm in diameter and a fraction of a mm thick. The feasibility of fabricating $\sim 1$-$\mu$m thick AT-cut and SC-cut quartz resonators by means of etching techniques has been demonstrated (Vig et al., 1977; Hunt and Smythe, 1985; Smythe and Angove, 1988), however, these resonators are not useful for precision frequency control applications. The primary reason is that such resonators are extremely sensitive to mass loading, which makes frequency adjustment difficult and degrades the stability, especially the aging. For example, a 1-$\mu$m-thick quartz resonator consists of about 2000 molecular layers. Therefore, a thickness change of a single molecular layer, for example, due to contamination adsorption, changes the frequency by one part in 2000, that is, by 500 ppm. This is a huge change relative to the stability of precision resonators.

For sensing, however, the absolute frequency and the long-term stability are not important. The important factors are the frequency change caused by a measurand, and how accurately that frequency change can be measured. In this chapter, the potential of microresonator arrays as IR sensors is explored. It is shown that, combining the low noise characteristics of quartz crystal oscillators, the steep frequency versus temperature characteristics of resonators made of certain cuts of quartz, and the small thermal mass and high thermal isolation capability of microresonators can result in high-performance IR sensors and sensor arrays.

## II. Quartz Microresonators as Infrared Sensors

The performance of any thermal detector (Kruse et al., 1962; Kruse, 1995; Keyes, 1980) is determined mainly by the following:

1. Steady-state response, given by

$$\Delta T = \frac{P}{G} \qquad (2)$$

where $P$ is the power absorbed from the heat source, $\Delta T$ is the temperature increase due to $P$, and $G$ is the thermal conductance from the sensitive element to a heat sink

2. Noise of the detector relative to the signal produced by $\Delta T$, which limits the degree to which the $\Delta T$ can be resolved
3. Time constant $\tau_T$ of the detector element, given by

$$\tau_T = \frac{C}{G} \qquad (3)$$

where $C$ is the heat capacity of the element

For a quartz microresonator IR sensor array, the parameters that determine the performance are the microresonators'

1. Temperature coefficient of frequency
2. Noise
3. Thermal conductance to a heat sink
4. Heat capacity
5. IR absorbance
6. Dimensions
7. Fill factor (i.e., the ratio of sensing area to the total area of the array)

### III. Quartz Thermometers and their Temperature Coefficients

Microresonators may be fabricated from a large variety of (bulk or thin-film) piezoelectric materials, especially since the materials need not possess temperature-compensated cuts. Although materials other than quartz (e.g. zinc oxide and gallium arsenide) may eventually prove to be more suitable than quartz for microresonator sensors, the discussion that follows shall be confined to quartz microresonators because the technology of such resonators is the most advanced and the performance is best known.

The temperature sensitivity of quartz resonators has been known for a long time (Heising, 1946). The possibilities of using quartz resonators as precision thermometers have been described since at least 1962 (Wade and Slutsky, 1962; Gorini and Sarto, 1962; Smith and Spencer, 1963; Hammond et al., 1965; Ziegler and Tiesmeyer, 1983; Ziegler, 1983; Gagnepain et al., 1983; Ralph et al., 1985; EerNisse et al., 1988; Spassov, 1992; Hamrour and Galliou, 1994). The ability of quartz thermometers to measure small temperature changes was discussed by Smith and Spencer in 1963. They pointed out that it is possible to sense temperature changes of microkelvins.

Quartz resonators have also been proposed for IR sensing (Ziegler and Tiesmeyer, 1983; Ziegler, 1983; Ralph et al., 1985; Spassov, 1992). The resonators in these quartz thermometers and IR sensors had typical dimensions of about 1 cm in diameter and a fraction of a millimeter in thickness, that is, they were not microresonators.

The fractional frequency of quartz resonators can vary monotonically with temperature, with a slope of about $10^{-4}$/K. Due to the low-noise capabilities of crystal oscillators, the fractional frequency noise limitation for resolving temperature changes is, for example, $10^{-12}$ or less for a low-noise 10-MHz resonator. Therefore, the noise of such a resonator corresponds to temperature fluctuations of 10 nanokelvins or less.

The crystal cut for microresonator sensors may be chosen from among a wide variety. Some of the possibilities are the AC-cut (20 ppm/°C (Heising, 1946)), LC-cut (35.4 ppm/°C) (Hammond et al., 1965), Y-cut ($\sim 90$ ppm/°C (Heising, 1946)), SC-cut (b-mode: $-25.5$ ppm/°C (Kusters et al., 1978) or dual-mode: 80 ppm/°C to $>100$ ppm/°C (Schodowski, 1989; Filler and Vig, 1989), NLSC-cut ($\sim 14$ ppm/°C) (Nakazawa et al., 1984), or any other cut that can be made thin and have a well-behaved temperature-sensitive mode. The NLSC-cut, the dual-mode SC-cut, and other thermometer cuts from the stress-compensated-cut locus (Sinha, 1981) are especially advantageous when the pixels are switched on and off for short periods, for example, when the array is divided into subarrays and one oscillator circuit per subarray is used to sequentially excite the resonators in the subarray, or when the microresonators are illuminated intermittently, as would be the case when the microresonators are used in a scanned array.

## IV. Oscillator Noise

The noise floor (i.e., the Allan deviation floor (Gerber and Ballato, 1985)) of a state-of-the-art oscillator that uses an AT- or a SC-cut resonator can be described by $\sigma_y(\tau) \approx (2 \times 10^{-7})Q^{-1} \approx 1.2 \times 10^{-20}$ Hz$^{-1} \cdot f_0$ (Parker, 1985, 1987; Vig and Walls, 1994), where $\sigma_y(\tau)$ is the Allan deviation floor and $f_0$ is the resonator frequency in Hz. The best (bulk acoustic wave (BAW)) noise performance has been achieved with overtone resonators. Microresonator sensor arrays seem most promising when the resonators are in the 200-MHz to 1.0-GHz range, as shown below. Noise data are available for 400- to 900-MHz surface acoustic wave (SAW) resonators, and for these, at the noise floor, $\sigma_y(\tau)$ is in the range of $1 \times 10^{-11}$ to $1 \times 10^{-10}$ (Montress and Parker, 1994, also private communication, 1995) (after converting the data from the frequency domain to the time domain). For 100-MHz third overtone and 160-MHz fifth overtone BAW (SC-cut) resonators, the noise floor corresponds to 1.2 to $1.9 \times 10^{-20}$ Hz$^{-1} \cdot f_0$ (Driscoll and Hanson, 1993).

The noise of microresonator sensors may be adversely affected by size effects (Walls, 1992; Parker and Andres, 1993) and by contamination adsorption and desorption (Yong and Vig, 1988). For purposes of calculating the properties of microresonator sensors, we assume that the noise floor will be 10 times worse than $1.2 \times 10^{-20}$ Hz$^{-1} \cdot f_0$, that is, we assume that the noise will be $1.2 \times 10^{-19}$ Hz$^{-1} \cdot f_0$. This predicts, for example, that the noise floor of an 800 MHz microresonator will be $1 \times 10^{-10}$.

A noise contribution that is negligibly small in ordinary resonators, but is the dominant noise contribution in microresonators below about 2 $\mu$m in thickness, is the temperature fluctuation noise. This noise is due to the quantum nature of heat exchange, that is, that heat exchange takes place by the absorption and emission of photons and phonons. The smaller the heat capacity (i.e., the higher the frequency) of a microresonator, the larger the temperature fluctuations due to this noise source. It can be shown that the mean temperature fluctuation $\Delta T$ of any object due to this noise is given by

$$\Delta T = \sqrt{\frac{\kappa_B T^2}{C}} \qquad (4)$$

where $C$ is the heat capacity and $\kappa_B$ is Boltzmann's constant (Kruse et al., 1962; Kruse, 1965). This temperature fluctuation noise manifests itself as frequency noise, via the temperature coefficient of the microresonator, and it adds to the other oscillator noise sources, as shown in the Appendix.

## V. Frequency Measurement

The power required to simultaneously excite all microresonators in a large array may be excessive in many applications. Dividing the array into subarrays and exciting the resonators sequentially, with one oscillator per subarray, can reduce the power requirement to manageable levels. One resonator that is shielded from the IR source can serve as a reference, and the frequency differences between that reference and the other resonators can be measured rapidly.

In a conventional crystal oscillator, how rapidly the frequencies may be measured is a function of the resonators' noise characteristics. The shortest measurement time $\tau$, which may be used for frequency measurement without loss of resolution, is the $\tau$ at the knee of the $\sigma_y(\tau)$ versus $\tau$ curve, that is, it is the lower $\tau$ limit of the noise floor. At this limit, other noise processes, such as white-phase noise or Johnson noise (which is white FM noise) begin to exceed the flicker of frequency noise, so that $\sigma_y(\tau)$ starts to increase as $\tau$ decreases. The $\tau$ at the knee of the $\sigma_y(\tau)$ versus $\tau$ curve is a function of the drive level, so that in the design of a microresonator array, it is necessary to balance the drive level, self-heating, and the measurement time.

For example, for 500-MHz SAW resonator-oscillator, the knee of the $\sigma_y(\tau)$ versus $\tau$ curve was at about $10^{-3}$ sec when a 40-MHz high-frequency cutoff ($f_c$) was used for measuring at small $\tau$'s (T. E. Parker, NIST, private communication, 1994). Assuming that with the same $f_c$, the knee of the $\sigma_y(\tau)$-versus-$\tau$ curve of a 500-MHz microresonator would be at the same $\tau$, and using an $f_c$ of $\sim 40$ kHz, one may lower the knee to a $\tau$ of about 30 $\mu$sec (because $\sigma_y(\tau)$ scales as $(f_c)^{1/2}$ in the white-phase noise region) (F. L. Walls, NIST, private communication, 1994). Therefore, except for the effect of dead time, which is described in the Appendix, in principle, one can sequentially excite each resonator in a 16 × 20 subarray with one oscillator in a 33-msec video frame time, so that 252 oscillators (and counters) would be required to excite (and measure) a 240 × 336 array. Measuring the 320 resonators in 33 msec allows 103 $\mu$sec per resonator for the switching, startup, and measurement of each resonator.

Whether one may lower the $\tau$ to 30 msec without increasing the noise depends on the resonator frequency, drive level, and measurement dead time. For example, measuring the frequency of a 500-MHz microresonator at a 1-$\mu$W drive level and with $\tau = 30$ $\mu$sec increases the oscillator noise by about a factor of 10 because, as is shown in the Appendix, the Johnson noise becomes significant at low drive levels and short measurement times. However, at 500 MHz (and higher frequencies), the temperature fluctuation noise is larger than the resonator flicker noise, and therefore the total noise increases by less than a factor of 10 over the case in which the Johnson noise is negligible (i.e., when flicker noise is the dominant oscillator noise source). Therefore, in designing a scanned array, one must balance the number of resonators per subarray with the drive level and the noise equivalent temperature difference (NETD). The larger the number of resonators in a subarray, the higher the NETD for a given drive level.

## VI. Thermal Isolation

From Eq. (2) it follows that to maximize the response of an IR detector to a given IR source, the thermal conductance $G$, from the sensitive element to a heat sink must be minimized. However, Eq. (3) shows that minimizing

$G$ for a given $C$ also maximizes $\tau$. Reducing $C$ reduces $\tau$, but increases the temperature fluctuation noise, as can be seen from Eq. (4). Therefore, in designing a microresonator array, one must find the proper balance between $G$ and $C$.

The thermal isolation of microresonators may be controlled in a variety of ways. The ultimate isolation can be achieved by levitating (Kumar *et al.*, 1992; Cho *et al.*, 1991; Pelrine, 1990) the microresonators in a vacuum and combining the levitation with electrodeless excitation (Vig and Walls, 1994). In such an arrangement, heat exchange would be by means of radiation only. The resulting long time constants and the extra complexity would make such an array unsuitable for most applications (although the time constant could be controlled by admitting a controlled amount of helium, for example).

A controlled amount of thermal isolation may also be produced during the etching of the microresonators. For example, one may form thin and narrow bridges or thin and narrow rings to surround the resonators, the rings being connected to the resonators (and to each other, for multiple rings) by bridges. Figure 1 illustrates square resonators isolated by means of bridges at each corner. Figure 2 illustrates circular resonators, each isolated by means of three evenly spaced bridges. A third way to produce the thermal isolation and mechanical supports is to use freestanding thin-film-strip supports (Lee, 1990). Because, for example, the mass of an 800-MHz microresonator is only about 30 nanograms, the supports need not be massive.

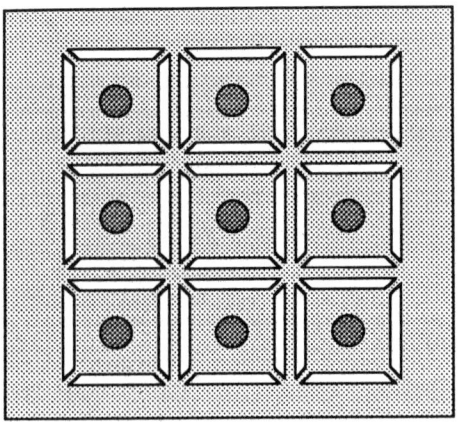

FIG. 1. Thermal isolation of square microresonators.

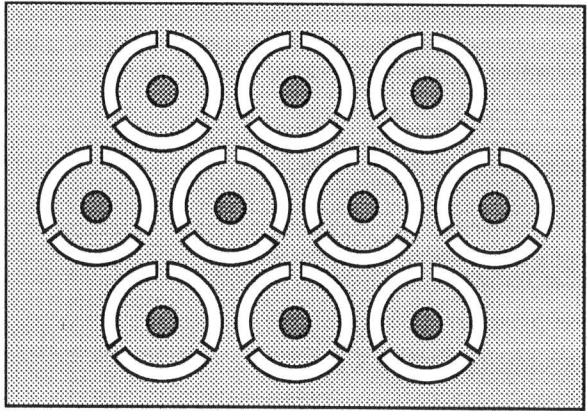

FIG. 2. Thermal isolation of circular microresonators.

For purposes of calculating the figures of merit of microresonator IR sensor arrays, bridge supports are assumed. The dimensions of the bridges are selected to provide a $G$ that results in a time constant of 10 msec, that is, $\tau$ is fixed at 10 msec, and $G$ is a variable that is a function of resonator frequency. The calculation of bridge dimensions is described in the Appendix.

## VII. Infrared Absorption of Microresonators

Quartz is a good absorber of IR near 9, 12, 20, and 26 $\mu$m (Spitzer and Kleiman, 1961). We performed a theoretical calculation over the 8- to 14-$\mu$m wavelength range using refractive indices given by Palik (1985). It indicates that 2.5-$\mu$m-thick quartz has about 50% absorption peaks near 9.5 and 12.5 $\mu$m. The average energy that would be absorbed by a 2.5-$\mu$m-thick quartz plate in the 8- to 14-$\mu$m IR band is about 17%. The overall IR absorption can be increased by depositing IR absorbing coatings onto the resonators.

A wide variety of IR absorbing coatings is available (Hadley and Dennison, 1947; Lang et al., 1991, 1992; Liddiard, 1993; Parsons and Pedder, 1988; Havens, 1955; Advena et al., 1993; McCarthy, 1963). Thin films that have a sheet resistance that is half the impedance of free space, that is, 188 $\Omega$/square, can absorb 50% of incident IR radiation (Hadley and Dennison, 1947). (The impedance of free space is 377 $\Omega$.) At such an optimum film thickness, 50% of the incident IR radiation is absorbed, 25% is reflected, and 25% is transmitted. The absorption can be increased to over 50% by creating multiple passes of the IR through the resonator, for example, by having an IR reflecting film on the back side. Making the enclosure into a resonant optical cavity is another way to obtain high absorption. According to Lang et al. (1991, 1992), at a critical thickness of 17 nm (34 $\mu g/cm^2$), a vacuum evaporated gold (Au) film has about a 0.5 absorption. An 8-nm nichrome film is also a good absorber and has been used on some thermal detectors.

A three-layer absorber stack consisting of a metal film of 377 $\Omega$/square on the side facing the IR, a dielectric layer of refractive index $n$ and thickness $d = \lambda_{max}/4n$, and an IR reflecting film on the back side has been shown to be capable of over 95% absorption at $\lambda_{max}$ (Liddiard, 1993). The quartz plate itself could become a quarter wave plate for approximately 10-$\mu$m IR, and thereby help produce a strong absorption in the 10-$\mu$m range.

The IR absorbing film can be deposited so as to contribute to energy trapping, which is required for optimum resonator performance (Bottom, 1982; Gerber and Ballato, 1985). Mass loading that is equivalent to about 1/50 of the plate thickness could provide the proper energy trapping (H. F. Tiersten, private communication, 1995). (For an 800-MHz resonator, for example, 1/50 of the plate thickness is approximately 40 nm of quartz-equivalent mass loading.) Additional absorbing film can be outside the active area of the resonator. Aluminum is suitable from the mass loading point of view, but it has high reflectance. However, its reflectance drops rapidly below 10 nm of film thickness; for 8-$\mu$m IR radiation, the transmission versus thickness and the absorption versus thickness curves cross at approximately 6 nm film thickness and the values are approximately 30% for each (Havens, 1955). If the back side of the enclosure is reflecting (e.g., an Al film more than 15-nm thick has approximately 100% reflectance to IR wavelengths up to at least 50 $\mu$m), then two passes of the IR result in approximately 64% absorption. For 1.6-GHz AT-cut resonators, approximately 40-nm-thick Al electrodes were about the lowest mass electrodes that could be used (Gerber and Ballato, 1985; R. C. Smythe, private communication, 1995). Below 40 nm, the resistivity of the electrodes was too high.

Simultaneously obtaining sufficient energy trapping and high IR absorption presents a challenging design and fabrication problem. Some possibilities are using a ring electrode (H. F. Tiersten, private communication, 1995; Belser and Hicklin, 1967; Nakamura et al., 1990), which surrounds an IR absorbing film on the side facing the IR radiation and either a ring or a solid (and reflecting) electrode on the back side; using other arrangements of electrodes with opening(s) for allowing the IR radiation to pass through; using lateral field excitation (Ballato et al., 1985; Warner and Goldfrank,

1985; Smythe and Tiersten, 1987), such that the exciting electrodes are on the side of the resonators facing away from the IR source and the IR absorbing film is on the front side; using electrodeless excitation (Vig and Walls, 1994; Hammond and Cutler, 1967, 1969; Besson, 1977), with the electrode facing the IR source being a ring electrode (H. F. Tiersten, private communication, 1995); and etching energy trapping mesas into the microresonators (H. F. Tiersten, private communication, 1995).

## VIII. Predicted Performance of Microresonator Arrays

The performance of microresonator arrays has been calculated for scanning array and staring array operations. In the first case, the microresonators are excited periodically, which implies dead time between measurements. In the second case, all the microresonators are excited and measured continually. Figure 3 shows NETD as a function of microresonator frequency for the two conditions: (a) scanning array, with 30-$\mu$sec $\tau$ and 30-Hz frame rate ($r \sim 1000$) and (b) staring array, 30-msec $\tau$ and 30-Hz frame rate ($r \sim 1$); $\tau$ is the frequency measurement time and $r$ is the ratio $(\tau + \tau_d)/\tau$, where $\tau_d$ is the dead time between measurements. (If there is no dead time, then $r = 1$. See the Appendix for discussion $r$, $\tau$, $\tau_d$ and the details of the calculations of $D^*$ and NETD.) Also shown in Fig. 3 are the individual contributions to NETD of the oscillator noise and the temperature fluctuation noise. Figure 4 shows NETD as functions of $G$ (and $\tau_T$) for a 600-MHz array, down to the radiation thermal conductance limit of $1.3 \times 10^{-4}$ K. Table 1 shows a summary of NETD, quartz wafer area needed for a 240 × 336 array, and the bridge length needed for a 10-msec time constant for 100-MHz, 200-MHz, 600-MHz, 1.0-GHz, 1.4-GHz, and 1.8-GHz arrays. In the calculations, the following assumptions were made: $T = 300$ K, $\tau_T = 10$ msec (except in Fig. 4, where $G$, and therefore $\tau_T$, is a variable), resonator diameter-to-thickness ratio equals 80, the frequency constant equals 1.6 MHz·mm, temperature coefficient equals 50 ppm/K, oscillator noise at the noise floor equals $1.2 \times 10^{-19}$ Hz$^{-1} \cdot f_0$, no chopping, transmittance of the atmosphere between the target and the sensor equals 0.8, 56% IR absorption, 80% fill factor, f/1 optics, and the wavelength range of 8 to 14 $\mu$m. Of course, cooling the array improves the performance (as it does in the ideal limit for all IR detectors).

For comparison, the best uncooled IR imaging arrays that have been reported are the Si microbolometer array (Wood, 1993), with NETD = 0.04 K, and the ferroelectric bolometer array (Hanson, 1993), with NETD = 0.05 K. The best available cryogenically cooled IR imaging arrays (mercury cadmium telluride (HgCdTe), indium antimonide (InSb), platinum silicide

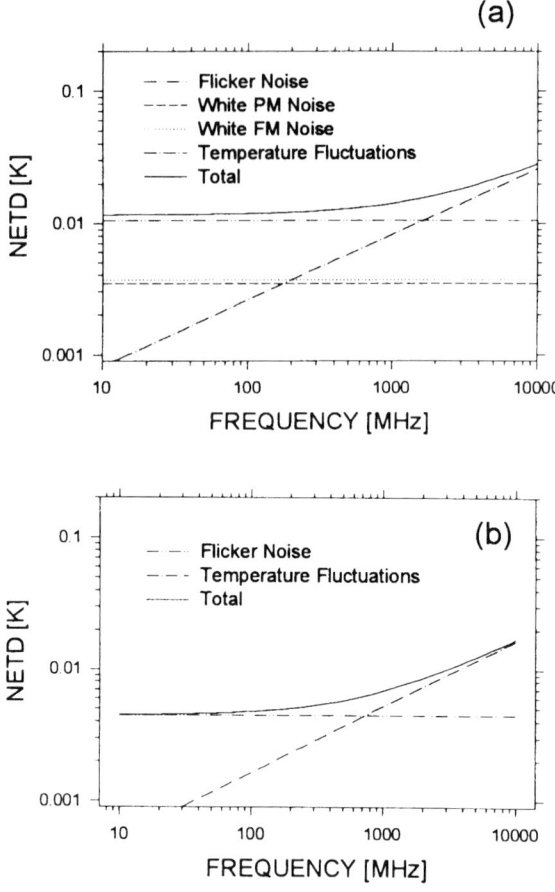

FIG. 3. Noise equivalent temperature difference (NETD) as a function of microresonator frequency at (a) $r = 1000$ and (b) $r = 1$. White PM and white FM noise contributions are negligibly small for $r = 1$.

(PtSi) and gallium arsenide–aluminum gallium arsenide (GaAs/AlGaAs)) exhibit NETD = 0.01 K (Wood and Foss, 1993). Typical thermal imaging cameras exhibit NETDs of 0.02 to 0.2 K (Demeis, 1995). (To be fair, it must be pointed out that the NETD values used for comparison are measured values, whereas the microresonator array NETD values are calculated ones, based on some known and some estimated properties of quartz microresonators, i.e., no microresonator arrays have yet been built. The calculated NETDs assume that the design and fabrication problems, discussed subsequently, can be solved without degrading the NETD.)

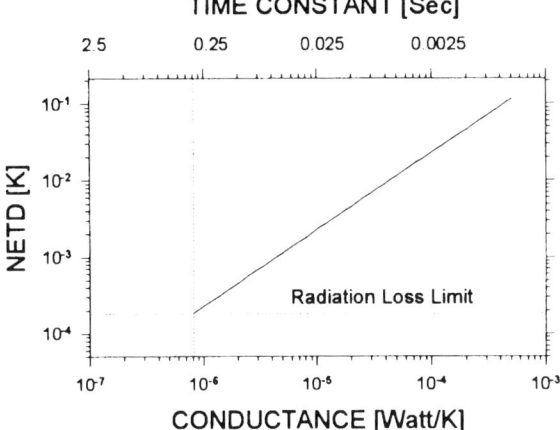

FIG. 4. Noise equivalent temperature difference (NETD) as a function of $G$ (and $\tau$) for a 600-MHz microresonator array at $r = 1$.

TABLE I

COMPARISON OF KEY PARAMETERS OF 100-MHz, 200-MHz, 600-MHz, 1.0-GHz, 1.4-GHz, AND 1.8-GHz ARRAYS WITH 240 × 336 PIXELS

| Frequency (MHz) | NETD (K) | | Wafer size [cm × cm] | Bridge length [$\mu$m] |
|---|---|---|---|---|
| | $r = 1000$ | $r = 1$ | | |
| 100 | 0.011 | 0.0047 | — | 1.3 |
| 200 | 0.012 | 0.0050 | — | 2.7 |
| 600 | 0.013 | 0.0059 | 6.4 × 9.0 | 8.1 |
| 1000 | 0.014 | 0.0068 | 3.9 × 5.4 | 13.6 |
| 1400 | 0.015 | 0.0076 | 2.8 × 3.9 | 19.4 |
| 1800 | 0.016 | 0.0083 | 2.2 × 3.0 | 25.3 |

NETD, noise equivalent temperature difference.

## IX. Producibility and Other Challenges

Small arrays are producible with currently available technology. For example, strip AT-cut resonators are currently being produced by photolithography and etching techniques at frequencies up to 250 MHz (T. Payne, Avance Technology Inc., private communication, Cedar City, UT. 1995; J. Standte, XECO, Inc., Cedar City, UT, 1995). Arrays of 250 MHz and even

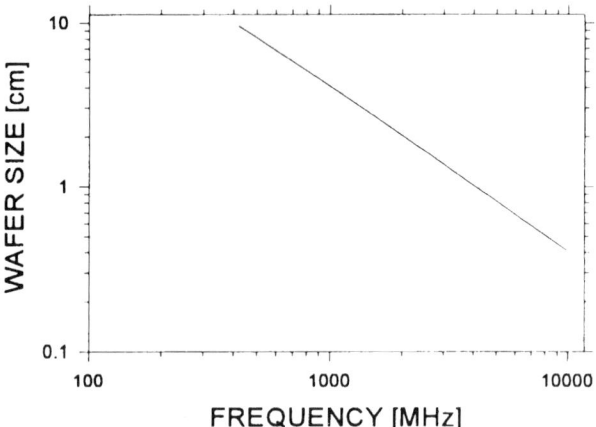

FIG. 5. Wafer size required for a 240 × 336 pixel array as a function of resonator frequency, assuming a square wafer, a diameter-to-thickness ratio of 80, an 80% fill factor, and a frequency constant of 1.6 MHz·mm.

lower frequency resonators can be useful for scanned IR imaging arrays. However, the lower the frequency, the larger is each resonator, and the fewer the resonators per wafer, as is illustrated in Fig. 5. The example in Fig. 5 shows the wafer size needed for a 240 × 336 array (i.e., 80,640 resonators per wafer). Because the largest quartz wafers currently available are 10 × 10 cm, it is clear that for 240 × 336 arrays to be on a single wafer, the frequencies of the microresonators must be currently above about 500 MHz.

Several problems must be overcome before large arrays of microresonators may be realized. The techniques being used for producing 250-MHz strip AT-cut resonators will need to be extended to higher frequencies, or new methods will need to be developed. Low defect density quartz will need to be used to achieve high yields. Maintaining the surface sufficiently parallel is just one of the fabrication challenges. The arrangement of IR absorbing films and resonator electrodes will require careful design, as was discussed earlier.

Plating the microresonators to frequency is not a problem because the exact frequencies are not important. Moreover, not having all the microresonators at the same frequency will be an advantage because it will reduce the potential cross talk among resonators. Additional challenging design and fabrication problems are packaging; making the necessary interconnections; dealing with the reactances of the structure; minimizing cross talk; and making, driving, and measuring the resonators in the array so that they exhibit sufficiently low noise.

## X. Summary and Conclusions

Calculations based on the properties of quartz resonators indicate that high-frequency microresonator arrays can potentially result in high-performance sensors. In optimizing the performance of a microresonator sensor array, one must find the proper balance between the thermal conductance from each microresonator to a heat sink and the heat capacity of the resonator (which is determined by the frequency of the resonator). Another important balance that needs to be found is between the drive level of the resonator (and the resulting self-heating) and the noise characteristics (i.e., the optimum frequency measurement method).

For IR imaging arrays, one must also find the balance among performance, wafer size, and ease of fabrication. The calculated NETD at 300 K is 0.01 K or less when the resonator frequency is 1 GHz or less. At lower frequencies the performance and ease of fabrication improve, but the number of microresonators per unit area decreases. The measured NETDs of uncooled IR imaging arrays made with other technologies are 0.04 K or more. (The calculated microresonator array NETDs assume that the design and fabrication problems can be solved without degrading the NETD.) Unlike some competing technologies that depend on precise temperature control near a transition temperature, the microresonators can be at any temperature, over a wide range, for example, from below $-55°C$ to above $85°C$.

Although the design and fabrication of microresonator arrays present many challenges, if the obstacles can be overcome, microresonator arrays can potentially result in high-performance sensors.

### Acknowledgments

The authors acknowledge and thank Paul Kruse for his helpful comments on preliminary reports on this work, and for his book (Kruse *et al.*, 1962) from which we learned some of the basics of IR detector technology; Fred Walls and Thomas Parker, National Institute of Standards and Technology (NIST); Charles Greenhall, Jet Propulsion Laboratory; and David Allan, Allan's TIME; for their helpful comments on the noise aspects; Professor Harry Tiersten of Renssalaer Polytechnic Institute for suggesting the use of ring electrodes and for analysis of the effects of displacement of levitated electrodeless resonators (Tiersten, 1995); and Al Benjaminson for suggesting circuitry that may be used to excite the microresonators.

## Appendix. Performance Calculations

The theoretical calculation for various figures of merit of IR detectors as a function of resonator frequency is shown in this Appendix.

The thermal time constant of the detector $\tau_T$ is given by (Kruse et al., 1962)

$$\tau_T = \frac{C}{G} \tag{A1}$$

where $C$ is the heat capacity of the detector and $G$ is the thermal conductance due to both conduction via the mounting structure and radiation from the surface of the resonator.

The heat capacity of the quartz microresonator $C$ is

$$C = c_V V \tag{A2}$$

where $c_V = 2.08 \ \text{W·sec/cm}^{-3}\text{·K}$ is the specific heat of quartz and $V$ is the volume of the resonator.

When the mounting structure consists of quartz bridges,

$$G = \frac{k_q n w t}{L} + 4\sigma T^3 (2A) \tag{A3}$$

where $k_q = 0.07 \ \text{W/cm·K}$ is the assumed thermal conductivity of the quartz bridges; $\sigma = 5.67 \times 10^{-10} \ \text{W/cm}^2\text{·K}^4$ is the Stefan-Boltzmann constant; $n$, $w$, $t$, and $L$ are the number, width, thickness, and length of the bridges; and $A$ and $T$ are the area and temperature of the resonator, respectively. The first term is the contribution due to conduction and the second term is due to radiation. The factor $2A$ is included in the second term because the resonator radiates from both the front and back surfaces.

The common practice in video application is to make the time constant be one third the frame time, that is, to make the thermal time constant approximately 10 msec for a 30-Hz video frame rate. (P. W. Kruse, private communication, 1994). The bridge length required to force the thermal time constant of the detector to be $\tau_T$ is, using Eqs. (A2) and (A3) in Eq. (A1) and rearranging,

$$L(\tau_T) = \frac{k_q n w t}{(c_V V / \tau_T) - 8\sigma T^3 A} \tag{A4}$$

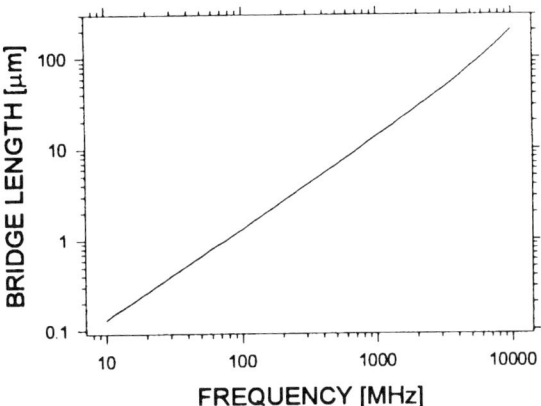

FIG. A.1. Bridge length for a response time of 10 msec as a function of resonator frequency.

The calculated value of $L$ for a 10-msec time constant is plotted in Fig. A1. For purposes of this calculation, it is assumed that the shape of the resonator is square, the frequency-thickness constant is 1.6 MHz·mm, the diameter-to-thickness ratio is 80, there are four identical bridges with square cross section (i.e., $w = t$ = resonator thickness), and $T = 300$ K.

In these temperature-sensitive microresonators, since the mass is extremely small and the temperature coefficient is large, background thermal fluctuation noise is significant. The mean square value of temperature fluctuations $\overline{\Delta T^2}$ of any system is given by (Kruse et al., 1962)

$$\overline{\Delta T^2} = \frac{\kappa_B T^2}{C} \tag{A5}$$

where $\kappa_B = 1.38 \times 10^{-23}$ W·sec/K is Boltzmann's constant.

The differential equation describing the heat transfer of the detector is given by (Kruse et al., 1962)

$$C \frac{d(\Delta T)}{dt} + G \Delta T = P(t) \tag{A6}$$

where $P(t)$ represents the power fluctuations from a source. We assume that the spectrum of $P(t)$ is independent of frequency (i.e., that the noise is white noise). The steady-state solution for the modulus of Eq. (A6) is

$$\overline{\Delta T_f^2} = \frac{S_P \Delta f}{G^2 + 4\pi^2 f^2 C^2} \tag{A7}$$

where $S_P$ is a constant. The mean square value of the fluctuations over all frequencies is obtained by integrating Eq. (A7):

$$\overline{\Delta T^2} = \int_0^\infty \frac{S_P\, df}{G^2 + 4\pi^2 f^2 C^2} = \frac{S_P}{4GC} \tag{A8}$$

Thus equating Eqs. (A5) and (A8) gives the value of $S_P$ to be $4\kappa_B G T^2$.

Substituting the value of $S_P$ into Eq. (A7) leads to the spectral density of the noise due to temperature fluctuations:

$$S_{\Delta T}(f) = \frac{4\kappa_B T^2/G}{1 + 4\pi^2 f^2 C^2/G^2} \tag{A9}$$

This can be regarded as a white-noise source, with spectral density of $4\kappa_B T^2/G$, filtered by a single-pole low-pass filter with cut-off frequency $f_{c0} = 1/(2\pi\tau_T)$, as follows:

$$S_{\Delta T}(f) = \frac{4\kappa_B T^2/G}{1 + f^2/f_{c0}^2} \tag{A10}$$

The power spectral density of thermal fluctuation–induced frequency noise is the power spectral density of temperature fluctuations multiplied by the square of the temperature coefficient $\alpha_T$, that is, $S_y(f) = S_{\Delta T}(f)\alpha_T^2$.

We ignore the noise contribution of the radiative exchange between the detector and its surroundings because, in the structure under consideration, the conduction is the dominant thermal transport mechanism.

In addition to the temperature fluctuation noise, two noise processes are typically observed (T. E. Parker, NIST, private communication, 1994):

1. Flicker of frequency, where the power spectral density of frequency fluctuations follows a $1/f$ characteristic, that is, $S_y(f) = h_{-1}/f$, where $h_{-1}$ is a constant. We assumed that $\sigma_y = 1.2 \times 10^{-19}\, \text{Hz}^{-1} \cdot f_0$ where $f_0$ is the resonator frequency in Hz (T. E. Parker, NIST, private communication, 1994), and it can be shown that

$$h_{-1} = \frac{\sigma_y^2}{2\ln(2)} = \frac{(1.2 \times 10^{-19}\, \text{Hz}^{-1} \cdot f_0)^2}{2\ln(2)} \tag{A11}$$

2. White phase modulation where the power spectral density of frequency fluctuations follows a $f^2$ characteristic, that is, $S_y(f) = h_2 f^2$, where $h_2$

is a constant taken to be the value necessary to make the knee of the Allan deviation curve occur at $\tau_{knee} = 1$ msec for a 500-MHz device with a 40-MHz cut-off filter $f_c$.

$$h_2 = \frac{(1.2 \times 10^{-19} \text{ Hz}^{-1} \cdot f_0)^2 (8\pi)(\tau_{knee})^2}{3 f_c} \quad \text{(A12)}$$

There is also white frequency noise (Johnson noise) that is independent of offset frequency, that is, $S_y(f) = h_0$. White frequency noise has a power spectral density that depends on the loaded $Q$, $Q_L$, and the power dissipated in the resonator $P$.

$$h_0 = \frac{\kappa_B T}{Q_L^2 P} \quad \text{(A13)}$$

This usually is not observed because its level is usually below the other two noise processes. The relatively low power dissipation and the need to make rapid measurements may make white frequency noise a factor for microresonator arrays.

Figure A2, shows the four contributions for a 500-MHz resonator using the noise floor and knee assumptions stated earlier. The sum of these four noise processes is shown in Fig. A3.

The specific detectivity $D^*$ being a function of the modulation frequency, is analogous to the noise spectrum. The frequency dependence of $D^*$ can be

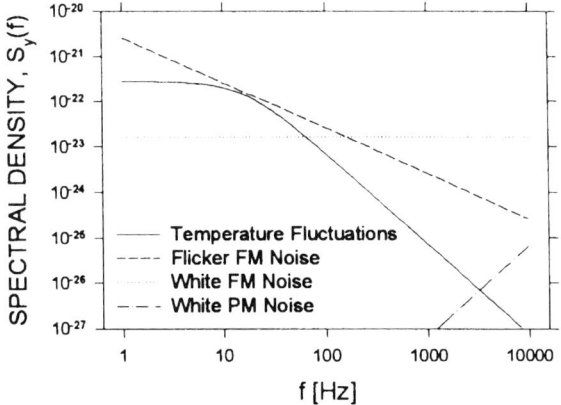

FIG. A.2. Power spectral density $S_y(f)$ versus offset frequency $f$ of white phase, white frequency, flicker of frequency, and thermal noise processes for a 500-MHz resonator.

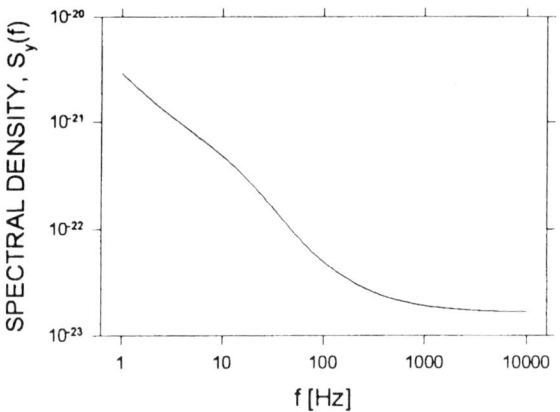

FIG. A.3. Sum of white phase, white frequency, flicker of frequency, and thermal noise processes.

expressed as

$$D^*(f) = \frac{R(f)A^{1/2}}{S_y(f)} \quad (A14)$$

where $S_y(f)$ is the power spectral density of frequency fluctuations in the detector. The responsivity $R$ is defined by (Kruse et al., 1962)

$$R(f) = \frac{\eta \alpha_T}{G(1 + (2\pi f)^2 \tau_T^2)^{1/2}} \quad (A15)$$

where $\eta$ is the optical absorptance multiplied by the fill factor, and $\alpha_T = 1/f_0 \cdot (df/dT)$ is the temperature coefficient of the resonator frequency. The units of $R$ are watt$^{-1}$.

Figure A4 is $D^*(f)$ for three resonator frequencies, all with $\tau_T = 100$ msec. Figure A5 is the thermal conductance required to obtain 10- and 100-msec time constants as a function of resonator frequency (with a constant diameter-to-thickness ratio of 80). The temperature increase due to self-heating caused by the drive level can be readily determined from Fig. A5 and Eq. (2).

Figure A6 is $D^*$ as a function of $G$ (and $\tau_T$) for 600 MHz at the frequency, where $D^*$ has a maximum value, $f = 1/(2\pi\tau_T)$ (Kruse et al., 1962). For comparison, the values of $D^*$ of other uncooled thermal detectors are

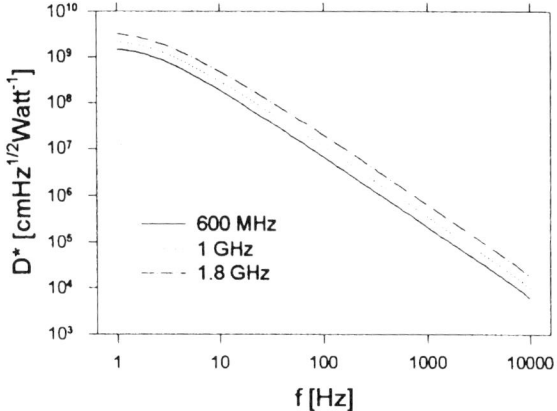

FIG. A.4. Detectivity $D^*$ as a function of modulation frequency $f$ and resonator frequency, with $\tau = 100$ msec.

quoted: $1 - 3 \times 10^8$ at 15 Hz for a thermistor bolometer, $1.1 \times 10^8$ at 100 Hz for a ferroelectric bolometer (Hudson, 1969), $10^9$ at 10 Hz for a pyroelectric detector, approximately $3 \times 10^9$ at 2 Hz for a spectroscopic thermopile, and approximately $6 \times 10^9$ at 2 Hz for an alanine-doped triglycine sulfate (TGS) detector (Keyes, 1980).

Although the noise spectrum is a useful analytical tool, when we make real measurements we actually integrate the spectrum over some band of

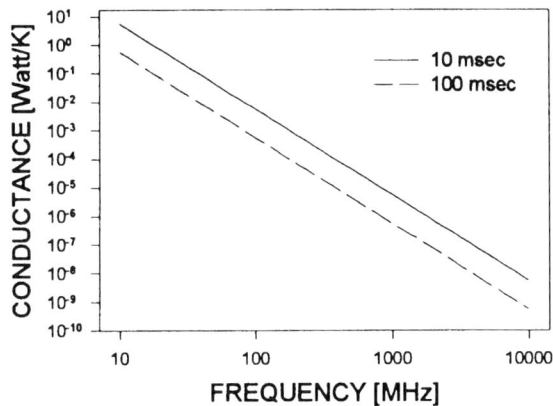

FIG. A.5. Thermal conductance for a response time of 10 msec, and 100 msec as a function of resonator frequency.

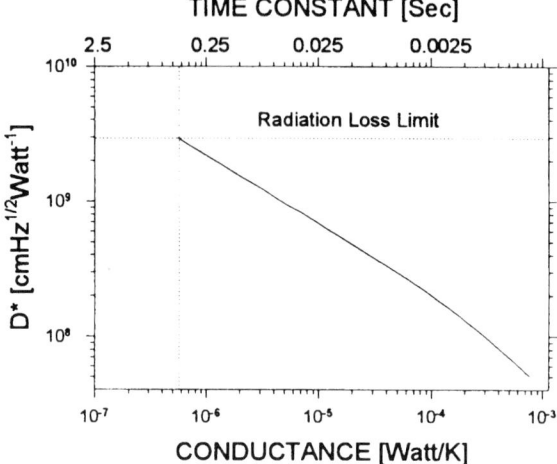

FIG. A.6. $D^*$ versus $G$ (and $\tau_T$) at $f = 1/(2\pi\tau_T)$ for 600 MHz.

frequencies. A measure of the dispersion, that is, noise, of the measurements is the variance (or its square root, the deviation).

In general, the variance of a noise process (or the sum of processes) is given by (Lesage and Audoin, 1979)

$$\langle \sigma_y^2(N, r, \tau) \rangle = \frac{N}{N-1} \int_0^\infty S_y(f) \frac{\sin^2(\pi f \tau)}{(\pi f \tau)^2} \left[ 1 - \frac{\sin^2(\pi r f N \tau)}{N^2 \sin^2(\pi r f \tau)} \right] df \quad (A16)$$

where $N$ is the number of samples, $\tau$ is the measurement time, and $r$ is the ratio $(\tau + \tau_d)/\tau$, where $\tau_d$ is the dead time between measurements. If there is no dead time, then $r = 1$.

If $N = \infty$ and $\tau = 0$, we have

$$\langle \sigma_y^2(\infty, r, 0) \rangle = \int_0^\infty S_y(f) \, df \quad (A17)$$

independent of $r$, which is the standard definition of the variance of a continuous variable.

If $N = \infty$ and $\tau$ is finite, we have the standard variance of the averaged frequency fluctuations

$$\langle \sigma_y^2(\infty, r, \tau) \rangle = \int_0^\infty S_y(f) \frac{\sin^2(\pi f \tau)}{(\pi f \tau)^2} \, df \quad (A18)$$

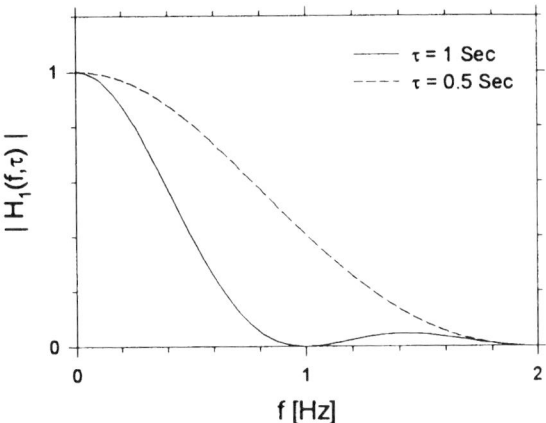

FIG. A.7. Filter shape for the ∞-sample variance for $\tau = 1$ sec and $\tau = 0.5$ sec.

which is the total power output from a filter with $S_y(f)$ as input and transfer function $H_1(f, \tau)$ where

$$H_1(f, \tau) = \frac{\sin^2(\pi f \tau)}{(\pi f \tau)^2} \tag{A19}$$

On a linear plot, the response of this filter is shown in Fig. A7. The total area under the curve is $1/(2\tau)$; therefore, a measurement of $\tau$ sec looks like a bandwidth of $1/(2\tau)$ (in Hz).

A problem arises in noise processes such as flicker, where the noise power increases as the frequency decreases like $1/f^n$. In that case the integral depends on $N$ and may not converge.

If $N$ and $\tau$ are finite and there is no dead time, the filter looks like a bandpass filter whose lower band edge varies as $1/(N\tau)$.

If $N = 2$ and there is no dead time, we have the definition of the Allan variance, which looks like a filter with transfer function $H_2(f, \tau)$:

$$H_2(f, \tau) = 2 \frac{\sin^2(\pi f \tau)}{(\pi f \tau)^2} \left[ 1 - \frac{\sin^2(\pi f 2\tau)}{2^2 \sin^2(\pi f \tau)} \right] = 2 \frac{\sin^4(\pi f \tau)}{(\pi f \tau)^2} \tag{A20}$$

This filter looks like Fig. A8. Both the bandwidth and center frequency of the bandpass filter move with averaging time. The decrease in the bandwidth, as the frequency decreases, is by exactly the right amount to cancel out the $1/f$ dependence of the flicker noise. This is why flicker of frequency is the only noise process whose Allan variance is independent of averaging time. This integral also converges for random walk ($1/f^2$) frequency noise.

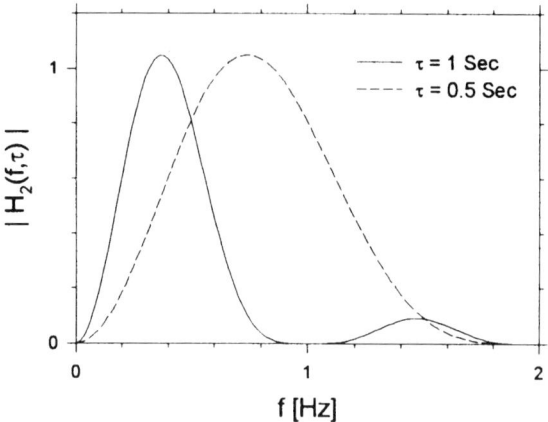

FIG. A.8. Filter shape for the two-sample variance for $\tau = 1$ sec and $\tau = 0.5$ sec.

Physically, the Allan variance is a measure of the fluctuations from measurement to measurement when the measurement intervals are contiguous.

The filter function of the two-sample variance when the measurements are not contiguous depends on the frame interval $T$ and the measurement interval $\tau$. The function is

$$H_3(f, T, \tau) = 2\frac{\sin^2(\pi f \tau)}{(\pi f \tau)^2}\left[1 - \frac{\sin^2(\pi f N T)}{N^2 \sin^2(\pi f T)}\right] \quad \text{(A21)}$$

Figure A9 shows the filter functions for a $T/\tau$ ratio of 1, and a ratio of 10 for $\tau = 30$ μsec. The effect of the dead time is to increase the contribution of low frequencies.

The two-sample variance of the spectrum of thermal fluctuations is given by (Lesage and Audoin, 1979)

$$\sigma_{\Delta T}^2(T, \tau) = \int_0^\infty (4\kappa_B T^2/G)\frac{1}{1 + (f/f_{c0})^2}|H_3(f, T, \tau)|^2\, df \quad \text{(A22)}$$

The two-sample deviation with no dead time for the noise discussed previously is shown in Fig. A10 for a 500-MHz resonator frequency. The effect of increasing the dead time is shown in Fig. A11. A frame rate of 30 Hz is equivalent to a repetition ratio of 1100 for $\tau = 30$ μsec. Both thermal and flicker noise increase significantly when dead time is introduced.

## 9 QUARTZ MICRORESONATORS

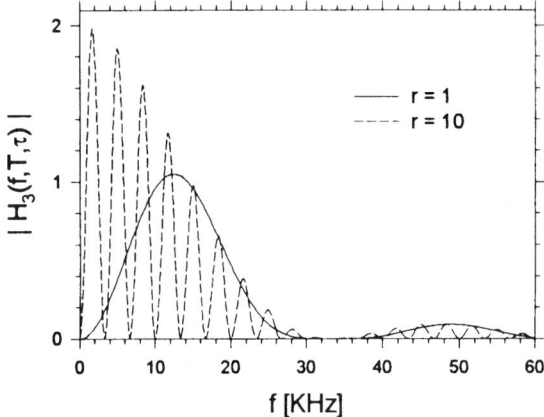

FIG. A.9. Filter shape for the two-sample variance for $r = 1$ and $r = 10$.

The total frequency noise $\sigma_N$ is given by a root mean square of the temperature fluctuations times the temperature coefficient and the oscillator noise:

$$\sigma_N = \sqrt{(\sigma_{\Delta T}\alpha_T)^2 + (\sigma_y)^2} \qquad (A23)$$

The noise equivalent temperature difference (NETD) is defined as (Johnson

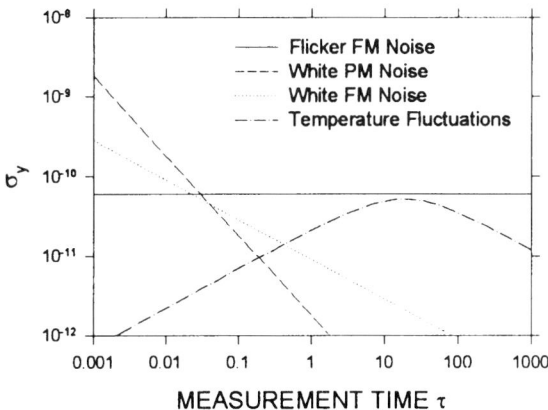

FIG. A.10. Two-sample variance of noise processes as a function of measurement time at $r = 1$ for the resonator frequency of 500 MHz.

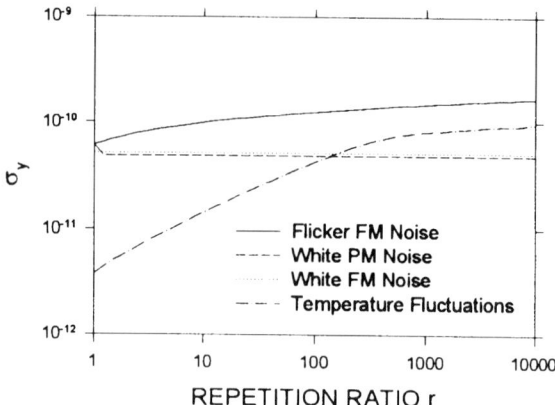

FIG. A.11. Two-sample variance of noise processes as a function of repetition ratio $r$.

and Kruse, 1993)

$$\text{NETD} = \frac{(4F^2 + 1)\sigma_N}{\tau_0 A(\Delta P/\Delta T)_{\lambda_1 - \lambda_2} R} \quad \text{(A24)}$$

where $F$ is the numerical aperture of the optics, $\tau_0$ is the transmittance of the atmosphere between the target and the sensor, and $(\Delta P/\Delta T)_{\lambda_1 - \lambda_2}$ is the temperature dependence of the blackbody function over the wavelength interval from $\lambda_1$ to $\lambda_2$, respectively. For thermal imaging systems operating in 8- to 14-$\mu$m atmospheric window, the value of $(\Delta P/\Delta T)_{\lambda_1 - \lambda_2}$ is 2.62 W/m$^2 \cdot$K (P. W. Kruse, private communication, 1994).

## References

Advena, D. J., Bly, V. T., and Cox, T. J. (1993). *Appl. Opt.* **32**, 1136–1144.

Ballato, A. D., Hatch, E. R., Mizan, M., Lukaszek, T. J., and Tilton, R. (1985). *Proc. 39th Annu. Symp. Freq. Control*, pp. 462–472.

Belser, R. B., and Hicklin, W. H. (1967). *Proc. 21st Annu. Symp. Freq. Control*, pp. 211–223.

Besson, R. J. (1977). *Proc. 31st Annu. Symp. Freq. Control*, pp. 147–152.

Bottom, V. E. (1982). "Introduction to Quartz Crystal Unit Design." Van Nostrand-Reinhold, New York.

Cho, D., Kumar, S., and Carr, W. (1991). "Electrostatic Levitation Control System for Micromechanical Devices." U.S. Pat. 5,015,906.

Demeis, R. (1995). *Laser Focus World*, July, pp. 105–112.

Driscoll, M. M., and Hanson, W. P. (1993). *Proc. IEEE Int. Freq. Control Symp., 1993*, pp. 186–192.

EerNisse, E. P., Ward, R. W., and Wiggins, R. B. (1988). *IEEE Trans. Ultrason. Ferroelect. Freq. Control* **35**, 323–330.
Filler, R. L., and Vig. J. R. (1989). *Proc. 43rd Annu. Symp. Freq. Control*, pp. 8–15.
Gagnepain, J.-J., Hauden, D., Coquerel, R., and Pegeot, C. (1983). U.S. Pat. 4,398,115.
Gerber, E. A., and Ballato, A., eds. (1985). "Precision Frequency Control." Academic Press, New York.
Gorini, I., and Sartori, S. (1962). *Rev. Sci. Instrum.* **33**, 883–884.
Hadley, L. N., and Dennison, D. M. (1947). *J. Opt. Soc. Am.* **37**, 451–465.
Hammond, D. L., and Cutler, L. S. (1967). U.S. Pat. 3,339,091; reissue 26,707 (1969).
Hammond, D. L., Adams, C. A., and Schmidt, P. (1965). *ISA Trans.* **4**, 349–354.
Hamrour, M. R., and Galliou, S. (1994). *Proc. Ultrason. Symp., 1994*, pp. 513–516.
Hanson, C. M. (1993). *Proc. SPIE* **2020** (Infrared Technol. XIX), 330–339.
Havens, O. S. (1955). "Optical Properties of Thin Solid Films," Chapters 4 and 6. Butterworths, Washington, D.C.
Heising, R. A. (1946). "Quartz Crystals for Electrical Circuits," pp. 26–27, 32. Van Nostrand, New York.
Hudson, R. D. (1969). "Infrared System Engineering." Wiley, New York.
Hunt, J. R., and Smythe, R. C. (1985). *Proc. 39th Annu. Freq. Control Symp.*, pp. 292–300.
Johnson, B. R., and Kruse, P. W. (1993). *Proc. SPIE* **2020** (Infrared Technol. XIX), 2–11.
Keyes, R. J. ed. (1980). "Topics in Applied Physics," Vol. 19, esp. Chapter 3. Springer-Verlag, Berlin.
Kruse, P. W. (1995). *Infrared Phys. Technol.* **36**, 869–882.
Kruse, P. W., McGlauchlin, L. D., and McQuistan R. B. (1962). "Elements of Infrared Technology." Wiley, New York.
Kumar, S., Cho, D., and Carr, W. N. (1992). *J. Microelectromechan. Syst.* **1**, 23–30.
Kusters, J. A., Fisher, M. C., and Leach, J. G. (1978). *Proc. 32nd Annu. Symp. Freq. Control*, pp. 389–397.
Lang, W., Kühl, K., and Sandmaier, H. (1991). *Transducers '91, Int. Conf. Solid States Sens. Actuators, Dig. Tech. Pap.*, IEEE Cat. No. 91CH2817-5, pp. 635–638.
Lang, W., Kühl, K., and Sandmaier, H. (1992). *Sens. Actuators, A: Phys.* **34**, 243–248.
Lee, K. C. (1990). *J. Electrochem. Soc.* **137**, 2556–2574.
Lesage, P., and Audoin, C. (1979). *Radio Sci.* **14**, 521–539.
Liddiard, K. C. (1993). *Infrared Phys.* **34**, 379–387.
McCarthy, D. E. (1963). *Appl. Opt.* **2**, 591–595.
Montress, G. K., and Parker, T. E. (1994). *Proc. IEEE Freq. Control Symp., 1994*, pp. 365–373.
Nakamura, K., Yasuike, R., Hirama, K., and Shimizu, H. (1990). *Proc. 44th Annu. Symp. Freq. Control*, pp. 372–377.
Nakazawa, M., Yamaguchi, H., Ballato, A. D., and Lukaszek, T. J. (1984). *Proc. 38th Annu. Freq. Control Symp.*, pp. 240–244.
Palik, E. D. (1985). "Handbook of Optical Constants of Solids," Vol. 1. Academic Press, Orlando, FL.
Parker, T. E. (1985). *Appl. Phys. Lett.* **46**, 246–248.
Parker, T. E. (1987). *Proc. 41st Annu. Symp. Freq. Control*, pp. 99–110.
Parker, T. E., and Andres, D. (1993). *Proc. IEEE Int. Freq. Control Symp., 1993*, pp. 178–185.
Parsons, A. D., and Pedder, D. J. (1988). *J. Vac. Sci. Technol., A* **6**(3), 1686–1689.
Pelrine, R. E. (1990). *Proc. IEEE MicroElectroMech. Sys.*, IEEE Cat. No. 90CH2832-4, pp. 35–37.
Ralph, J. E., King, R. C., Curran, J. E., and Page, J. S. (1985). *Proc. Ultrason. Symp., 1985*, pp. 362–364.
Schodowski, S. S. (1989). *Proc. 43rd Annu. Symp. Freq. Control*, pp. 2–7.

Sinha, B. (1981). *Proc. 35th Annu. Freq. Control Symp.*, pp. 213–221.
Smith, W. L., and Spencer, W. J. (1963). *Rev. Sci. Instrum.* **34**, 268–70.
Smythe, R. C., and Angove, R. B. (1988). *Proc. 42nd Annu. Freq. Control Symp.*, pp. 73–77.
Smythe, R. C., and Tiersten, H. F. (1987). *Proc. 41st Annu. Symp. Freq. Control*, pp. 311–313.
Spassov, L. (1992). *Sens. Actuators A: Phys.* **30**, 67–72.
Spitzer, W. G., and Kleiman, D. A. (1961). *Phys. Rev.* **121**, 1324–1335.
Tiersten, H. F. (1995). *IEEE Int. Freq. Control Symp.*, pp. 740–745.
Vig, J. R., and Walls, F. L. (1994). *Proc. IEEE Int. Freq. Control Symp., 1994*, pp. 506–523.
Vig. J. R., LeBus, J. W., and Filler, R. L. (1977). *Proc. 31st Annu. Symp. Freq. Control*, pp. 131–143.
Wade, W. H., and Slutsky, L. J. (1962). *Rev. Sci. Instrum.* **33**, 212–213.
Walls, F. L. (1992). *Proc. IEEE Int. Freq. Control Symp., 1992*, pp. 327–333.
Walls, F. L., and Vig, J. R. (1995). *IEEE Trans. Ultrason. Ferroelect., Freq. Control* **42**, 576–589.
Warner, A. W., Jr., and Goldfrank, B. (1985). *Proc. 39th Annu. Symp. Freq. Control*, pp. 473–474.
Wood, R. A. (1993). *Proc. SPIE* **2020** (Infrared Technol. XIX), 322–329.
Wood, R. A., and Foss, N. A. (1993). *Laser Focus World*, June, pp. 101–106.
Yong, Y. K., and Vig, J. R. (1988). *Proc. 42nd Annu. Freq. Control Symp.*, pp. 397–403.
Ziegler, H. (1983). *Sens. Actuators* **5**, 169–178.
Ziegler, H., and Tiesmeyer, J. (1983). *Sens. Actuators* **4**, 363–367.

CHAPTER 10

# Application of Uncooled Monolithic Thermoelectric Linear Arrays to Imaging Radiometers

*Paul W. Kruse*

Infrared Solutions, Inc.
Minneapolis, Minnesota

I. INTRODUCTION . . . . . . . . . . . . . . . . . . . . . . . . . . . 297
II. IDENTIFICATION OF INCIPIENT FAILURE OF RAILCAR WHEELS . . . . . . . . . . 298
   1. *Technical Description of the Model IR 1000 Imaging Radiometer* . . . . . . . 298
   2. *Performance of the Model IR 1000 Imaging Radiometer* . . . . . . . . . . 300
   3. *Initial Application* . . . . . . . . . . . . . . . . . . . . . . . . 304
   4. *Summary* . . . . . . . . . . . . . . . . . . . . . . . . . . . . 309
III. IMAGING RADIOMETER FOR PREDICTIVE AND PREVENTIVE MAINTENANCE . . . . 309
   1. *Description* . . . . . . . . . . . . . . . . . . . . . . . . . . . 310
   2. *Operation* . . . . . . . . . . . . . . . . . . . . . . . . . . . . 312
   3. *Specifications* . . . . . . . . . . . . . . . . . . . . . . . . . . 317
   4. *Summary* . . . . . . . . . . . . . . . . . . . . . . . . . . . . 317
   *References* . . . . . . . . . . . . . . . . . . . . . . . . . . . . 318

## I. Introduction

This chapter describes the application of uncooled silicon microstructure thermoelectric linear arrays to imaging radiometers. Of all the types of uncooled arrays, the monolithic silicon thermoelectric arrays are the easiest to fabricate and the easiest to implement in imaging radiometers. Imaging radiometers require measurement of the temperature of every pixel of the scene being imaged. Thermoelectric arrays, which require one or more pairs of junctions for each pixel, automatically provide only the signal due to the absorbed radiant power, that is, there is no "pedestal" due to Joulean heating to subtract. Their responsivity is low, nevertheless their noise equivalent temperature difference (NETD) is adequate for many applications. Two such applications are described below.

## II. Identification of Incipient Failure of Railcar Wheels*

If one were to specify the properties of a system that would find widespread application in determining the temperature of fast-moving discrete objects, such as railroad train wheels or continuous material flow on a production or process line, the desirable attributes would include the following:

- High transverse spatial resolution
- High longitudinal spatial resolution
- Accurate temperature measurement
- Wide dynamic range of target temperatures
- Low system power

Infrared Solutions, Inc. (ISI) has developed the Model IR 1000 high-speed infrared imaging radiometer we believe meets these attributes. Details of the imaging radiometer, developed by ISI, and its thermoelectric linear array, developed by the Honeywell Technology Center (HTC; Plymouth, MN), are presented below. Excellent results have been achieved by the Science Applications International Corporation (SAIC), which is using the imaging radiometer in a new track-side system to read the temperature and temperature distribution of wheels and bearings on fast-moving trains.

### 1. TECHNICAL DESCRIPTION OF THE MODEL IR 1000 IMAGING RADIOMETER

The Model IR 1000 imaging radiometer employs an uncooled thermoelectric linear array consisting of 96 pixels, each 150 by 150 μm. Each of the 96 pixels consists of 30 series-connected chromel–constantan thermoelectric junctions deposited on a silicon nitride ($Si_3N_4$) thermal isolation structure, (Fig. 1). This structure, based on Si micromachining technology, was first discussed by HTC in 1991 (Listvan et al., 1991; Wilson et al., 1991). A similar structure is employed in the ISI IR SnapShot® camera (Wood et al., 1995). The use of Si micromachining technology makes possible changes in pixel size, shape, and speed of response by means of mask changes, while the process for preparing the arrays remains unchanged. The linear arrays are packaged in ceramic multipin packages containing a germanium window and evacuated to prevent sensitivity reduction owing to heat conduction through air (Fig. 2).

*Reprinted from Kruse, P. W., Schmidt, R. N., and Danyluk, J. D. (1996). High speed imaging radiometer employing uncooled thermoelectric linear array. Presented at the SPIE Thermosense Meeting, Orlando, FL; April; with permission of the authors and the SPIE.

10 APPLICATION OF LINEAR ARRAYS TO IMAGING RADIOMETERS 299

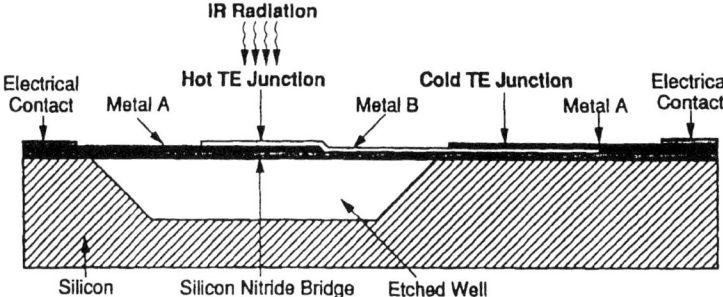

FIG. 1. Thermoelectric pixel structure. IR = infrared; TE = thermoelectric. Reprinted from Kruse et al. (1996) with permission from the authors and the SPIE.

FIG. 2. The Model IR 1000 electronics board with an evacuated array package. Reprinted from Kruse et al. (1996) with permission from the authors and the SPIE.

The spectral response of the pixels, determined by their optical absorption properties, is largely wavelength-independent. Thus, for room temperature targets, most of the radiant power is in the 8 to 12 μm spectral region. The pixel thermal response time, that is, the ratio of pixel heat capacity to the thermal conductance of the pixel legs, is designed to be 0.75 msec to provide high-speed operation. Because of the tradeoff in responsivity with speed (the inverse of the response time) at short response times (Kruse, 1995), it is necessary to reduce the system noise by limiting the system bandwidth. Accordingly, each pixel has its own preamplifier providing a gain of 4000, followed by an integrate and dump filter, which integrates the signal for 0.8 msec. Each pixel is scanned for 1 msec. The signals from the pixels are multiplexed and digitized. A proprietary algorithm translates the digitized signal into absolute temperature. Because of the parallel nature of the pixel output, the scan time of an entire linear array of 96 pixels is 1 msec. Translating this into more familiar terms, the Model IR 1000 imaging radiometer can image objects or processes moving at speeds of up to 80 mph.

The spatial resolution of the Model IR 1000 is determined by the pixel size and the focal length of the optics. Most frequently employed is an f/0.8, 1.8-cm focal length germanium (Ge) lens. This lens provides a pixel instantaneous field of view (IFOV) of 8.33 mrad (0.477 degree) square, and a total field of view for the 96 pixels of 800 by 8.33 mrad, or 45.8 by 0.477 degrees. With this lens, the NETD for room temperature targets is 0.7°C. The target temperature dynamic range is factory-adjustable. Units described here have a dynamic range from less than 30°C to greater than 430°C. The system power requirement is about 25 W.

Figure 3 is a photograph of the ISI Model IR 1000 imaging radiometer. The circular object above the lens is a motorized shutter, employed for temperature calibration. Figure 4 shows the unit disassembled.

## 2. Performance of the Model IR 1000 Imaging Radiometer

The prototype Model IR 1000 imaging radiometer described here is used with a host computer that contains the calibration coefficients and converts the scanner output of digital counts to temperature. Future production imaging radiometers will output calibrated temperatures directly.

System performance of individual pixels was evaluated for a variety of target temperatures with the scanner at different environmental temperatures. Data for target temperatures from 30°C to 390°C and environmental temperatures at room temperature and 60°C are presented in Figs. 5 and 6 for the prototype imaging radiometer, Model IR 1000 serial number

10 APPLICATION OF LINEAR ARRAYS TO IMAGING RADIOMETERS 301

FIG. 3. The Model IR 1000 high-speed imaging radiometer. Reprinted from Kruse *et al.* (1996) with permission from the authors and the SPIE.

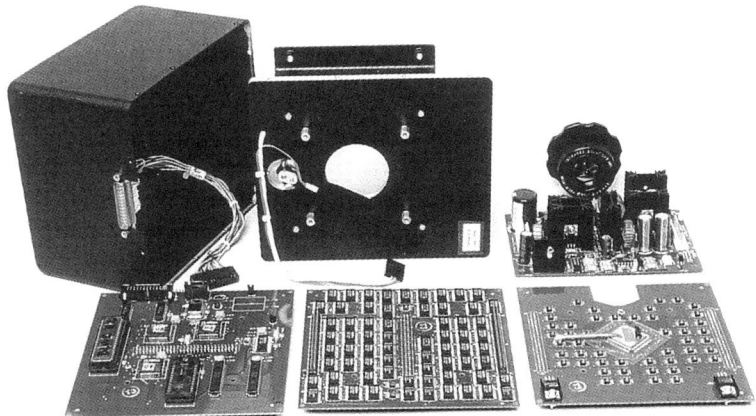

FIG. 4. The Model IR 1000 high-speed imaging radiometer disassembled. Clockwise: enclosure with wire harness, front cover and mounting bracket with shutter and shutter motor lens, 24-V power supply, analog board 1 with uncooled linear array, analog board 2, and digital board. Reprinted from Kruse *et al.* (1996) with permission from the authors and the SPIE.

K. The system's responsivity, noise, and NETD for each pixel are shown in Figs. 5 and 6.

System responsivity, in counts per degree celcius, was calculated by first obtaining 95 independent test readings, $V_t$, for each of the 96 pixels with the imaging radiometer shutter open. Data were taken for a fixed temperature blackbody target, with the imaging radiometer at a fixed environmental temperature. The process was repeated with the shutter closed to yield $V_s$ readings. The $V_t$ and $V_s$ readings were averaged over the 95 independent readings and the differences in the averages ($\bar{V}_t - \bar{V}_s$) were computed. The system responsivity was calculated by subtracting the average difference values [$\bar{V}_t - \bar{V}_s$), for two consecutive target temperatures and dividing by the difference in the target temperatures.

For example, the system responsivity at 60°C was computed by subtrac-

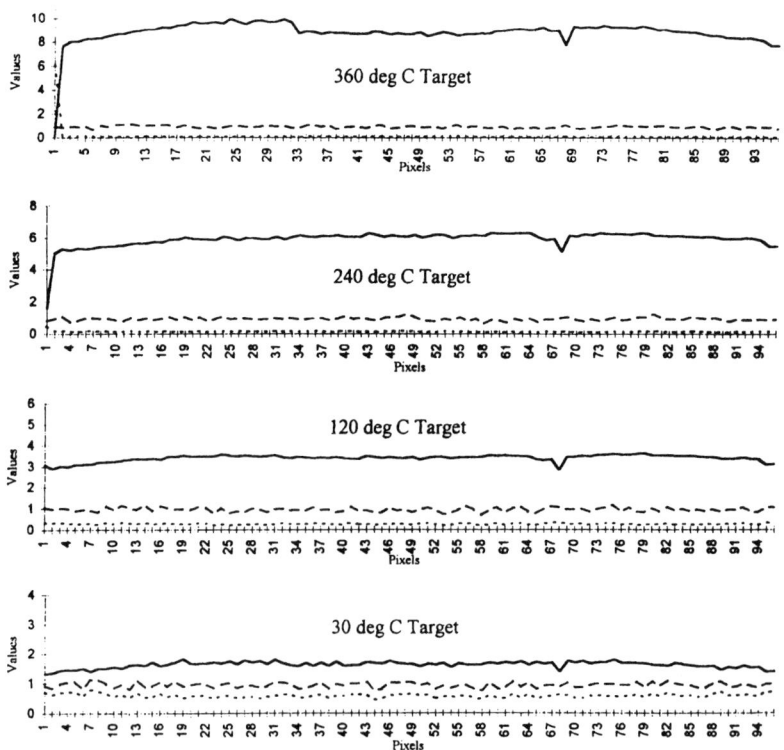

FIG. 5. Performance of the prototype serial number K at room temperature. —— = responsivity (counts/°C); ---- = noise (counts); ---- = NETD (°C). Reprinted from Kruse et al. (1996) with permission from the authors and the SPIE.

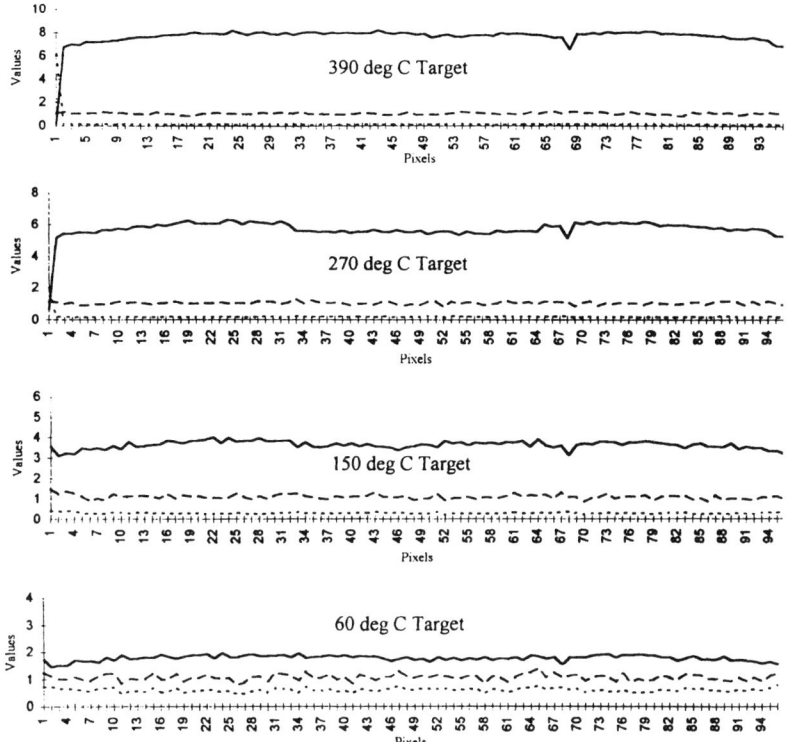

FIG. 6. Performance of the prototype serial number K at 60°C. —— = responsivity (counts/°C); ---- = noise (counts); ···· = NETD (°C). Reprinted from Kruse et al. (1996) with permission from the authors and the SPIE.

ting the $(\bar{V}_t - \bar{V}_s)$ values at 60°C from the values at 90°C and dividing by the temperature difference, 30°C. The resulting system responsivity values are plotted on the graphs. This process was repeated for different target temperatures with the environmental temperature remaining relatively constant and then repeated for different environmental temperatures.

System noise, in counts, was calculated as the standard deviation of $V_s$ over the 95 samples for each of the 96 pixels. It, too, is plotted in the graphs.

System NETD, in degrees Celcius, was calculated by dividing the system noise value for each pixel by the system responsivity of that pixel at the same target temperature for which the noise data were recorded. These data are also plotted in Figs. 5 and 6.

Note that some of the plots in Figs. 5 and 6 show poor data for pixel number 1. This has been indentified as a multiplexer problem and has been corrected. The prototype IR 1000 imaging radiometer meets or exceeds our design requirements.

3. INITIAL APPLICATION

The initial application of the IR 1000 infrared imaging radiometer is to detect defective bearings and brakes on railroad equipment traveling at speeds of up to 80 mph. Potential advantages over existing technology include more accurate defect detection, reduced maintenance, and increased reliability.

Excessive heat in a railcar wheel bearing (called a *hot box*) is an indication of a defective bearing that requires immediate corrective action. Railcar roller bearings can—and do—burn off, in as few as 2 miles, creating the potential for derailment. Railroads use bearing defect detectors to provide, in real time, early warning of impending bearing failure. Railroads also use hot wheel detectors to identify railcars with defective brakes.

Currently available sensing technologies, using carefully aligned single-element thermistor bolometers or pyroelectric detectors, have limitations. They generally have poor reliability, poor repeatability, and high maintenance, and they cause many false alarms. Railroads report false-alarm rates exceeding 50%, resulting in thousands of unnecessary train delays.

SAIC has designed and built a prototype system, known as Thermalview, which uses the Model IR 1000 imaging radiometer to detect defective railcar bearings and brake systems. Thermalview consists of two Model IR 1000 infrared imaging radiometers interfaced to a ruggedized STD bus, a card cage-based microcomputer suited for operation in the harsh railroad environment. One imaging radiometer is mounted on each side of the tracks to allow scanning of both sides of a passing train. The systems use dial access modem links for remote support. Train speed is calculated based on inputs from rail-mounted wheel detectors. This allows the use of pushbroom scanning by varying the line scan acquisition rate with train speed. An application-specific interface board is used to handle the high-speed acquisition of data from two Model IR 1000 imaging radiometers. Up to 32 MB of data is acquired by each imaging radiometer for a train sampled at 1-msec intervals. The Model IR 1000 imaging radiometers are mounted in ruggedized environmental housings for protection.

A prototype installation of the Thermalview system for in-transit inspection of train wheels is shown in Fig. 7. For this application, the imaging radiometer enclosures are mounted at a distance of 48 in. from the nearest rail, at a height of 18 in. above the top of the rail. The wide vertical field of

FIG. 7. Prototype Thermalview installation. Reprinted from Kruse et al. (1996) with permission from the authors and the SPIE.

view (48 degrees) provided by the Model IR 1000 imaging radiometer allows the thermal inspection of the entire wheel. An installed Thermalview enclosure, with railcar wheel in the field of view, is shown in Fig. 8.

Railcars with defective air brakes are identified by the large thermal gradient in the wheels. Two specific groups of defects have been identified to date:

1. Brakes that fail to properly release or apply
2. Brakes with excessive force that cause wheels to "slide"

Railroads are interested in identifying railcars with brakes that fail to release or apply in order to route defective equipment for maintenance. Failed brakes can be identified with the Thermalview system by monitoring temperatures of the wheel rim. A circular ring of heat as shown in Fig. 9 is clear evidence of a potential air brake defect on a train that would not normally be operating with its brakes applied. Similarly, the system could be used to identify railcars with cooler wheels, indicating railcars with brakes that fail to properly apply.

A more serious situation was detected during the initial testing of the prototype system. Wheels were identified that had stopped rotating because of excessive braking force from a defective braking system. This situation

Fig. 8. Line scan camera enclosure with railcar wheel in the field of view. Reprinted from Kruse et al. (1996) with permission from the authors and the SPIE.

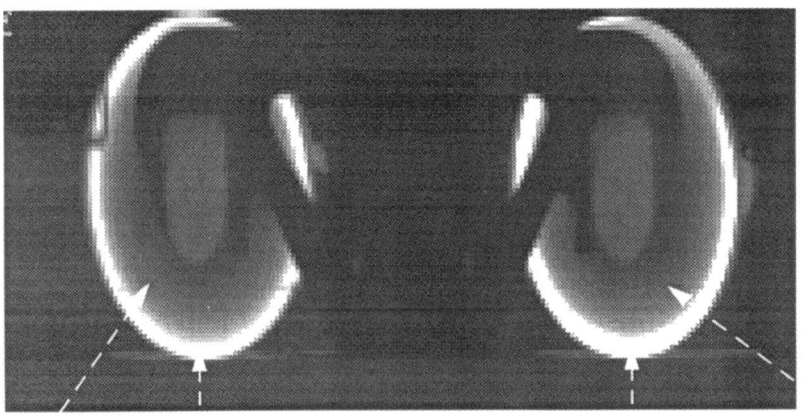

140 F     248 F            266 F     149 F

Fig. 9. Thermal profile of railcar wheels with brakes applied. Reprinted from Kruse et al. (1996) with permission from the authors and the SPIE.

10 APPLICATION OF LINEAR ARRAYS TO IMAGING RADIOMETERS 307

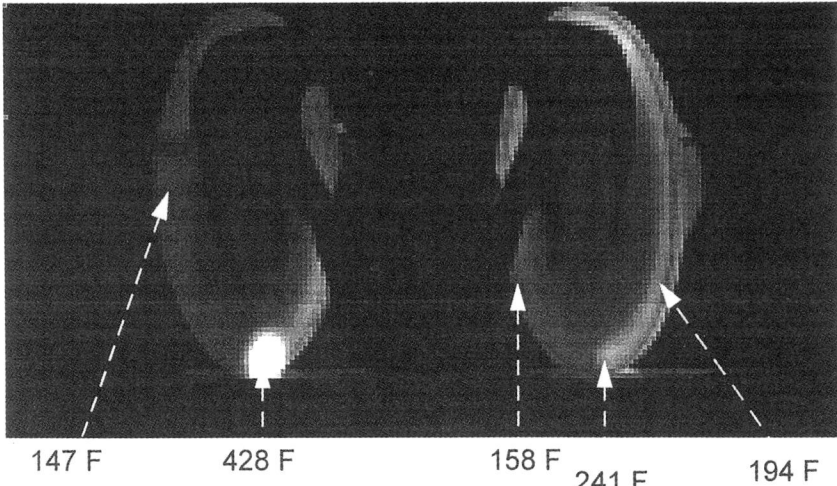

FIG. 10. Thermal profile of "sliding" railcar wheels. Reprinted from Kruse *et al.* (1996) with permission from the authors and the SPIE.

FIG. 11. A railcar wheel with a "flat" spot caused by "sliding." Reprinted from Kruse *et al.* (1996) with permission from the authors and the SPIE.

TABLE I
MODEL IR 1000 LINE SCANNER SPECIFICATIONS

|  | Design | Measured |
|---|---|---|
| Scan rate | 1000 Hz | 1000 Hz |
| Scan angle (field of view) | 46 degrees | 48 degrees |
| Spot size (instantaneous field of view) | 0.48 degree | 0.51 degree |
| Noise equivalent temperature difference (NETD)* | 0.7°C | 0.7°C |
| Accuracy | ±4°C or 2% | Calibrated to ±2.5°C or 2% |
| Focus |  |  |
|   Standard | 4 in. to infinity | — |
|   Option | Lens dependent |  |
| Target temperature range |  |  |
|   Standard | 0 to 500°C | <30°C to 400°C |
|   Option | Factory adjustable |  |
| Lens |  |  |
|   Standard | 18 mm, f/0.8 | — |
|   Option | Application dependent |  |
| Array |  |  |
|   Type | Uncooled thermoelectric | — |
|   Size | 96 pixel, 150 $\mu$m square |  |
| Spectral band |  |  |
|   Standard | 8 to 12 $\mu$m | — |
|   Option | Optics-dependent |  |
| Operating temperature | 0 to 50°C | −20°C to 60°C |
| Power | 25 W | 19.2 W with external power supply |
| Size (without lens and mount) | 7.5 × 6.5 × 4.5 in. | — |
| Weight (with lens and mount) | <10 lb | 5 lb |

*NETD for room temperature targets.

had occurred on a downgrade, in which heavy braking is normal operating procedure. Once the brakes are released, the flat spot created by the "sliding" wheel can cause damage to the rails and bearings, and can possibly lead to derailment. Figure 10 shows the thermal profile of "sliding" wheels. Figure 11 shows the flat spot on the wheel that resulted from the "sliding."

The system has also demonstrated the ability to improve the identification of overheated railroad wheel bearings. For this application, the linescan camera enclosures are mounted at rail level, approximately 32 in. from the nearest rail. The camera is aimed upward at approximately a 45-degree angle to provide a profile view of the outer surface of the bearing cup. The resulting temperature profiles combined with digital analysis give the system a far more accurate detection capability than existing bearing defect detectors that use a single-element sensor.

### 4. Summary

The performance of the Model IR 1000 imaging radiometer is summarized in Table I. We believe that the Model IR 1000 imaging radiometer addresses very well the desirable attributes listed in the introduction of this chapter. The initial application of identifying hot wheel bearings (about to fail) and hot wheels (stuck brakes) on fast-moving trains has been successfully demonstrated.

### III. Imaging Radiometer for Predictive and Preventive Maintenance*

There is a large and rapidly growing commercial market for *imaging radiometers*, which are infrared cameras that can display a thermal image of a scene and determine the temperature of any point in that scene. The largest market segment is predictive and preventive maintenance. Here, potential problems with the equipment in the scene are identified by the higher-than-normal temperature owing to failure mechanisms. Examples might be a hot bearing on a motor, a hot connection in an electrical distribution system, or a warmer-than-normal spot on a thermally insulated device. Other segments include, but are not limited to, nondestructive testing, industrial process monitoring and control, integrated circuit inspection, and quality assurance.

Nearly all the imaging radiometers available today employ cryogenic photon detectors that scan or stare at the scene. The scanning approaches usually employ mercury cadmium telluride [(Hg, Cd)Te] detectors operating in the 8 to 12 $\mu$m atmospheric window or indium antimonide (InSb) detectors operating in the 3 to 5 $\mu$m atmospheric window. The staring approaches employ two-dimensional focal planes of (Hg, Cd)Te, InSb, and also platinum silicide (PtSi), which operates in the 3 to 5 $\mu$m atmospheric window (Stetson and Landry, 1994). Nearly all systems available today operate at a frame rate

---

*Most of this section is reprinted from Wood, R. A. Rozachek, T. M. Kruse, P. W., and Schmidt, R. N. (1995). IR SnapShot® Camera. *Proceedings of SPIE Conference on Infrared Technology 21st* **2552**, 654; with the permission of the authors and the SPIE.

of 30 Hz. Their sensitivity, expressed as NETD, ranges from 0.2 to 0.02°C. Their angular resolution is typically 1 to 3 mrad, and their total field of view is 15 to 20 degrees. Absolute temperature measurements employ a "dialed-in" emissivity.

The exception to the cryogenic photon detector approach is the pyroelectric vidicon. The vidicon system employs the pyroelectric effect and a thermal detection mechanism, and it operates at ambient temperature. Although the cost of the pyroelectric vidicon system is roughly half that of the cryogenic photon detector system, its sensitivity as a function of spatial resolution is poorer.

A new approach to imaging radiometry employing a low-cost uncooled scanned linear array of thermoelectric thermal sensors is described herein. System tradeoffs have been made to provide the lowest possible cost with performance meeting nearly all the requirements for the predictive and preventive maintenance market and the other segments identified previously herein.

1. DESCRIPTION

The handheld IR SnapShot® camera, originally developed by the Honeywell Technology Center and optimized by Infrared Solutions, Inc., is an uncooled imaging radiometer somewhat larger in size and weight than is a 35-mm photographic camera, about 8.0 in. long, 4.25 in. high, and 3.5 in. deep, with a Ge lens projecting from the front of the body. It weighs 1.5 kg and requires 8 W. A removable, rechargeable battery, similar to that used

FIG. 12. The IR SnapShot® imaging radiometer (Infrared Solutions Inc., Minneapolis, MN). Reprinted from Wood *et al.* (1995) with permission from the authors and the SPIE.

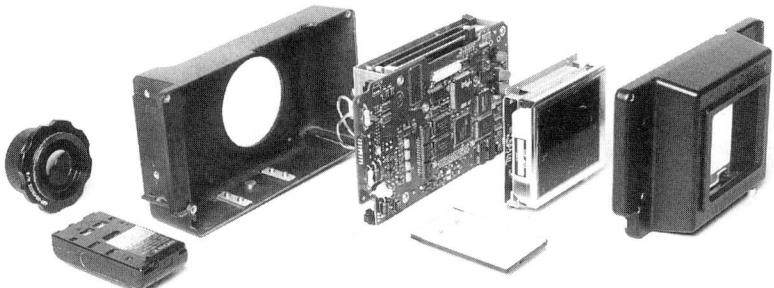

FIG. 13. Photograph of the IR SnapShot® imaging radiometer disassembled. Left to right: lens, battery, front cover, electronics, Personal Computer Memory Card International Association (PCMCIA) card, liquid crystal display, and back cover. Reprinted from Wood *et al.* (1995) with permission from the authors and the SPIE.

with camcorders, is attached to the left side of the body of the camera. Figures 12 and 13 show the IR SnapShot® assembled and disassembled.

Depressing a push-button on the top-front of the camera body causes a 120-element linear thermoelectric array to be scanned across the focal plane of the Ge lens in 1.44 sec. The scan provides a 120-by-120-pixel image in normal operation, although the number of pixels in the direction of the scan can be adjusted by way of a menu function.

The 120-pixel thermoelectric linear array is based on Si microstructure technology (Fig. 1). A $Si_3N_4$ membrane extends across an etch pit in the Si substrate. Two metals, chromel and constantan, are deposited on the membrane in thin-film form, providing the infrared-absorbing "hot" junction. A second junction on the Si substrate, and thereby heat sunk to the substrate, provides the reference or "cold" junction. An absorbing film (not

TABLE II
PIXEL PROPERTIES OF SILICON MICROSTRUCTURE THERMOELECTRIC INFRARED LINEAR ARRAYS

|  | Type 1 | Type 2 |
|---|---|---|
| Number of pixels | 128 | 128 |
| Number of pixels accessed | 120 | 120 |
| Pixel size ($\mu$m) | 50 | 75 |
| Number of junctions per pixel | 3 | 3 |
| Resistance at 300 K (ohms) | 2380 | 1800 |
| Thermal response time (msec) | 12 | 12 |
| $D^*(cmHz^{1/2}/W)$ | $1.7 \times 10^8$ | $1.4 \times 10^8$ |

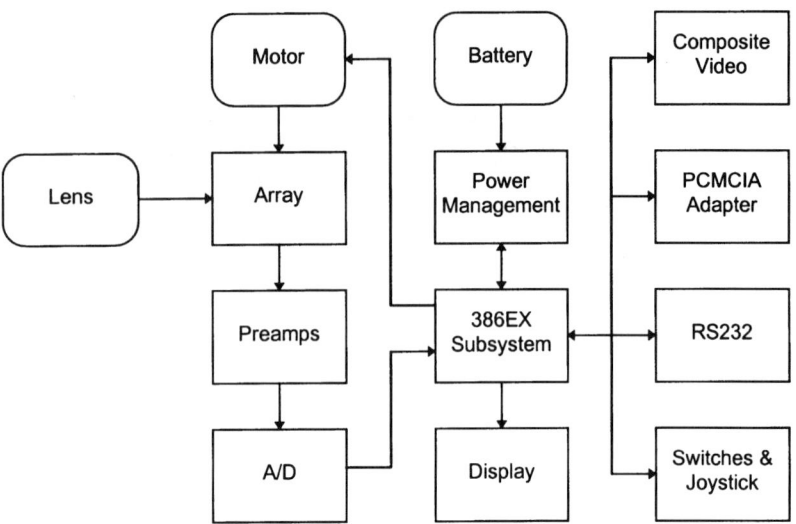

FIG. 14. Block diagram of the IR SnapShot® imaging radiometer employing uncooled thermoelectric linear array. PCMCIA = Personal Computer Memory Card International Association. Reprinted from Wood et al. (1995) with permission from the authors and the SPIE.

shown) is deposited on the hot junction. Two pixel sizes are available: 50 by 50 μm and 75 by 75 μm. Each has three pairs of junctions per pixel. The pixel properties are listed in Table II.

The 120 pixels of the array are electrically combined six at a time in 20 multiplexers whose output is fed to 20 preamplifiers. Their outputs are multiplexed and fed to a 16-bit analog-to-digital converter. The digital signal is processed by a 386EX microprocessor that, in turn, drives a 4-in. diagonal color liquid crystal display (LCD) on the back of the camera (Fig. 14).

2. OPERATION

All IR SnapShot® camera setup parameters and special IR camera functions normally provided by imaging radiometers are controlled by way of menus displayed by the LCD. The menus are activated by a miniature joystick on the back of the camera. Menu options include selection of temperature units, color palette, image polarity, panning functions, audible functions, temperature threshold, date, background temperature, and emissivity.

10 APPLICATION OF LINEAR ARRAYS TO IMAGING RADIOMETERS 313

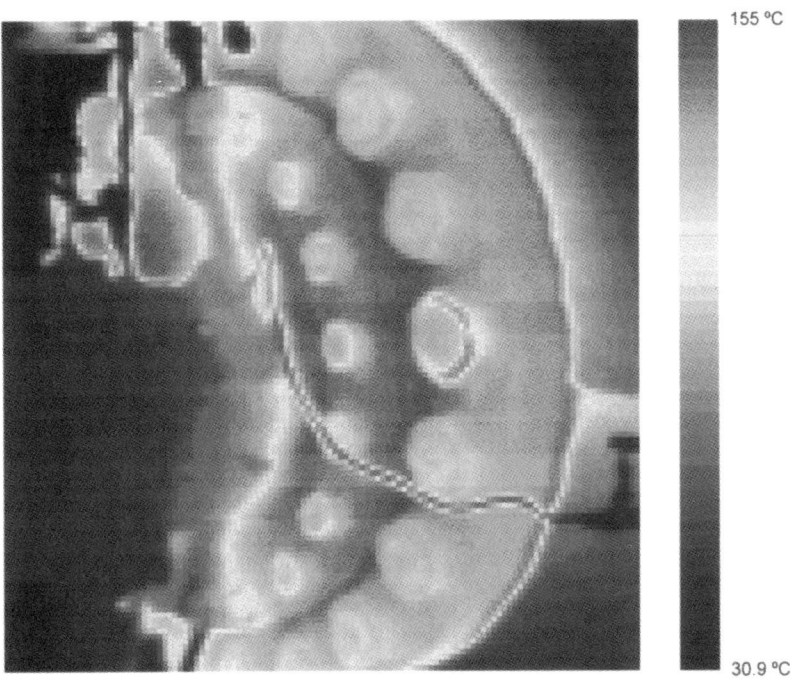

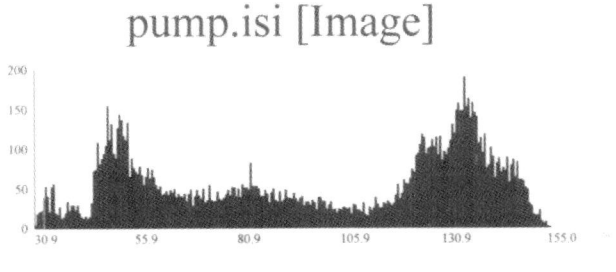

FIG. 15. Thermal image of an industrial pump and a histogram of the temperature distribution. Reprinted from Wood *et al.* (1995) with permission from the authors and the SPIE.

Focusing and framing of the thermal image is accomplished by holding down the push-button and observing the display as the scan continues back and forth across the display. The lens focus is adjusted manually during this procedure. The picture is completed, fixed on the display, and stored on the memory card (see below) when the push-button is released.

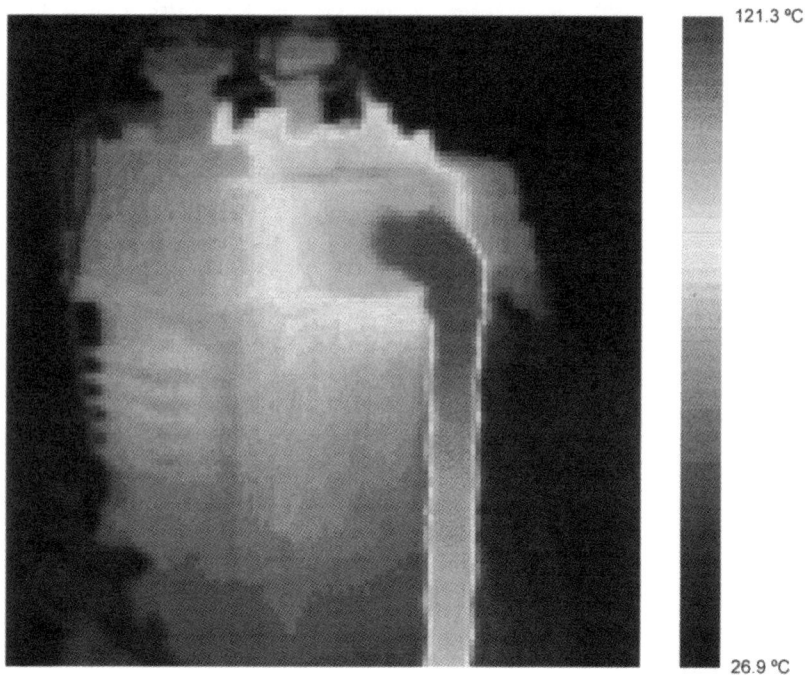

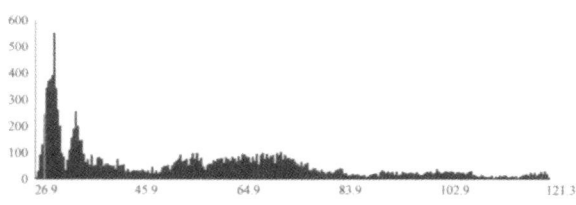

Temperature Range

Fig. 16. Thermal image of an industrial compressor and a histogram of the temperature distribution. Reprinted from Wood et al. (1995) with permission from the authors and the SPIE.

The temperature range of the displayed scene is automatically adjusted (called *autoscaling*), and the thermal image and a histogram illustrating the temperature distribution are displayed. To emphasize temperature distribution over a more limited temperature range, autoscaling can be replaced by

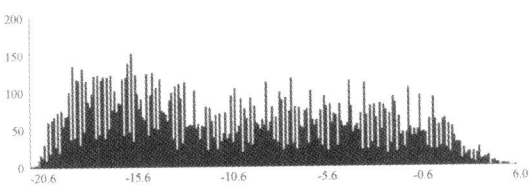

FIG. 17. Thermal image of an electric power distribution substation and a histogram of the temperature distribution. Reprinted from Wood *et al.* (1995) with permission from the authors and the SPIE.

manually setting the upper and lower temperatures. The temperature of the central pixel is shown numerically on the display. The temperature at any other point in the displayed image can be obtained by moving a cursor by way of a menu function and joystick to the desired point and reading the displayed temperature. An emissivity value must be dialed-in by way of a menu function. A threshold temperature can be selected to activate an audible alarm whenever a preset value is exceeded. Figures 15 to 17 provide

TABLE III
IR SNAPSHOT® IMAGING RADIOMETER SPECIFICATIONS

| | |
|---|---|
| Detector | 120-element linear array of uncooled thermoelectric detectors |
| Infrared dynamic range | 12 bits |
| Digitizing resolution | 16 bits |
| Field of view | 20 degrees with 50-$\mu$m square pixels and 18-mm lens |
| Instantaneous field of view | 2.8 mrad |
| Spectral band | 8 to 12 $\mu$m |
| Noise equivalent temperature difference | <0.2°C at 30°C |
| Temperature accuracy | 2°C or 2% |
| Temperature measurement range | −20 to 450°C |
| Operating temperature Range | 0 to 40°C |
| Power | NP-98 compatible camcorder battery, 3 Ah |
| Image scan time | 1.68 sec |
| Turn-on-time | <15 sec |
| Battery operating time | >75 min with 3 Ah battery |
| Image storage | 16-bit ISI format, 29 KB/image, 512 KB SRAM PC card standard |
| Communication | RS-232, 300 to 115 Kbaud selectable |
| Display | 102-mm/4-in. active matrix high-luminance color liquid crystal display |
| Video output | National Television System Committee (NTSC) |
| Functions | Focus, snap, acquisition, store, recall, temperature measurement, and setup menu |
| Setup menu options | Emissivity, background temperature, temperature units, color palette, autoscale and fixed ranges, histogram, acquisition threshold temperature focus width, time, date RS-232 baud rate |
| Size | 240 × 100 × 130 mm/9.4 × 3.9 × 5.1 in.; length × width × height, not including lens |
| Weight | <2.0 kg/4.4 lbs with lens and battery |
| Standard accessories | 3-Ah camcorder battery, battery charger, video cable, padded neck strap, manual, SnapView® image viewer for Microsoft® *Windows*® 95, 512 KB PC card, 18-mm f/0.7 lens |
| Optional accessories | SnapView Pro® for Microsoft® *Windows 95*®, PC cards, batteries and chargers, AC adapter and interface, padded hard case, lenses, video and RS-232 cables, tripod |

Ah, ampere-hour; ISI, Infrared Solutions, Inc.; PC, personal computer; AC, alternating current.

examples of thermal imagery of industrial equipment and an electric power distribution substation. Although these images are shown in gray scale, color palettes are available.

A removable Personal Computer Memory Card International Association (PCMCIA) memory card with storage capacity for about 30 images in Targeted Image File Format (TIFF) format is built into the camera. Stored images can be recalled for viewing and analysis on the camera display. The PCMCIA memory card can be removed and the stored images read, displayed, and analyzed in a personal computer. An RS232 port on the camera provides an interface to a personal computer for software input and diagnostics. An RS170 port provides an interface to a personal computer for software input and diagnostics. An RS170 port provides an interface to a television monitor for indirect viewing of the imagery on the LCD display.

IR SnapView® software installed in a personal computer is employed for additional processing of the IR SnapShot® imagery. The user can sort images into different directories, view them, enlarge them, export them into different graphical formats, and use any of the several report writing features. The image can also be exported in spreadsheet format, which gives the user a pixel-by-pixel breakdown of the temperatures within the image and allows for statistical analysis. IR SnapView® also provides detailed image-analysis and report-writing capabilities.

## 3. Specifications

The camera specifications are listed in Table III.

## 4. Summary

The IR SnapShot® is an entirely new low-cost approach to an imaging radiometer. The design emphasis has been to exploit uncooled thermoelectric arrays made by Si microstructure technology to provide excellent thermal sensitivity, resolution, and field of view by trading-off frame time. A very large fraction of the imaging radiometry market, including predictive and preventive maintenance, is accessible, yet the cost of the system is far less than that of cryogenic photon detector or pyroelectric vidicon imaging radiometers.

### Acknowledgments

This chapter could not have been written without input from many individuals from Honeywell, SAIC, and Infrared Solutions, Inc. Among them are Andrew Wood, Tom Rezachek, John Danyluk, and Roger Schmidt, contributors to the initial publications from which this chapter was prepared.

## References

Kruse, P. W. (1995). Uncooled IR focal plane arrays. *Proc. SPIE Infrared Technology, 21st* **2552**, 654.

Kruse, P. W., Schmidt, R. N., and Danyluk, J. D. (1996). High-speed imaging radiometer employing uncooled thermoelectric linear array. Presented at the SPIE Thermosense Mtg., Orlando, FL, April.

Listvan, M., Rhodes, M., and Wilson, M. L. (1991). On-line thermal profiling for industrial process control. *Proc. Instr. Soc. Am. Symp. Innovation in Measurement Science*, Geneva, NY, August.

Stetson, N., and Landry, J. (1994). Handheld imaging using PtSi, InSb, and HgCdTe focal plane technology. *Proc. SPIE Infrared Technology, 20th* **2269**, 498.

Wilson, M. L., Kubisiak, D., Wood, R. A., Ridley, J. A., and Listvan, M. (1991). An uncooled thermo-electric microthermopile camera developed using silicon microstructure sensors. *Proc. IRIS Spec. Group Infrared Detect.*, Boulder, CO, August.

Wood, R. A., Rezachek, T. M., Kruse, P. W., and Schmidt, R. N. (1995). IR SnapShot® camera. *Proc. SPIE Infrared Technology, 21st* **2552**, 654.

# Index

## A

Applications, infrared imaging systems, overview, 4

## B

Background fluctuation noise limit, uncooled infrared focal plane arrays, 33–37
Bolometers, noise in, 75–85
 bolometer resistance noise, 75–79
 noise from bias resistors, 79–80
 preamplifier noise, 85
 radiation noise, 81–82
 thermal conductance noise, 80–81
 total electrical noise, 83–84
Bolometric arrays, resistive, 11–12
Brewster, D., first usage of term, pyroelectricity, 6

## E

Electrical noise, in bolometers, 83–84
Electronic materials research, uncooled imagers, 6–9
 ferroelectric-pyroelectric materials, 6–8
 Langley, S.P., experimentation by, 8–9
 pyroelectricity, first usage of term, 6
 resistive bolometric arrays, 11–12
 resistive materials, 8–9
 uncooled image sensor, by Texas Instruments, Inc., 8

## F

Ferroelectric ceramic barium strontium titanate, uncooled infrared focal plane arrays, 28
Ferroelectric-pyroelectric bolometer arrays, hybrid, 123–74
 choppers, 162–69
 designs, 154–69
 ferroelectric material selection, 154–56
 modes of operation, 139–43
 modulation transfer function, 158–59
 principles, 124–54
 pyroelectricity, ferroelectric materials, 124–39
 read-out electronics, 159–61
 signal, noise and, 144–54
 system electronics, 161–62
 systems implementation, 169–73
 thermal isolation, 156–58
Ferroelectric-pyroelectric materials
 electronic uncooled imagers, 6–8
 uncooled imagers, 6–8
Focal plane arrays, uncooled, infrared, 17–42
 ferroelectric ceramic barium strontium titanate, 28
 fundamental limits, 31–37
 Honeywell thermoelectric detector, structure of, 30
 microbolometer pixel structure, 20
 overview, 37–40
 radiation thermocouple, 29
 resistive bolometer, 31
 responsivity, pyroelectric detector, as function of frequency, 27
 structures, pyroelectric detector device, 19
 thermal detection mechanisms, 23–31

Focal plane arrays, uncooled, infrared (*continued*)
  bolometer, usage of term, 23
  ferroelectric bolometers, pyroelectric detectors and, 25–29
  resistive bolometers, 23–25
  superconducting bolometers, 23
  thermoelectric voltage, 29
  thermal infrared sensors, principle of, 18
  thermal isolation structure, 17–23
  uncooled thermal arrays, comparison of principal types of, 31
Focusing, thermal radiation, 2

## G

Gallium arsenide-based integrated pyroelectric detectors, 197–99

## H

Hadni, A., use of triglycine sulfate, in camera tube in dielectric mode, 219
Heat balance, with applied bias, microbolometer, 59–61
Historical overview, thermal imaging, 1–16
  applications, infrared imaging systems, 4
  Brewster, D., first usage of term, pyroelectricity, 6
  display of resulting physical effect, 2
  electronic materials research, uncooled imagers, 6–9
    ferroelectric-pyroelectric materials, 6–8
    Langley, S.P., experimentation by, 8–9
    pyroelectricity, first usage of term, 6
    resistive bolometric arrays, 11–12
    resistive materials, 8–9
    uncooled image sensor, by Texas Instruments, Inc., 8
  future developments, 12–13
  industrial applications, 4
  infrared imaging, steps, 2
  Langley, S.P., experimentation by, 8–9
  law enforcement applications, 4
  medical applications, 4
  military applications, 4
  reading of resulting physical effect, 2
  rescue applications, 4
  scientific applications, 4
  silicon read-out, uncooled imaging arrays using, 9–12
    ferroelectric-pyroelectric arrays, 9–11
  thermal radiation focusing, 2
Hybrid pyroelectric-ferroelectric bolometer arrays, 123–74
  choppers, 162–69
  designs, 154–69
  ferroelectric material selection, 154–56
  modes of operation, 139–43
    induced pyroelectric mode, 141–43
    inherent pyroelectric mode, 140–41
    pulsed pyroelectric mode, 143
  modulation transfer function, 158–59
  principles, 124–54
  pyroelectricity, ferroelectric materials, 124–39
    first-order ferroelectrics, 132–39
    second-order ferroelectric, 128–32
  read-out electronics, 159–61
  signal, noise and, 144–54
    noise, 148–54
    responsivity, 144–48
  system electronics, 161–62
  systems implementation, 169–73
  thermal isolation, 156–58

## I

Imagers, electronic materials research
  pyroelectricity, first usage of term, 6
  resistive bolometric arrays, 11–12
  resistive materials, 8–9
  uncooled image sensor, by Texas Instruments, Inc., 8
Industrial applications, thermal imaging, 4
IR. *See* Infrared

## J

Johnson noise approximation, microbolometer signal-to-noise, 91

## L

Langley, S.P., experimentation by, 8–9
Law enforcement, thermal imaging

applications, 4
Load line, microbolometer responsivity, 64–68
Long wavelength limit, dependence on, uncooled infrared focal plane arrays, 37

## M

Maintenance, preventive, uncooled monolithic thermoelectric linear arrays, imaging radiometers for, 310–12
Materials research, electronic, uncooled imagers, 6–9
    ferroelectric-pyroelectric materials, 6–8
    Langley, S.P., experimentation by, 8–9
    pyroelectricity, first usage of term, 6
    resistive bolometric arrays, 11–12
    resistive materials, 8–9
    uncooled image sensor, by Texas Instruments, Inc., 8
Medical applications, thermal imaging, 4
Microbolometer pixel structure, uncooled infrared focal plane arrays, 20
Microbolometer signal-to-noise, 86–91
    comparison with ideal bolometer, 89–91
    detectivity, 87–88
Military applications, thermal imaging, 4
Monolithic pyroelectric bolometer arrays, 175–201
    gallium arsenide-based integrated pyroelectric detectors, 197–99
    methodology, design, 176–77
    overview, 199
    process, design, 187–89
    silicon-based integrated pyroelectric detector arrays, 189–97
Monolithic silicon microbolometer arrays, 43–121
    absorber structure, microbolometer, 107
    absorption coefficient, two-level microbolometer, 106
    bias circuit, load lines of, 65
    camera, compact, infrared images, 117
    camera block diagram, microbolometer, 116
    circuit, serial read-out, 92
    collection optics, 49
    construction, microbolometer, single-level, 103
    design, 98–115
    electronic read-out circuits, two-dimensional microbolometer arrays, 91–95
    fabrication, and packaging, 98–115
    gain correction, 97
    Johnson noise equivalent circuit, resistor, 76
    light cone geometry, 49
    load line
        with internal low-frequency noise, 69
        with pulsed bias, microbolometer, 72
    materials in fabrication
        one-level microbolometers, 100–102
        packaging, 109–15
        parameters of, 99
        two-level microbolometers, 102–9
    measured responsivity, microbolometer arrays compared, 111
    modulation transfer function, 98
    noise
        in bolometers, 75–85
            bias resistors, noise from, 79–80
            preamplifier noise, 85
            radiation noise, 81–82
            resistance noise, 75–79
            thermal conductance noise, 80–81
            total electrical noise, 83–84
        microbolometer arrays compared, 111
    noise equivalent temperature difference, microbolometer arrays compared, 111
    offset compensation schemes, 95–97
    one-level microbolometer design, 99
    packaging, 98–115
    parameters of materials, in microbolometer fabrication, 99
    portable microbolometer camera, 117
    practical camera development, 116
    prototype single-level microbolometer, microscope photograph, 100
    radiation, pulse of, response of microbolometer to, 58
    read-out circuit
        larger arrays, 94
        microbolometer, general form of, 92
    resistance parameters of microbolometer, 60

Monolithic silicon microbolometer arrays (*continued*)
  responsivity, 47–75
    heat balance
      with applied bias, 59–61
      equation, 56, 57
      with no applied bias, 57–59
    k\load line, constant voltage bias, 68
    load line, 64–68
      metal-temperature coefficient, 67
    low-frequency noise, 68–70
    microbolometer model, 47–51
    numerical calculation, microbolometer performance, 71–75
    with pulsed bias, large radiation signals, 70–71
    resistance changes, 51–55
    V-I curves calculations, 61–63
    variation, with gas pressure, composition, 115
  signal-to-noise, 86–91
    comparison with ideal bolometer, 89–91
    detectivity, 87–88
    Johnson noise approximation, 91
    noise equivalent power, 86
    noise equivalent temperature difference, 86–87
  silicone wafer, with two-level microbolometer arrays, 110
  surroundings, thermal conductances between, 57
  two-level microbolometer, 99, 112
    fabrication sequence for, 104
  vacuum package, sealed, for microbolometer arrays, 115
  vanadium dioxide, resistance, *vs.* temperature for, 53–55
Monolithic thermoelectric linear arrays, imaging radiometers, uncooled for predictive, preventive maintenance, 309–17
  description, 310–12
  operation, 312–17
  specifications, 317

## N

NEP. *See* Noise equivalent power

NETD. *See* Noise equivalent temperature difference
Noise equivalent power, microbolometer signal-to-noise, 86
Noise equivalent temperature difference, microbolometer signal-to-noise, 86–87
Noise limit, background fluctuation, uncooled infrared focal plane arrays, 33–37

## P

Polysilicon thermopile infrared detector, 210–11
Preamplifier noise, in bolometers, 85
Preventive maintenance, uncooled monolithic thermoelectric linear arrays, imaging radiometers for, 310–12
Pyroelectric bolometer arrays, monolithic, 175–201
  gallium arsenide-based integrated pyroelectric detectors, 197–99
  methodology, design, 176–77
    integrated circuits, 186–87
    materials characterization, 181–83
    materials processing, 178–81
    micromachined sensor process design, 184–86
    thermal isolation structures, 183–84
  overview, 199
  process, design, 187–89
  silicon-based integrated pyroelectric detector arrays, 189–97
Pyroelectric materials
  electronic uncooled imagers, 6–8
  uncooled imagers, 6–8
Pyroelectric vidicon, 219–25
  ferroelectric camera tubes, history of, 219
  Hadni, A., use of triglycine sulfate, in camera tube in dielectric mode, 219
  history, 219–23
  performance analysis, 223–25
  triglycine sulfate, use of, in camera tube, dielectric mode, 219
Pyroelectricity
  ferroelectric materials, 124–39
    ferroelectric-pyroelectric bolometer arrays, hybrid, 124–39

first-order ferroelectrics, 132–39
first usage of term, 6
Brewster, D., 6

## Q

Quartz microresonators, uncooled infrared imaging arrays, 269–96
  frequency measurement, 274–75
  infrared absorption of, 277–79
  oscillator noise, 273–74
  overview, 283
  performance, microresonator arrays, 279–80
  performance calculations, 284–94
  producibility, 281–82
  quartz microresonators, as infrared sensors, 271–72
  temperature coefficients, quartz thermometers, 272–73
  thermal isolation, 275–77

## R

Radiation noise, in bolometers, 81–82
Radiation signals, large, pulsed bias, microbolometer responsivity, 70–71
Read-out electronics, ferroelectric-pyroelectric bolometer arrays, hybrid, 159–61
Rescue applications, thermal imaging, 4
Resistance noise, bolometer, 75–79
Resistive bolometric arrays, 11–12
Resistive materials, 8–9
  imagers, electronic materials research, 8–9
Responsivity, microbolometers, 47–75
  koad line, constant voltage bias, 68
  load line, 64–68
    metal-temperature coefficient, 67
  low-frequency noise, 68–70
  microbolometer model, 47–51
  numerical calculation, microbolometer performance, 71–75
  with pulsed bias, large radiation signals, 70–71
  resistance changes, 51–55

V-I curves calculations, 61–63
  temperature coefficient of resistance
    metal-like, 62
    semiconductor-like, 62–63

## S

Scientific applications, thermal imaging, 4
Seebeck effect, thermopile infrared detector, 206–9
Silicon-based integrated pyroelectric detector arrays, 189–97
  cell structure, 190–91
  circuit operation, 191–94
Silicon microbolometer arrays, monolithic, 43–121
  absorber structure, microbolometer, 107
  absorption coefficient, two-level microbolometer, 106
  bias circuit, load lines of, 65
  camera, compact, infrared images, 117
  camera block diagram, microbolometer, 116
  circuit, serial read-out, 92
  collection optics, 49
  construction, microbolometer, single-level, 103
  design, 98–115
  electronic read-out circuits, two-dimensional microbolometer arrays, 91–95
  fabrication, and packaging, 98–115
  gain correction, 97
  Johnson noise equivalent circuit, resistor, 76
  light cone geometry, 49
  load line
    with internal low-frequency noise, 69
    with pulsed bias, microbolometer, 72
  materials in fabrication
    one-level microbolometers, 100–102
    packaging, 109–15
    parameters of, 99
    two-level microbolometers, 102–9
  measured responsivity, microbolometer arrays compared, 111
  modulation transfer function, 98
  noise
    in bolometers, 75–85

Silicon microbolometer arrays, monolithic
    (*continued*)
  bias resistors, noise from, 79–80
  preamplifier noise, 85
  radiation noise, 81–82
  resistance noise, 75–79
  thermal conductance noise, 80–81
  total electrical noise, 83–84
  microbolometer arrays compared, 111
  noise equivalent temperature difference,
      microbolometer arrays compared,
      111
  offset compensation schemes, 95–97
  one-level microbolometer design, 99
  packaging, 98–115
  parameters of materials, in
      microbolometer fabrication, 99
  portable microbolometer camera, 117
  practical camera development, 116
  prototype single-level microbolometer,
      microscope photograph, 100
  radiation, pulse of, response of
      microbolometer to, 58
  read-out circuit
    general form of, 92
    for larger arrays, 94
  resistance parameters of microbolometer,
      60
  responsivity, 47–75
    heat balance
      with applied bias, 59–61
      equation, 56, 57
      with no applied bias, 57–59
    k\load line, constant voltage bias, 68
    load line, 64–68
      metal-temperature coefficient, 67
    low-frequency noise, 68–70
    microbolometer model, 47–51
    numerical calculation, microbolometer
      performance, 71–75
    with pulsed bias, large radiation
      signals, 70–71
    resistance changes, 51–55
    V-I curves calculations, 61–63
    temperature coefficient of resistance
      metal-like, 62
      semiconductor-like, 62–63
    variation, with gas pressure,
      composition, 115
  signal-to-noise, 86–91
    comparison with ideal bolometer,
      89–91
    detectivity, 87–88
    Johnson noise approximation, 91
    noise equivalent power, 86
    noise equivalent temperature
      difference, 86–87
  silicone wafer, with two-level
      microbolometer arrays, 110
  surroundings, thermal conductances
      between, 57
  two-level microbolometer, 112
    design, 99
    fabrication sequence for, 104
  vacuum package, sealed, for
      microbolometer arrays, 115
  vanadium dioxide, resistance, *vs.*
      temperature for, 53–55
Silicon read-out, uncooled imaging arrays
    using, 9–12
  ferroelectric-pyroelectric arrays, 9–11

## T

Texas Instruments, Inc., imagers, electronic
    materials research, uncooled image
    sensor, 8
TGS. *See* Triglycine sulfate
Thermal conductance noise, in bolometers,
    80–81
Thermal isolation, ferroelectric-pyroelectric
    bolometer arrays, hybrid, 156–58
Thermal radiation, focusing, 2
Triglycine sulfate, use of, in camera tube,
    dielectric mode, 219
Tunneling infrared sensors, 227–67
  assembly process, 249
  corrugation process, 247
  design, 245–53
    nitride membranes, 246
    pinholes, 246
  fabrication, 245–53
  future prospects, 264–65
  infrared absorber process, 248–49
  infrared sensor operation, testing, 259–64
  membrane release, 246
  modeling, 229–39
    mechanical, electrical model, 232–35

INDEX 325

noise model, 236–39
sensor thermal model, 229–32
overview, 266
substrate metal process, 247–48
tip formation process, 247
transducer
  background, 239–45
    tunneling design, 243–45
  comparisons, 241–43
  tunneling transducer operation, 253–59

## U

Uncooled imagers, electronic materials research, 6–9
  ferroelectric-pyroelectric materials, 6–8
  Langley, S.P., experimentation by, 8–9
  pyroelectricity, first usage of term, 6
  resistive bolometric arrays, 11–12
  resistive materials, 8–9
  uncooled image sensor, by Texas Instruments, Inc., 8
Uncooled infrared focal plane arrays, 17–42
  ferroelectric ceramic barium strontium titanate, 28
  fundamental limits, 31–37
    background fluctuation noise limit, 33–37
    long wavelength limit, dependence on, 37
    temperature fluctuation noise limit, 31–33
  Honeywell thermoelectric detector, structure of, 30
  microbolometer pixel structure, 20
  overview, 37–40
  radiation thermocouple, 29
  resistive bolometer, 31
  responsivity, pyroelectric detector, as function of frequency, 27
  structures, pyroelectric detector device, 19
  thermal detection mechanisms, 23–31
    bolometer, usage of term, 23
    ferroelectric bolometers, pyroelectric detectors and, 25–29
    resistive bolometers, 23–25
    superconducting bolometers, 23
    thermoelectric voltage, 29
  thermal infrared sensors, principle of, 18

  thermal isolation structure, 17–23
  thermoelectric, 203–18
    overview, 217–18
    thermopile infrared detector, 204–9
      mechanism for, 204–5
      Seebeck effect, 206–9
      uncooled infrared detector schemes, compared, 205–6
    thermopile infrared focal plane array, 210–17
      characteristics of, 211
      circuit, signal read-out, 211–12
      future improvements, 217
      package, 214–15
      performance, 215–16
      polysilicon thermopile infrared detector, 210–11
      scanner, charge-coupled device, 213–14
  uncooled thermal arrays, comparison of principal types of, 31
    ferroelectric bolometer, 31
    pyroelectric detector, 31
    thermoelectric detector, 31
Uncooled infrared imaging arrays, quartz microresonators, 269–96
  frequency measurement, 274–75
  infrared absorption of, 277–79
  oscillator noise, 273–74
  overview, 283
  performance
    calculations, 284–94
    microresonator arrays, 279–80
  producibility, 281–82
  quartz microresonators, as infrared sensors, 271–72
  temperature coefficients, quartz thermometers, 272–73
  thermal isolation, 275–77
Uncooled monolithic thermoelectric linear arrays, imaging radiometers, 297–318
  Model IR 1000 imaging radiometer
    performance, 300–304
    technical description, 298–300
  for predictive, preventive maintenance, 309–17
    description, 310–12
    operation, 312–17
    specifications, 317

railcar wheels, incipient failure of, identification of, 298–309

## V

V-I curves calculations, microbolometer responsivity, 61–63
temperature coefficient of resistance
  metal-like, 62
  semiconductor-like, 62–63

## W

Wavelength limit, long, dependence on, uncooled infrared focal plane arrays, 37

# Contents of Volumes in This Series

### Volume 1  Physics of III–V Compounds

*C. Hilsum*, Some Key Features of III–V Compounds
*Franco Bassani*, Methods of Band Calculations Applicable to III–V Compounds
*E. O. Kane*, The k-p Method
*V. L. Bonch-Bruevich*, Effect of Heavy Doping on the Semiconductor Band Structure
*Donald Long*, Energy Band Structures of Mixed Crystals of III–V Compounds
*Laura M. Roth and Petros N. Argyres*, Magnetic Quantum Effects
*S. M. Puri and T. H. Geballe*, Thermomagnetic Effects in the Quantum Region
*W. M. Becker*, Band Characteristics near Principal Minima from Magnetoresistance
*E. H. Putley*, Freeze-Out Effects, Hot Electron Effects, and Submillimeter Photoconductivity in InSb
*H. Weiss*, Magnetoresistance
*Betsy Ancker-Johnson*, Plasma in Semiconductors and Semimetals

### Volume 2  Physics of III–V Compounds

*M. G. Holland*, Thermal Conductivity
*S. I. Novkova*, Thermal Expansion
*U. Piesbergen*, Heat Capacity and Debye Temperatures
*G. Giesecke*, Lattice Constants
*J. R. Drabble*, Elastic Properties
*A. U. Mac Rae and G. W. Gobeli*, Low Energy Electron Diffraction Studies
*Robert Lee Mieher*, Nuclear Magnetic Resonance
*Bernard Goldstein*, Electron Paramagnetic Resonance
*T. S. Moss*, Photoconduction in III–V Compounds
*E. Antoncik ad J. Tauc*, Quantum Efficiency of the Internal Photoelectric Effect in InSb
*G. W. Gobeli and I. G. Allen*, Photoelectric Threshold and Work Function
*P. S. Pershan*, Nonlinear Optics in III–V Compounds
*M. Gershenzon*, Radiative Recombination in the III–V Compounds
*Frank Stern*, Stimulated Emission in Semiconductors

## Volume 3  Optical of Properties III–V Compounds

*Marvin Hass*, Lattice Reflection
*William G. Spitzer*, Multiphonon Lattice Absorption
*D. L. Stierwalt and R. F. Potter*, Emittance Studies
*H. R. Philipp and H. Ehrenveich*, Ultraviolet Optical Properties
*Manuel Cardona*, Optical Absorption above the Fundamental Edge
*Earnest J. Johnson*, Absorption near the Fundamental Edge
*John O. Dimmock*, Introduction to the Theory of Exciton States in Semiconductors
*B. Lax and J. G. Mavroides*, Interband Magnetooptical Effects
*H. Y. Fan*, Effects of Free Carries on Optical Properties
*Edward D. Palik and George B. Wright*, Free-Carrier Magnetooptical Effects
*Richard H. Bube*, Photoelectronic Analysis
*B. O. Seraphin and H. E. Bennett*, Optical Constants

## Volume 4  Physics of III–V Compounds

*N. A. Goryunova, A. S. Borschevskii, and D. N. Tretiakov*, Hardness
*N. N. Sirota*, Heats of Formation and Temperatures and Heats of Fusion of Compounds $A^{III}B^V$
*Don L. Kendall*, Diffusion
*A. G. Chynoweth*, Charge Multiplication Phenomena
*Robert W. Keyes*, The Effects of Hydrostatic Pressure on the Properties of III–V Semiconductors
*L. W. Aukerman*, Radiation Effects
*N. A. Goryunova, F. P. Kesamanly, and D. N. Nasledov*, Phenomena in Solid Solutions
*R. T. Bate*, Electrical Properties of Nonuniform Crystals

## Volume 5  Infrared Detectors

*Henry Levinstein*, Characterization of Infrared Detectors
*Paul W. Kruse*, Indium Antimonide Photoconductive and Photoelectromagnetic Detectors
*M. B. Prince*, Narrowband Self-Filtering Detectors
*Ivars Melngalis and T. C. Harman*, Single-Crystal Lead-Tin Chalcogenides
*Donald Long and Joseph L. Schmidt*, Mercury-Cadmium Telluride and Closely Related Alloys
*E. H. Putley*, The Pyroelectric Detector
*Norman B. Stevens*, Radiation Thermopiles
*R. J. Keyes and T. M. Quist*, Low Level Coherent and Incoherent Detection in the Infrared
*M. C. Teich*, Coherent Detection in the Infrared
*F. R. Arams, E. W. Sard, B. J. Peyton, and F. P. Pace*, Infrared Heterodyne Detection with Gigahertz IF Response
*H. S. Sommers, Jr.*, Macrowave-Based Photoconductive Detector
*Robert Sehr and Rainer Zuleeg*, Imaging and Display

## Volume 6  Injection Phenomena

*Murray A. Lampert and Ronald B. Schilling*, Current Injection in Solids: The Regional Approximation Method
*Richard Williams*, Injection by Internal Photoemission
*Allen M. Barnett*, Current Filament Formation

*R. Baron and J. W. Mayer*, Double Injection in Semiconductors
*W. Ruppel*, The Photoconductor-Metal Contact

## Volume 7  Application and Devices
## Part A

*John A. Copeland and Stephen Knight*, Applications Utilizing Bulk Negative Resistance
*F. A. Padovani*, The Voltage-Current Characteristics of Metal-Semiconductor Contacts
*P. L. Hower, W. W. Hooper, B. R. Cairns, R. D. Fairman, and D. A. Tremere*, The GaAs Field-Effect Transistor
*Marvin H. White*, MOS Transistors
*G. R. Antell*, Gallium Arsenide Transistors
*T. L. Tansley*, Heterojunction Properties

## Part B

*T. Misawa*, IMPATT Diodes
*H. C. Okean*, Tunnel Diodes
*Robert B. Campbell and Hung-Chi Chang*, Silicon Junction Carbide Devices
*R. E. Enstrom, H. Kressel, and L. Krassner*, High-Temperature Power Rectifiers of $GaAs_{1-x}P_x$

## Volume 8  Transport and Optical Phenomena

*Richard J. Stirn*, Band Structure and Galvanomagnetic Effects in III–V Compounds with Indirect Band Gaps
*Roland W. Ure, Jr.*, Thermoelectric Effects in III–V Compounds
*Herbert Piller*, Faraday Rotation
*H. Barry Bebb and E. W. Williams*, Photoluminescence I: Theory
*E. W. Williams and H. Barry Bebb*, Photoluminescence II: Gallium Arsenide

## Volume 9  Modulation Techniques

*B. O. Seraphin*, Electroreflectance
*R. L. Aggarwal*, Modulated Interband Magnetooptics
*Daniel F. Blossey and Paul Handler*, Electroabsorption
*Bruno Batz*, Thermal and Wavelength Modulation Spectroscopy
*Ivar Balslev*, Piezopptical Effects
*D. E. Aspnes and N. Bottka*, Electric-Field Effects on the Dielectric Function of Semiconductors and Insulators

## Volume 10  Transport Phenomena

*R. L. Rhode*, Low-Field Electron Transport
*J. D. Wiley*, Mobility of Holes in III–V Compounds
*C. M. Wolfe and G. E. Stillman*, Apparent Mobility Enhancement in Inhomogeneous Crystals
*Robert L. Petersen*, The Magnetophonon Effect

## Volume 11 Solar Cells

*Harold J. Hovel*, Introduction; Carrier Collection, Spectral Response, and Photocurrent; Solar Cell Electrical Characteristics; Efficiency; Thickness; Other Solar Cell Devices; Radiation Effects; Temperature and Intensity; Solar Cell Technology

## Volume 12 Infrared Detectors (II)

*W. L. Eiseman, J. D. Merriam, and R. F. Potter*, Operational Characteristics of Infrared Photodetectors
*Peter R. Bratt*, Impurity Germanium and Silicon Infrared Detectors
*E. H. Putley*, InSb Submillimeter Photoconductive Detectors
*G. E. Stillman, C. M. Wolfe, and J. O. Dimmock*, Far-Infrared Photoconductivity in High Purity GaAs
*G. E. Stillman and C. M. Wolfe*, Avalanche Photodiodes
*P. L. Richards*, The Josephson Junction as a Detector of Microwave and Far-Infrared Radiation
*E. H. Putley*, The Pyroelectric Detector–An Update

## Volume 13 Cadmium Telluride

*Kenneth Zanio*, Materials Preparations; Physics; Defects; Applications

## Volume 14 Lasers, Junctions, Transport

*N. Holonyak, Jr. and M. H. Lee*, Photopumped III–V Semiconductor Lasers
*Henry Kressel and Jerome K. Butler*, Heterojunction Laser Diodes
*A Van der Ziel*, Space-Charge-Limited Solid-State Diodes
*Peter J. Price*, Monte Carlo Calculation of Electron Transport in Solids

## Volume 15 Contacts, Junctions, Emitters

*B. L. Sharma*, Ohmic Contacts to III–V Compounds Semiconductors
*Allen Nussbaum*, The Theory of Semiconducting Junctions
*John S. Escher*, NEA Semiconductor Photoemitters

## Volume 16 Defects, (HgCd)Se, (HgCd)Te

*Henry Kressel*, The Effect of Crystal Defects on Optoelectronic Devices
*C. R. Whitsett, J. G. Broerman, and C. J. Summers*, Crystal Growth and Properties of $Hg_{1-x}Cd_xSe$ alloys
*M. H. Weiler*, Magnetooptical Properties of $Hg_{1-x}Cd_xTe$ Alloys
*Paul W. Kruse and John G. Ready*, Nonlinear Optical Effects in $Hg_{1-x}Cd_xTe$

## Volume 17 CW Processing of Silicon and Other Semiconductors

*James F. Gibbons*, Beam Processing of Silicon
*Arto Lietoila, Richard B. Gold, James F. Gibbons, and Lee A. Christel*, Temperature Distribu-

tions and Solid Phase Reaction Rates Produced by Scanning CW Beams
*Arto Leitoila and James F. Gibbons*, Applications of CW Beam Processing to Ion Implanted Crystalline Silicon
*N. M. Johnson*, Electronic Defects in CW Transient Thermal Processed Silicon
*K. F. Lee, T. J. Stultz, and James F. Gibbons*, Beam Recrystallized Polycrystalline Silicon: Properties, Applications, and Techniques
*T. Shibata, A. Wakita, T. W. Sigmon, and James F. Gibbons*, Metal-Silicon Reactions and Silicide
*Yves I. Nissim and James F. Gibbons*, CW Beam Processing of Gallium Arsenide

## Volume 18 Mercury Cadmium Telluride

*Paul W. Kruse*, The Emergence of $(Hg_{1-x}Cd_x)Te$ as a Modern Infrared Sensitive Material
*H. E. Hirsch, S. C. Liang, and A. G. White*, Preparation of High-Purity Cadmium, Mercury, and Tellurium
*W. F. H. Micklethwaite*, The Crystal Growth of Cadmium Mercury Telluride
*Paul E. Petersen*, Auger Recombination in Mercury Cadmium Telluride
*R. M. Broudy and V. J. Mazurczyck*, (HgCd)Te Photoconductive Detectors
*M. B. Reine, A. K. Soad, and T. J. Tredwell*, Photovoltaic Infrared Detectors
*M. A. Kinch*, Metal-Insulator-Semiconductor Infrared Detectors

## Volume 19 Deep Levels, GaAs, Alloys, Photochemistry

*G. F. Neumark and K. Kosai*, Deep Levels in Wide Band-Gap III–V Semiconductors
*David C. Look*, The Electrical and Photoelectronic Properties of Semi-Insulating GaAs
*R. F. Brebrick, Ching-Hua Su, and Pok-Kai Liao*, Associated Solution Model for Ga-In-Sb and Hg-Cd-Te
*Yu. Ya. Gurevich and Yu. V. Pleskon*, Photoelectrochemistry of Semiconductors

## Volume 20 Semi-Insulating GaAs

*R. N. Thomas, H. M. Hobgood, G. W. Eldridge, D. L. Barrett, T. T. Braggins, L. B. Ta, and S. K. Wang*, High-Purity LEC Growth and Direct Implantation of GaAs for Monolithic Microwave Circuits
*C. A. Stolte*, Ion Implantation and Materials for GaAs Integrated Circuits
*C. G. Kirkpatrick, R. T. Chen, D. E. Holmes, P. M. Asbeck, K. R. Elliott, R. D. Fairman, and J. R. Oliver*, LEC GaAs for Integrated Circuit Applications
*J. S. Blakemore and S. Rahimi*, Models for Mid-Gap Centers in Gallium Arsenide

## Volume 21 Hydrogenated Amorphous Silicon
## Part A

*Jacques I. Pankove*, Introduction
*Masataka Hirose*, Glow Discharge; Chemical Vapor Deposition
*Yoshiyuki Uchida*, di Glow Discharge
*T. D. Moustakas*, Sputtering
*Isao Yamada*, Ionized-Cluster Beam Deposition
*Bruce A. Scott*, Homogeneous Chemical Vapor Deposition

*Frank J. Kampas*, Chemical Reactions in Plasma Deposition
*Paul A. Longeway*, Plasma Kinetics
*Herbert A. Weakliem*, Diagnostics of Silane Glow Discharges Using Probes and Mass Spectroscopy
*Lester Gluttman*, Relation between the Atomic and the Electronic Structures
*A. Chenevas-Paule*, Experiment Determination of Structure
*S. Minomura*, Pressure Effects on the Local Atomic Structure
*David Adler*, Defects and Density of Localized States

## Part B

*Jacques I. Pankove*, Introduction
*G. D. Cody*, The Optical Absorption Edge of a-Si:H
*Nabil M. Amer and Warren B. Jackson*, Optical Properties of Defect States in a-Si:H
*P. J. Zanzucchi*, The Vibrational Spectra of a-Si:H
*Yoshihiro Hamakawa*, Electroreflectance and Electroabsorption
*Jeffrey S. Lannin*, Raman Scattering of Amorphous Si, Ge, and Their Alloys
*R. A. Street*, Luminescence in a-Si:H
*Richard S. Crandall*, Photoconductivity
*J. Tauc*, Time-Resolved Spectroscopy of Electronic Relaxation Processes
*P. E. Vanier*, IR-Induced Quenching and Enhancement of Photoconductivity and Photoluminescence
*H. Schade*, Irradiation-Induced Metastable Effects
*L. Ley*, Photoelectron Emission Studies

## Part C

*Jacques I. Pankove*, Introduction
*J. David Cohen*, Density of States from Junction Measurements in Hydrogenated Amorphous Silicon
*P. C. Taylor*, Magnetic Resonance Measurements in a-Si:H
*K. Morigaki*, Optically Detected Magnetic Resonance
*J. Dresner*, Carrier Mobility in a-Si:H
*T. Tiedje*, Information about band-Tail States from Time-of-Flight Experiments
*Arnold R. Moore*, Diffusion Length in Undoped a-Si:H
*W. Beyer and J. Overhof*, Doping Effects in a-Si:H
*H. Fritzche*, Electronic Properties of Surfaces in a-Si:H
*C. R. Wronski*, The Staebler-Wronski Effect
*R. J. Nemanich*, Schottky Barriers on a-Si:H
*B. Abeles and T. Tiedje*, Amorphous Semiconductor Superlattices

## Part D

*Jacques I. Pankove*, Introduction
*D. E. Carlson*, Solar Cells
*G. A. Swartz*, Closed-Form Solution of I-V Characteristic for a a-Si:H Solar Cells
*Isamu Shimizu*, Electrophotography
*Sachio Ishioka*, Image Pickup Tubes

*P. G. LeComber and W. E. Spear*, The Development of the a-Si:H Field-Effect Transistor and Its Possible Applications
*D. G. Ast*, a-Si:H FET-Addressed LCD Panel
*S. Kaneko*, Solid-State Image Sensor
*Masakiyo Matsumura*, Charge-Coupled Devices
*M. A. Bosch*, Optical Recording
*A. D'Amico and G. Fortunato*, Ambient Sensors
*Hiroshi Kukimoto*, Amorphous Light-Emitting Devices
*Robert J. Phelan, Jr.*, Fast Detectors and Modulators
*Jacques I. Pankove*, Hybrid Structures
*P. G. LeComber, A. E. Owen, W. E. Spear, J. Hajto, and W. K. Choi*, Electronic Switching in Amorphous Silicon Junction Devices

## Volume 22 Lightwave Communications Technology
### Part A

*Kazuo Nakajima*, The Liquid-Phase Epitaxial Growth of IngaAsp
*W. T. Tsang*, Molecular Beam Epitaxy for III–V Compound Semiconductors
*G. B. Stringfellow*, Organometallic Vapor-Phase Epitaxial Growth of III–V Semiconductors
*G. Beuchet*, Halide and Chloride Transport Vapor-Phase Deposition of InGaAsP and GaAs
*Manijeh Razeghi*, Low-Pressure Metallo-Organic Chemical Vapor Deposition of $Ga_x In_{1-x} As P_{1-y}$ Alloys
*P. M. Petroff*, Defects in III–V Compound Semiconductors

### Part B

*J. P. van der Ziel*, Mode Locking of Semiconductor Lasers
*Kam Y. Lau and Ammon Yariv*, High-Frequency Current Modulation of Semiconductor Injection Lasers
*Charles H. Henry*, Special Properties of Semiconductor Lasers
*Yasuharu Suematsu, Katsumi Kishino, Shigehisa Arai, and Fumio Koyama*. Dynamic Single-Mode Semiconductor Lasers with a Distributed Reflector
*W. T. Tsang*, The Cleaved-Coupled-Cavity ($C^3$) Laser

### Part C

*R. J. Nelson and N. K. Dutta*, Review of InGaAsP InP Laser Structures and Comparison of Their Performance
*N. Chinone and M. Nakamura*, Mode-Stabilized Semiconductor Lasers for 0.7–0.8- and 1.1–1.6-$\mu$m Regions
*Yoshiji Horikoshi*, Semiconductor Lasers with Wavelengths Exceeding $2\,\mu m$
*B. A. Dean and M. Dixon*, The Functional Reliability of Semiconductor Lasers as Optical Transmitters
*R. H. Saul, T. P. Lee, and C. A. Burus*, Light-Emitting Device Design
*C. L. Zipfel*, Light-Emitting Diode-Reliability
*Tien Pei Lee and Tingye Li*, LED-Based Multimode Lightwave Systems
*Kinichiro Ogawa*, Semiconductor Noise-Mode Partition Noise

## Part D

*Federico Capasso*, The Physics of Avalanche Photodiodes
*T. P. Pearsall and M. A. Pollack*, Compound Semiconductor Photodiodes
*Takao Kaneda*, Silicon and Germanium Avalanche Photodiodes
*S. R. Forrest*, Sensitivity of Avalanche Photodetector Receivers for High-Bit-Rate Long-Wavelength Optical Communication Systems
*J. C. Campbell*, Phototransistors for Lightwave Communications

## Part E

*Shyh Wang*, Principles and Characteristics of Integrable Active and Passive Optical Devices
*Shlomo Margalit and Amnon Yariv*, Integrated Electronic and Photonic Devices
*Takaoki Mukai, Yoshihisa Yamamoto, and Tatsuya Kimura*, Optical Amplification by Semiconductor Lasers

## Volume 23  Pulsed Laser Processing of Semiconductors

*R. F. Wood, C. W. White, and R. T. Young*, Laser Processing of Semiconductors: An Overview
*C. W. White*, Segregation, Solute Trapping, and Supersaturated Alloys
*G. E. Jellison, Jr.*, Optical and Electrical Properties of Pulsed Laser-Annealed Silicon
*R. F. Wood and G. E. Jellison, Jr.*, Melting Model of Pulsed Laser Processing
*R. F. Wood and F. W. Young, Jr.*, Nonequilibrium Solidification Following Pulsed Laser Melting
*D. H. Lowndes and G. E. Jellison, Jr.*, Time-Resolved Measurement During Pulsed Laser Irradiation of Silicon
*D. M. Zebner*, Surface Studies of Pulsed Laser Irradiated Semiconductors
*D. H. Lowndes*, Pulsed Beam Processing of Gallium Arsenide
*R. B. James*, Pulsed $CO_2$ Laser Annealing of Semiconductors
*R. T. Young and R. F. Wood*, Applications of Pulsed Laser Processing

## Volume 24  Applications of Multiquantum Wells, Selective Doping, and Superlattices

*C. Weisbuch*, Fundamental Properties of III–V Semiconductor Two-Dimensional Quantized Structures: The Basis for Optical and Electronic Device Applications
*H. Morkoc and H. Unlu*, Factors Affecting the Performance of (Al,Ga)As/GaAs and (Al,Ga)As/InGaAs Modulation-Doped Field-Effect Transistors: Microwave and Digital Applications
*N. T. Linh*, Two-Dimensional Electron Gas FETs: Microwave Applications
*M. Abe et al.*, Ultra-High-Speed HEMT Integrated Circuits
*D. S. Chemla, D. A. B. Miller, and P. W. Smith*, Nonlinear Optical Properties of Multiple Quantum Well Structures for Optical Signal Processing
*F. Capasso*, Graded-Gap and Superlattice Devices by Band-Gap Engineering
*W. T. Tsang*, Quantum Confinement Heterostructure Semiconductor Lasers
*G. C. Osbourn et al.*, Principles and Applications of Semiconductor Strained-Layer Superlattices

## Volume 25  Diluted Magnetic Semiconductors

*W. Giriat and J. K. Furdyna*, Crystal Structure, Composition, and Materials Preparation of Diluted Magnetic Semiconductors

*W. M. Becker*, Band Structure and Optical Properties of Wide-Gap $A_{1-x}^{II} Mn_x B^{IV}$ Alloys at Zero Magnetic Field

*Saul Oseroff and Pieter H. Keesom*, Magnetic Properties: Macroscopic Studies

*Giebultowicz and T. M. Holden*, Neutron Scattering Studies of the Magnetic Structure and Dynamics of Diluted Magnetic Semiconductors

*J. Kossut*, Band Structure and Quantum Transport Phenomena in Narrow-Gap Diluted Magnetic Semiconductors

*C. Riquaux*, Magnetooptical Properties of Large-Gap Diluted Magnetic Semiconductors

*J. A. Gaj*, Magnetooptical Properties of Large-Gap Diluted Magnetic Semiconductors

*J. Mycielski*, Shallow Acceptors in Diluted Magnetic Semiconductors: Splitting, Boil-off, Giant Negative Magnetoresistance

*A. K. Ramadas and R. Rodriquez*, Raman Scattering in Diluted Magnetic Semiconductors

*P. A. Wolff*, Theory of Bound Magnetic Polarons in Semimagnetic Semiconductors

## Volume 26  III–V Compound Semiconductors and Semiconductor Properties of Superionic Materials

*Zou Yuanxi*, III–V Compounds

*H. V. Winston, A. T. Hunter, H. Kimura, and R. E. Lee*, InAs-Alloyed GaAs Substrates for Direct Implantation

*P. K. Bhattachary and S. Dhar*, Deep Levels in III–V Compound Semiconductors Grown by MBE

*Yu. Yu. Gurevich and A. K. Ivanov-Shits*, Semiconductor Properties of Supersonic Materials

## Volume 27  High Conducting Quasi-One-Dimensional Organic Crystals

*E. M. Conwell*, Introduction to Highly Conducting Quasi-One-Dimensional Organic Crystals

*I. A. Howard*, A Reference Guide to the Conducting Quasi-One-Dimensional Organic Molecular Crystals

*J. P. Pouquet*, Structural Instabilities

*E. M. Conwell*, Transport Properties

*C. S. Jacobsen*, Optical Properties

*J. C. Scott*, Magnetic Properties

*L. Zuppiroli*, Irradiation Effects: Perfect Crystals and Real Crystals

## Volume 28  Measurement of High-Speed Signals in Solid State Devices

*J. Frey and D. Ioannou*, Materials and Devices for High-Speed and Optoelectronic Applications

*H. Schumacher and E. Strid*, Electronic Wafer Probing Techniques

*D. H. Auston*, Picosecond Photoconductivity: High-Speed Measurements of Devices and Materials

*J. A. Valdmanis*, Electro-Optic Measurement Techniques for Picosecond Materials, Devices, and Integrated Circuits

*J. M. Wiesenfeld and R. K. Jain*, Direct Optical Probing of Integrated Circuits and High-Speed Devices

*G. Plows*, Electron-Beam Probing

*A. M. Weiner and R. B. Marcus*, Photoemissive Probing

## Volume 29  Very High Speed Integrated Circuits: Gallium Arsenide LSI

M. Kuzuhara and T. Nazaki, Active Layer Formation by Ion Implantation
H. Hasimoto, Focused Ion Beam Implantation Technology
T. Nozaki and A. Higashisaka, Device Fabrication Process Technology
M. Ino and T. Takada, GaAs LSI Circuit Design
M. Hirayama, M. Ohmori, and K. Yamasaki, GaAs LSI Fabrication and Performance

## Volume 30  Very High Speed Integrated Circuits: Heterostructure

H. Watanabe, T. Mizutani, and A. Usui, Fundamentals of Epitaxial Growth and Atomic Layer Epitaxy
S. Hiyamizu, Characteristics of Two-Dimensional Electron Gas in III–V Compound Heterostructures Grown by MBE
T. Nakanisi, Metalorganic Vapor Phase Epitaxy for High-Quality Active Layers
T. Nimura, High Electron Mobility Transistor and LSI Applications
T. Sugeta and T. Ishibashi, Hetero-Bipolar Transistor and LSI Application
H. Matsueda, T. Tanaka, and M. Nakamura, Optoelectronic Integrated Circuits

## Volume 31  Indium Phosphide: Crystal Growth and Characterization

J. P. Farges, Growth of Discoloration-free InP
M. J. McCollum and G. E. Stillman, High Purity InP Grown by Hydride Vapor Phase Epitaxy
T. Inada and T. Fukuda, Direct Synthesis and Growth of Indium Phosphide by the Liquid Phosphorous Encapsulated Czochralski Method
O. Oda, K. Katagiri, K. Shinohara, S. Katsura, Y. Takahashi, K. Kainosho, K. Kohiro, and R. Hirano, InP Crystal Growth, Substrate Preparation and Evaluation
K. Tada, M. Tatsumi, M. Morioka, T. Araki, and T. Kawase, InP Substrates: Production and Quality Control
M. Razeghi, LP-MOCVD Growth, Characterization, and Application of InP Material
T. A. Kennedy and P. J. Lin-Chung, Stoichiometric Defects in InP

## Volme 32  Strained-Layer Superlattices: Physics

T. P. Pearsall, Strained-Layer Superlattices
Fred H. Pollack, Effects of Homogeneous Strain on the Electronic and Vibrational Levels in Semiconductors
J. Y. Marzin, J. M. Gerárd, P. Voisin, and J. A. Brum, Optical Studies of Strained III–V Heterolayers
R. People and S. A. Jackson, Structurally Induced States from Strain and Confinement
M. Jaros, Microscopic Phenomena in Ordered Suprlattices

## Volume 33  Strained-Layer Superlattices: Materials Science and Technology

R. Hull and J. C. Bean, Principles and Concepts of Strained-Layer Epitaxy
William J. Schaff, Paul J. Tasker, Marc C. Foisy, and Lester F. Eastman, Device Applications of Strained-Layer Epitaxy

*S. T. Picraux, B. L. Doyle, and J. Y. Tsao*, Structure and Characterization of Strained-Layer Superlattices
*E. Kasper and F. Schäffer*, Group IV Compounds
*Dale L. Martin*, Molecular Beam Epitaxy of IV–VI Compounds Heterojunction
*Robert L. Gunshor, Leslie A. Kolodziejski, Arto V. Nurmikko, and Nobuo Otsuka*, Molecular Beam Epitaxy of II–VI Semiconductor Microstructures

## Volume 34 Hydrogen in Semiconductors

*J. I. Pankove and N. M. Johnson*, Introduction to Hydrogen in Semiconductors
*C. H. Seager*, Hydrogenation Methods
*J. I. Pankove*, Hydrogenation of Defects in Crystalline Silicon
*J. W. Corbett, P. Deák, U. V. Desnica, and S. J. Pearton*, Hydrogen Passivation of Damage Centers in Semiconductors
*S. J. Pearton*, Neutralization of Deep Levels in Silicon
*J. I. Pankove*, Neutralization of Shallow Acceptors in Silicon
*N. M. Johnson*, Neutralization of Donor Dopants and Formation of Hydrogen-Induced Defects in $n$-Type Silicon
*M. Stavola and S. J. Pearton*, Vibrational Spectroscopy of Hydrogen-Related Defects in Silicon
*A. D. Marwick*, Hydrogen in Semiconductors: Ion Beam Techniques
*C. Herring and N. M. Johnson*, Hydrogen Migration and Solubility in Silicon
*E. E. Haller*, Hydrogen-Related Phenomena in Crystalline Germanium
*J. Kakalios*, Hydrogen Diffusion in Amorphous Silicon
*J. Chevalier, B. Clerjaud, and B. Pajot*, Neutralization of Defects and Dopants in III–V Semiconductors
*G. G. DeLeo and W. B. Fowler*, Computational Studies of Hydrogen-Containing Complexes in Semiconductors
*R. F. Kiefl and T. L. Estle*, Muonium in Semiconductors
*C. G. Van de Walle*, Theory of Isolated Interstitial Hydrogen and Muonium in Crystalline Semiconductors

## Volume 35 Nanostructured Systems

*Mark Reed*, Introduction
*H. van Houten, C. W. J. Beenakker, and B. J. van Wees*, Quantum Point Contacts
*G. Timp*, When Does a Wire Become an Electron Waveguide?
*M. Büttiker*, The Quantum Hall Effects in Open Conductors
*W. Hansen, J. P. Kotthaus, and U. Merkt*, Electrons in Laterally Periodic Nanostructures

## Volume 36 The Spectroscopy of Semiconductors

*D. Heiman*, Spectroscopy of Semiconductors at Low Temperatures and High Magnetic Fields
*Arto V. Nurmikko*, Transient Spectroscopy by Ultrashort Laser Pulse Techniques
*A. K. Ramdas and S. Rodriguez*, Piezospectroscopy of Semiconductors
*Orest J. Glembocki and Benjamin V. Shanabrook*, Photoreflectance Spectroscopy of Microstructures
*David G. Seiler, Christopher L. Littler, and Margaret H. Wiler*, One- and Two-Photon Magneto-Optical Spectroscopy of InSb and $Hg_{1-x}Cd_xTe$

## Volume 37  The Mechanical Properties of Semiconductors

*A.-B. Chen, Arden Sher and W. T. Yost*, Elastic Constants and Related Properties of Semiconductor Compounds and Their Alloys
*David R. Clarke*, Fracture of Silicon and Other Semiconductors
*Hans Siethoff*, The Plasticity of Elemental and Compound Semiconductors
*Sivaraman Guruswamy, Katherine T. Faber and John P. Hirth*, Mechanical Behavior of Compound Semiconductors
*Subhanh Mahajan*, Deformation Behavior of Compound Semiconductors
*John P. Hirth*, Injection of Dislocations into Strained Multilayer Structures
*Don Kendall, Charles B. Fleddermann, and Kevin J. Malloy*, Critical Technologies for the Micromachining of Silicon
*Ikuo Matsuba and Kinji Mokuya*, Processing and Semiconductor Thermoelastic Behavior

## Volume 38  Imperfections in III/V Materials

*Udo Scherz and Matthias Scheffler*, Density-Functional Theory of sp-Bonded Defects in III/V Semiconductors
*Maria Kaminska and Eicke R. Weber*, El2 Defect in GaAs
*David C. Look*, Defects Relevant for Compensation in Semi-Insulating GaAs
*R. C. Newman*, Local Vibrational Mode Spectroscopy of Defects in III/V Compounds
*Andrzej M. Hennel*, Transition Metals in III/V Compounds
*Kevin J. Malloy and Ken Khachaturyan*, DX and Related Defects in Semiconductors
*V. Swaminathan and Andrew S. Jordan*, Dislocations in III/V Compounds
*Krzysztof W. Nauka*, Deep Level Defects in the Epitaxial III/V Materials

## Volume 39  Minority Carriers in III–V Semiconductors: Physics and Applications

*Niloy K. Dutta*, Radiative Transitions in GaAs and Other III–V Compounds
*Richard K. Ahrenkiel*, Minority-Carrier Lifetime in III–V Semiconductors
*Tomofumi Furuta*, High Field Minority Electron Transport in p-GaAs
*Mark S. Lundstrom*, Minority-Carrier Transport in III–V Semiconductors
*Richard A. Abram*, Effects of Heavy Doping and High Excitation on the Band Structure of GaAs
*David Yevick and Witold Bardyszewski*, An Introduction to Non-Equilibrium Many-Body Analyses of Optical Processes in III–V Semiconductors

## Volume 40  Epitaxial Microstructures

*E. F. Schubert*, Delta-Doping of Semiconductors: Electronic, Optical, and Structural Properties of Materials and Devices
*A. Gossard, M. Sundaram, and P. Hopkins*, Wide Graded Potential Wells
*P. Petroff*, Direct Growth of Nanometer-Size Quantum Wire Superlattices
*E. Kapon*, Lateral Patterning of Quantum Well Heterostructures by Growth of Nonplanar Substrates
*H. Temkin, D. Gershoni, and M. Panish*, Optical Properties of Ga1-$_x$In$_x$As/InP Quantum Wells

## Volume 41    High Speed Heterostructure Devices

F. Capasso, F. Beltram, S. Sen, A. Pahlevi, and A. Y. Cho, Quantum Electron Devices: Physics and Applications
P. Solomon, D. J. Frank, S. L. Wright, and F. Canora, GaAs-Gate Semiconductor–Insulator–Semiconductor FET
M. H. Hashemi and U. K. Mishra, Unipolar InP-Based Transistors
R. Kiehl, Complementary Heterostructure FET Integrated Circuits
T. Ishibashi, GaAs-Based and InP-Based Heterostructure Bipolar Transistors
H. C. Liu and T. C. L. G. Sollner, High-Frequency-Tunneling Devices
H. Ohnishi, T. More, M. Takatsu, K. Imamura, and N. Yokoyama, Resonant-Tunneling Hot-Electron Transistors and Circuits

## Volume 42    Oxygen in Silicon

F. Shimura, Introduction to Oxygen in Silicon
W. Lin, The Incorporation of Oxygen into Silicon Crystals
T. J. Schaffner and D. K. Schroder, Characterization Techniques for Oxygen in Silicon
W. M. Bullis, Oxygen Concentration Measurement
S. M. Hu, Intrinsic Point Defects in Silicon
B. Pajot, Some Atomic Configurations of Oxygen
J. Michel and L. C. Kimerling, Electical Properties of Oxygen in Silicon
R. C. Newman and R. Jones, Diffusion of Oxygen in Silicon
T. Y. Tan and W. J. Taylor, Mechanisms of Oxygen Precipitation: Some Quantitative Aspects
M. Schrems, Simulation of Oxygen Precipitation
K. Simino and I. Yonenaga, Oxygen Effect on Mechanical Properties
W. Bergholz, Grown-in and Process-Induced Effects
F. Shimura, Intrinsic/Internal Gettering
H. Tsuya, Oxygen Effect on Electronic Device Performance

## Volume 43    Semiconductors for Room Temperature Nuclear Detector Applications

R. B. James and T. E. Schlesinger, Introduction and Overview
L. S. Darken and C. E. Cox, High-Purity Germanium Detectors
A. Burger, D. Nason, L. Van den Berg, and M. Schieber, Growth of Mercuric Iodide
X. J. Bao, T. E. Schlesinger, and R. B. James, Electrical Properties of Mercuric Iodide
X. J. Bao, R. B. James, and T. E. Schlesinger, Optical Properties of Red Mercuric Iodide
M. Hage-Ali and P. Siffert, Growth Methods of CdTe Nuclear Detector Materials
M. Hage-Ali and P Siffert, Characterization of CdTe Nuclear Detector Materials
M. Hage-Ali and P. Siffert, CdTe Nuclear Detectors and Applications
R. B. James, T. E. Schlesinger, J. Lund, and M. Schieber, $Cd_{1-x}Zn_xTe$ Spectrometers for Gamma and X-Ray Applications
D. S. McGregor, J. E. Kammeraad, Gallium Arsenide Radiation Detectors and Spectrometers
J. C. Lund, F. Olschner, and A. Burger, Lead Iodide
M. R. Squillante, and K. S. Shah, Other Materials: Status and Prospects
V. M. Gerrish, Characterization and Quantification of Detector Performance
J. S. Iwanczyk and B. E. Patt, Electronics for X-ray and Gamma Ray Spectrometers
M. Schieber, R. B. James, and T. E. Schlesinger, Summary and Remaining Issues for Room Temperature Radiation Spectrometers

## Volume 44  II–IV Blue/Green Light Emitters: Device Physics and Epitaxial Growth

J. Han and R. L. Gunshor, MBE Growth and Electrical Properties of Wide Bandgap ZnSe-based II–VI Semiconductors
Shizuo Fujita and Shigeo Fujita, Growth and Characterization of ZnSe-based II–VI Semiconductors by MOVPE
Easen Ho and Leslie A. Kolodziejski, Gaseous Source UHV Epitaxy Technologies for Wide Bandgap II–VI Semiconductors
Chris G. Van de Walle, Doping of Wide-Band-Gap II–VI Compounds — Theory
Roberto Cingolani, Optical Properties of Excitons in ZnSe-Based Quantum Well Heterostructures
A. Ishibashi and A. V. Nurmikko, II–VI Diode Lasers: A Current View of Device Performance and Issues
Supratik Guha and John Petruzello, Defects and Degradation in Wide-Gap II–VI-based Structures and Light Emitting Devices

## Volume 45  Effect of Disorder and Defects in Ion-Implanted Semiconductors: Electrical and Physiochemical Characterization

Heiner Ryssel, Ion Implantation into Semiconductors: Historical Perspectives
You-Nian Wang and Teng-Cai Ma, Electronic Stopping Power for Energetic Ions in Solids
Sachiko T. Nakagawa, Solid Effect on the Electronic Stopping of Crystalline Target and Application to Range Estimation
G. Müller, S. Kalbitzer and G. N. Greaves, Ion Beams in Amorphous Semiconductor Research
Jumana Boussey-Said, Sheet and Spreading Resistance Analysis of Ion Implanted and Annealed Semiconductors
M. L. Polignano and G. Queirolo, Studies of the Stripping Hall Effect in Ion-Implanted Silicon
J. Stoemenos, Transmission Electron Microscopy Analyses
Roberta Nipoti and Marco Servidori, Rutherford Backscattering Studies of Ion Implanted Semiconductors
P. Zaumseil, X-ray Diffraction Techniques

## Volume 46  Effect of Disorder and Defects in Ion-Implanted Semiconductors: Optical and Photothermal Characterization

M. Fried, T. Lohner and J. Gyulai, Ellipsometric Analysis
Antonios Seas and Constantinos Christofides, Transmission and Reflection Spectroscopy on Ion Implanted Semiconductors
Andreas Othonos and Constantinos Christofides, Photoluminescence and Raman Scattering of Ion Implanted Semiconductors. Influence of Annealing
Constantinos Christofides, Photomodulated Thermoreflectance Investigation of Implanted Wafers. Annealing Kinetics of Defects
U. Zammit, Photothermal Deflection Spectroscopy Characterization of Ion-Implanted and Annealed Silicon Films
Andreas Mandelis, Arief Budiman and Miguel Vargas, Photothermal Deep-Level Transient Spectroscopy of Impurities and Defects in Semiconductors
R. Kalish and S. Charbonneau, Ion Implantation into Quantum-Well Structures
Alexandre M. Myasnikov and Nikolay N. Gerasimenko, Ion Implantation and Thermal Annealing of III-V Compound Semiconducting Systems: Some Problems of III-V Narrow Gap Semiconductors

## Volume 47  Uncooled Infrared Imaging Arrays and Systems

*R. G. Buser and M. F. Tompsett*, Historical Overview
*P. W. Kruse*, Principles of Uncooled Infrared Focal Plane Arrays
*R. A. Wood*, Monolithic Silicon Microbolometer Arrays
*C. M. Hanson*, Hybrid Pyroelectric–Ferroelectric Bolometer Arrays
*D. L. Polla and J. R. Choi*, Monolithic Pyroelectric Bolometer Arrays
*N. Teranishi*, Thermoelectric Uncooled Infrared Focal Plane Arrays
*M. F. Tompsett*, Pyroelectric Vidicon
*T. W. Kenny*, Tunneling Infrared Sensors
*J. R. Vig, R. L. Filler and Y. Kim*, Application of Quartz Microresonators to Uncooled Infrared Imaging Arrays
*P. W. Kruse*, Application of Uncooled Monolithic Thermoelectric Linear Arrays to Imaging Radiometers

ISBN 0-12-752155-0